AF597971

Methods in Molecular Biology™

Series Editor
John M. Walker
School of Life Sciences
University of Hertfordshire
Hatfield, Hertfordshire, AL10 9AB, UK

For further volumes:
http://www.springer.com/series/7651

Heart Proteomics

Methods and Protocols

Edited by

Fernando Vivanco

Department of Immunology, IIS-Fundacion Jimenez Diaz, Madrid, Spain
Department of Biochemistry and Molecular Biology I, Universidad Complutense, Madrid, Spain

Editor
Fernando Vivanco
Department of Immunology
IIS-Fundacion Jimenez Diaz
Madrid, Spain

Department of Biochemistry and Molecular Biology I
Universidad Complutense
Madrid, Spain

ISSN 1064-3745 ISSN 1940-6029 (electronic)
ISBN 978-1-62703-385-5 ISBN 978-1-62703-386-2 (eBook)
DOI 10.1007/978-1-62703-386-2
Springer New York Heidelberg Dordrecht London

Library of Congress Control Number: 2013936072

Printed on acid-free paper

Humana Press is a brand of Springer
Springer is part of Springer Science+Business Media (www.springer.com)

Preface

The use of proteomics for the study of complex diseases is increasing and is particularly applicable to cardiovascular diseases, the leading cause of death in developed countries. Proteomics is a rapidly expanding investigation platform in cardiovascular medicine and is becoming integrated and incorporated into cardiovascular research. Driven by major improvements in mass spectrometry instrumentation, methodology, and data analysis, the proteomics field has exponentially grown in recent years. These advancements serve not only to identify proteins with high sensitivity but increasingly to use proteomic technologies to assay the dynamic properties of proteins in high throughput and to characterize the structure and composition of large multiprotein complexes. These new approaches and techniques are characterized by the development of quantitative MS-based methods that moved the field on from primarily identifying proteins to also providing measurements of relative changes in protein levels between different cell states, typically normal controls versus diseased samples. The application of proteomic techniques to the heart pathology is a true reflection of this progress. Overall, the recent development of MS-based methods and advanced analytical tools are transforming our ability to profile proteins in the cardiovascular system. This book encompasses a selection of techniques and methods that target the numerous processes implicated in the pathophysiology of heart. Thus, we provide protocols and updated methods in the heart proteomic area with a particular focus on MS-based methods of protein and peptide quantification and the analysis of posttranslational modifications. The development of LC-MS/MS approaches has been a significant advancement in many areas of biomedical research, and heart proteomics is not an exception. Several chapters describe system biology approaches, which provide a better understanding of normal and pathological processes. We have followed a hierarchical order in the different chapters starting with methods dedicated to cardiac tissue (Large-scale characterization of the murine cardiac proteome; Determining protein concentration of the human ventricular proteome; Multidimensional protein identification technology for direct-tissue proteomic of heart; Global protein quantification of mouse heart tissue based on the SILAC mouse; Global proteomic profiling and enrichment maps of dilated cardiomyopathy; Characterization of the human myocardial proteome in dilated cardiomyopathy by label-free quantitative shotgun proteomics of heart biopsies), aortic valves (Differential protein expression analysis of degenerative aortic stenosis by iTRAQ labeling; Proteomic analysis of interstitial aortic valve cells acquiring a pro-calcific profile), organelle subproteomes (Proteomic analysis of brain mitochondrial proteome and mitochondrial complexes; Oxidative modifications of mitochondria complex II), posttranslational modifications (Detection of O-GlcNAc modifications on cardiac myofilament proteins; Quantification of mitochondrial S-Nitrosylation by CysTMT6 switch assay), analysis of secretomes (Optimized method for identification of the proteomes secreted by cardiac cells; Secretome of human aortic valves), and quantitation of specific proteins (Using pure protein to build a multiple reaction monitoring mass spectrometry assay for targeted detection and quantitation; Identification of Thioredoxin target protein networks in cardiac tissues of a transgenic mouse).

This book does not attempt to describe exhaustively all the techniques used in the field of heart proteomics, rather is a representative selection of methods that can be a useful resource for experienced proteomics practitioners and, specially, for newcomers, in order to become acquainted with the practice of a selective group of proteomic techniques for cardiovascular research. The editors are especially grateful to all contributing authors for the time and effort they have put into writing their chapters and particularly to the Methods in Molecular Biology series editor, John Walker, for his continuous advice and support through the editorial process.

Madrid, Spain *Fernando Vivanco*
Toledo, Spain *María G. Barderas*

Contents

Contributors

SERGIO ALONSO-ORGAZ • *Department of Vascular Physiopathology, Hospital Nacional de Paraplejicos, SESCAM, Toledo, Spain*

GLORIA ALVAREZ-LLAMAS • *Department of Immunology, Instituto de Investigacion Sanitaria Fundacion Jimenez Diaz (IIS-FJD), Madrid, Spain*

GIORGIO ARRIGONI • *Proteomics Center of Padova University, VIMM, Padova, Italy; Padova University Hospital, Padova, Italy; Department of Biological Chemistry, Padova University, Padova, Italy*

JAVIER BARALLOBRE-BARREIRO • *King's British Heart Foundation Centre, London, UK*

MARÍA G. BARDERAS • *Department of Vascular Physiopathology, Hospital Nacional de Paraplejicos, SESCAM, Toledo, Spain*

ELISA BERTACCO • *Department of Clinical and Experimental Medicine, Medical School, University of Padova, Padova, Italy*

FRANCESCA BRAMBILLA • *Laboratory of Food Chemistry and Mass Spectrometry, Department of Endocrinology, University of Milan, Milan, Italy*

THOMAS BRAUN • *Biomolecular Mass Spectrometry, Max-Planck-Institute for Heart and Lung Research, Bad Nauheim, Germany*

ENRIQUE CALVO • *Unidad de Proteómica, Centro Nacional de Investigaciones Cardiovasculares, CNIC, Madrid, Spain*

CHWEN-LIH CHEN • *Department of Integrative Medical Sciences, College of Medicine, Northeast Ohio Medical University, Rootstown, OH, USA*

HAODONG CHEN • *Department of Anesthesiology, David Geffen School of Medicine, UCLA, Los Angeles, CA, USA; Department of Medicine, David Geffen School of Medicine, UCLA, Los Angeles, CA, USA; Department of Physiology, David Geffen School of Medicine, UCLA, Los Angeles, CA, USA*

YEONG-RENN CHEN • *Department of Integrative Medical Sciences, Northeastern Ohio Universities, Colleges of Medicine and Pharmacy, Rootstown, OH, USA*

HEA SEUNG CHUNG • *Department of Biological Chemistry, Johns Hopkins University, Baltimore, MD, USA*

JAKE COSME • *Department of Physiology, University of Toronto, Toronto, ON, Canada*

VERÓNICA M. DARDE • *Proteomic Unit, Hospital Nacional de Paraplejicos, Toledo, Spain*

KATRIN DARM • *Abteilung für Funktionelle Genomforschung, Interfakultäres Institut für Genetik und Funktionelle Genomforschung, Universitätsmedizin Greifswald, Greifswald, Germany*

FERNANDO DE LA CUESTA • *Laboratorio de Fisiopatología Vascular, Hospital Nacional de Parapléjicos, SESCAM, Toledo, Spain*

ATHANASIOS DIDANGELOS • *King's British Heart Foundation Centre, London, UK*

Dario Di Silvestre • *Proteomics and Metabolomics Unit, Institute for Biomedical Technologies, CNR, Milan, Italy*
Nieves Doménech • *Instituto de Investigación Biomédica de A Coruña (INIBIC), Coruña, Spain*
Andrew Emili • *The Donnelly Centre for Cellular and Biomolecular Research, University of Toronto, Toronto, ON, Canada; Banting and Best Department of Medical Research, University of Toronto, Toronto, ON, Canada; Department of Molecular Genetics, University of Toronto, Toronto, ON, Canada*
Carlos Fernandez-Patron • *University of Alberta, Edmonton, AB, Canada*
Isabel Martinez Ferrando • *Department of Pharmacology and Molecular Sciences, Johns Hopkins University, School of Medicine, Baltimore, MD, USA*
Cinzia Franchin • *Proteomics Center of Padova University, VIMM, Padova, Italy; Padova University Hospital, Padova, Italy; Department of Biological Chemistry, Padova University, Padova, Italy*
Sarah Franklin • *Department of Anesthesiology, David Geffen School of Medicine, UCLA, Los Angeles, CA, USA; Department of Medicine, David Geffen School of Medicine, UCLA, Los Angeles, CA, USA; Department of Physiology, David Geffen School of Medicine, UCLA, Los Angeles, CA, USA*
Cexiong Fu • *Pharmacokinetics, Dynamics and Metabolism, Pfizer Global Research and Development, Pfizer Inc., Groton, CT, USA*
Qin Fu • *Division of Cardiology, Department of Medicine, Johns Hopkins Bayview Proteomics Center, School of Medicine, Johns Hopkins University, Baltimore, MD, USA*
Felix Gil-Dones • *Department of Vascular Physiopathology, Hospital Nacional de Paraplejicos, SESCAM, Toledo, Spain*
Anthony O. Gramolini • *Department of Physiology, University of Toronto, Toronto, ON, Canada*
Kari B. Green • *Proteomics and Mass Spectrometry Facility, Campus Chemical Instrument Center, The Ohio State University, Columbus, OH, USA*
Eric Grote • *Division of Cardiology, Department of Medicine, Johns Hopkins Bayview Proteomics Center, School of Medicine, Johns Hopkins University, Baltimore, MD, USA*
Elke Hammer • *Abteilung für Funktionelle Genomforschung, Interfakultäres Institut für Genetik und Funktionelle Genomforschung, Universitätsmedizin Greifswald, Greifswald, Germany*
Gerald Hart • *Department of Biological Chemistry, Johns Hopkins University School of Medicine, Baltimore, MD, USA*
Albert J.R. Heck • *Biomolecular Mass Spectrometry and Proteomics Group, Bijvoet Center for Biomolecular Research, Utrecht University, Utrecht, The Netherlands; Utrecht Institute for Pharmaceutical Sciences, Netherlands Proteomics Centre, Utrecht University, Utrecht, The Netherlands*
Warren Heideman • *Department of Biomolecular Chemistry, School of Pharmacy, University of Wisconsin, Madison, WI, USA*
Ruth Isserlin • *The Donnelly Centre for Cellular and Biomolecular Research, University of Toronto, Toronto, ON, Canada; Banting and Best Department of Medical Research, University of Toronto, Toronto, ON, Canada; Department of Molecular Genetics, University of Toronto, Toronto, ON, Canada*

WEIHUA JI • *Division of Cardiology, Department of Medicine, Johns Hopkins Bayview Proteomics Center, School of Medicine, Johns Hopkins University, Baltimore, MD, USA*

PATRICK T. KANG • *Department of Integrative Medical Sciences, College of Medicine, Northeast Ohio Medical University, Rootstown, OH, USA*

ANNE KONZER • *Biomolecular Mass Spectrometry, Max-Planck-Institute for Heart and Lung Research, Bad Nauheim, Germany*

MARCUS KRÜGER • *Biomolecular Mass Spectrometry, Max-Planck-Institute for Heart and Lung Research, Bad Nauheim, Germany*

KEVIN A. LANHAM • *Department of Biomolecular Chemistry, University of Wisconsin, Madison, WI, USA*

HONG LI • *Department of Biochemistry and Molecular Biology, UMDNJ-NJMS Cancer Center, Newark, NJ, USA*

LINGJUN LI • *School of Pharmacy, University of Wisconsin, Madison, WI, USA*

TONG LIU • *Center for Advanced Proteomics Research, New Jersey Medical School Cancer Center, UMDNJ, Newark, NJ, USA; Department of Biochemistry and Molecular Biology, New Jersey Medical School Cancer Center, UMDNJ, Newark, NJ, USA*

XIAOQIAN LIU • *Division of Cardiology, Department of Medicine, Johns Hopkins Bayview Proteomics Center, School of Medicine, Johns Hopkins University, Baltimore, MD, USA*

JUAN ANTONIO LÓPEZ • *Unidad de Proteómica, Centro Nacional de Investigaciones Cardiovasculares, CNIC, Madrid, Spain*

ANA LOPEZ-CAMPISTROUS • *Department of Biochemistry, Institute for Biomolecular Design, University of Alberta, Edmonton, AB, Canada*

TATIANA MARTIN-ROJAS • *Department of Vascular Physiopathology, Hospital Nacional de Parapléjicos, SESCAM, Toledo, Spain*

PIER LUIGI MAURI • *Proteomics and Metabolomics Unit, Institute for Biomedical Technologies – CNR, Milan, Italy*

MANUEL MAYR • *King's British Heart Foundation Centre, London, UK*

DANIELE MERICO • *The Donnelly Centre for Cellular and Biomolecular Research, University of Toronto, Toronto, ON, Canada; Banting and Best Department of Medical Research, University of Toronto, Toronto, ON, Canada; Department of Molecular Genetics, University of Toronto, Toronto, ON, Canada*

EMMA MONTE • *Department of Anesthesiology, David Geffen School of Medicine at UCLA, Los Angeles, CA, USA; Department of Medicine, David Geffen School of Medicine at UCLA, Los Angeles, CA, USA; Department of Physiology, David Geffen School of Medicine at UCLA, Los Angeles, CA, USA*

ANNE MURPHY • *Department of Biological Chemistry, Johns Hopkins University School of Medicine, Baltimore, MD, USA*

CHRISTOPHER I. MURRAY • *Department of Biological Chemistry, Johns Hopkins University, Baltimore, MD, USA*

ANDREW M. PARROTT • *Center for Advanced Proteomics Research, New Jersey Medical School Cancer Center, UMDNJ, Newark, NJ, USA; Department of Biochemistry and Molecular Biology, New Jersey Medical School Cancer Center, UMDNJ, Newark, NJ, USA*

RICHARD E. PETERSON • *School of Pharmacy, University of Wisconsin, Madison, WI, USA*

GENARO A. RAMIREZ-CORREA • *Division of Cardiology, Department of Pediatrics, Johns Hopkins University School of Medicine, Baltimore, MD, USA*

MARCELLO RATTAZZI • *Dipartimento di Medicina Clinica e Sperimentale, Medicina Interna I, Padova University, Ospedale Ca' Foncello, Università degli Studi di Padova, Treviso, Italy*
MILLIONI RENATO • *Department of Medicine, Padova University, Padova, Italy; Proteomics Center of Padova University, VIMM, Padova, Italy; Padova University Hospital, Padova, Italy*
AARON RUHS • *Biomolecular Mass Spectrometry, Max-Planck-Institute for Heart and Lung Research, Bad Nauheim, Germany*
ARJEN SCHOLTEN • *Biomolecular Mass Spectrometry and Proteomics Group, Bijvoet Center for Biomolecular Research, Utrecht University, Utrecht, The Netherlands; Utrecht Institute for Pharmaceutical Sciences, Utrecht University, Utrecht, The Netherlands; Netherlands Proteomics Centre, Utrecht, The Netherlands*
MIROSLAVA STASTNA • *Division of Cardiology, Department of Medicine, School of Medicine, Johns Hopkins Bayview Proteomics Center, Johns Hopkins University, Baltimore, MD, USA; Institute of Analytical Chemistry of the ASCR, v.v.i., Brno, Czech Republic*
HELGE UHRIGSHARDT • *Division of Cardiology, Department of Medicine, Johns Hopkins University, Baltimore, MD, USA*
JENNIFER E. VAN EYK • *Division of Cardiology, Department of Medicine, Johns Hopkins Bayview Proteomics Research Center, School of Medicine, Johns Hopkins University, Baltimore, MD, USA*
FERNANDO VIVANCO • *Department of Immunology, IIIS-Fundacion Jimenez Diaz, Madrid, Spain; Department of Biochemistry and Molecular Biology I, Universidad Complutense, Madrid, Spain*
UWE VÖLKER • *Abteilung für Funktionelle Genomforschung, Interfakultäres Institut für Genetik und Funktionelle Genomforschung, Universitätsmedizin Greifswald, Greifswald, Germany*
THOMAS M. VONDRISKA • *Department of Anesthesiology, David Geffen School of Medicine, UCLA, Los Angeles, CA, USA; Department of Medicine, David Geffen School of Medicine, UCLA, Los Angeles, CA, USA; Department of Physiology, David Geffen School of Medicine, UCLA, Los Angeles, CA, USA*
XIAOKE YIN • *King's British Heart Foundation Centre, London, UK*
JIANG ZHANG • *School of Pharmacy, University of Wisconsin, Madison, WI, USA*
LIWEN ZHANG • *Proteomics and Mass Spectrometry Facility, Campus Chemical Instrument Center, The Ohio State University, Columbus, OH, USA*

Chapter 1

Large-Scale Characterization of the Murine Cardiac Proteome

Jake Cosme, Andrew Emili, and Anthony O. Gramolini

Abstract

Cardiomyopathies are diseases of the heart that result in impaired cardiac muscle function. This dysfunction can progress to an inability to supply blood to the body. Cardiovascular diseases play a large role in overall global morbidity. Investigating the protein changes in the heart during disease can uncover pathophysiological mechanisms and potential therapeutic targets. Establishing a global protein expression "footprint" can facilitate more targeted studies of diseases of the heart.

In the technical review presented here, we present methods to elucidate the heart's proteome through subfractionation of the cellular compartments to reduce sample complexity and improve detection of lower abundant proteins during multidimensional protein identification technology analysis. Analysis of the cytosolic, microsomal, and mitochondrial subproteomes separately in order to characterize the murine cardiac proteome is advantageous by simplifying complex cardiac protein mixtures. In combination with bioinformatic analysis and genome correlation, large-scale protein changes can be identified at the cellular compartment level in this animal model.

Key words Cardiac tissue subfractionation, Multidimensional protein identification technology (MuDPIT), Mass spectrometry

1 Introduction

Characterization of the murine cardiac proteome has allowed for the identification of differential expressed proteins in disease animal models (1) that can lead to identification of human analogues for diagnosis of disease. Establishing the murine cardiac proteome provides a basis for future studies in the heart. With the large dynamic range of protein expression in the heart (2), subsets need to be separated for analysis of lower abundant subproteomes that may offer novel differences between healthy and diseased cardiac tissue. Using differential centrifugation and ultracentrifugation, the cardiac proteome can be subdivided into cytosolic, mitochondrial, and microsomal fractions and analyzed separately on a tandem mass spectrometer. This approach reduces the complexity of each analyzed sample, allowing for the better detection of lower abundant proteins (3–5).

Fernando Vivanco (ed.), *Heart Proteomics: Methods and Protocols*, Methods in Molecular Biology, vol. 1005,
DOI 10.1007/978-1-62703-386-2_1, © Springer Science+Business Media New York 2013

The method outlined combines the approaches used in two of our large-scale proteomic and bioinformatic studies (1, 6). Multidimensional protein identification technology (MuDPIT) allows for better resolution of peptide separation compared to traditional two-dimensional gel separation (7). With the vast amount of data produced during proteomic analyses, comprehensive bioinformatic and network analysis (see Chapter 5 by Isserlin et al.) allows for efficient interpretation of protein identification that can translate to improved understanding of the cardiac proteome.

2 Materials

2.1 Ventricular Subfractionation

1. Cardiac lysis buffer: 250 mM sucrose, 50 mM Tris–HCl, pH 7.4, 5 mM $MgCl_2$, 1 mM dithiothreitol (DTT), and 1 mM phenylsulfonyl fluoride (PMSF). DTT and PMSF should be prepared fresh.
2. Sucrose cushion I: 0.9 M sucrose, 50 mM Tris–HCl, pH 7.4, 5 mM $MgCl_2$, 1 mM DTT, and 1 mM PMSF.
3. Sucrose cushion II: 2 M sucrose, 50 mM Tris–HCl, pH 7.4, 5 mM $MgCl_2$, 1 mM DTT, and 1 mM PMSF.
4. Nuclear extraction buffer I: 20 mM HEPES, pH 7.8, 1.5 mM $MgCl_2$, 450 mM NaCl, 0.2 mM EDTA, and 25 % glycerol.
5. Nuclear extraction buffer II: 1 % Triton X-100 added to nuclear extraction buffer I.
6. Mitochondrial extraction buffer I: 10 mM HEPES, pH 7.8.
7. Mitochondrial extraction buffer II: 1.5 % Triton X-100 added to mitochondrial extraction buffer I.
8. Beckman Ultraclear centrifuge tubes for SW40.1 rotor (cat. no. 344060).

2.2 Protein Precipitation and Digestion of Cardiac Proteins

1. 100–150 μg of protein in aqueous or detergent solution.
2. Acetone at −20 °C.
3. 8 M urea, 50 mM Tris–HCl, pH 8.5, 1 mM $CaCl_2$.
4. 50 mM ammonium bicarbonate.
5. Mass spectrometry-grade Trypsin Gold (Promega; Cat. No. V528A).

2.3 MuDPIT Analysis

1. MacroSpin reverse phase columns (The Nest Group; SMM SS18V).
2. 75- and 150-μm inner diameter capillary columns.
3. Magic C-18 reverse phase (RP) resin (Michrom Bioresources, Auburn, CA).

4. Luna strong cation exchange (SCX) resin (Phenomenex, Torrance, CA).
5. HPLC buffers: A—0.1 % formic acid in HPLC grade water, B—100 % acetonitrile with 0.1 % formic acid.
6. Salt buffers: C—250 mM ammonium acetate in buffer A, D—500 mM ammonium acetate in buffer A.

2.4 Bioinformatics

1. Cluster 3.0 software (http://rana.lbl.gov).
2. SEQUEST database search software (Thermo Fisher).
3. STATQUEST (in-house developed).
4. Swiss-Prot annotation (http://www.expasy.org/sprot).
5. Gene Ontology (GO) database (http://www.geneontology.org).
6. GoMiner (http://discover.nci.nih.gov/gominer).
7. TreeView (http://rana.lbl.govdownloads/TreeView).
8. Affymetrix GeneChip Mouse Genome 430 2.0 Array.
9. GEO DataSets (http://www.ncbi.nlm.nih.gov/gds).

3 Methods

3.1 Ventricular Subfractionation

1. Healthy adult mice are euthanized by CO_2 and cervical dislocation. Extracted hearts are rinsed in ice-cold PBS and removed of atria. The hearts are washed three times in ice-cold PBS and minced finely using a razor blade. Minced tissue is homogenized using a loose-fitting dounce homogenizer in cardiac lysis buffer. All following steps are done at 4 °C. The lysate is spun at 800 × *g* for 15 min. The supernatant is kept for later subfractionation. The pellet containing nuclei and contractile proteins is resuspended in cardiac lysis buffer, overlaid atop sucrose cushion I, and centrifuged at 800 × *g* for 15 min. The resulting pellet is resuspended in 8 ml of sucrose cushion I, overlaid atop 4 ml of sucrose cushion II in an Ultraclear centrifuge tube, and ultracentrifuged at 150,000 × *g* for 60 min. The nuclear pellet is washed in PBS and resuspended in nuclear extraction buffer I on ice for 15 min before pelleting at 8,000 × *g* for 20 min. The supernatant is collected as ***nuclear extract I***. The pellet is resuspended in nuclear extraction buffer II and kept on ice for 30 min. The supernatant after centrifuging for 20 min at 8,000 × *g* is taken as the ***nuclear extract II***. During sucrose cushion centrifugations, the interface formed will contain ***contractile proteins*** than can be collected after two washes in PBS, spinning at 14,000 × *g* for 10 min, and resuspended in mitochondrial extraction buffer II.
2. Centrifuging the initial 800 × *g* supernatant at 8,000 × *g* for 20 min pelleted the mitochondria. After resuspension and incubation on ice for 30 min in mitochondrial extraction

buffer I, the mixture is sonicated at maximal settings with brief pulses. Samples centrifuged at 8,000 × *g* for 20 min produced *mitochondria extract I* as the supernatant. The pellet is incubated for 30 min in mitochondrial extraction buffer II and centrifuged at 8,000 × *g* for 20 min. The supernatant is collected as the *mitochondrial extract II.*

3. Ultracentrifugation of the supernatant obtained after pelleting the mitochondria at 100,000 × *g* for 60 min on a SW40.1 rotor pelleted the *microsomal fraction* (see Note 1). The supernatant is taken as the *cytosolic fraction*. The microsomal pellet is incubated on ice for 30 min with mitochondrial extraction buffer II and centrifuged at 8,000 × *g* for 30 min.

3.2 Protein Digestion for MuDPIT Analysis

1. One hundred micrograms of protein from each fraction are precipitated overnight in five volumes of acetone at −20 °C. The precipitate is pelleted at 21,000 × *g* for 20 min (see Note 2).
2. Precipitated proteins are denatured and solubilized in 8 M urea, 50 mM Tris–HCl, pH 8.5, which is then diluted to 1.5 M with 50 mM ammonium bicarbonate. $CaCl_2$ is added to a final concentration of 1 mM.
3. Denatured proteins are rotated overnight in Promega Trypsin Gold at 37 °C.
4. The tryptic peptides are solid phase extracted on a MacroSpin column according to manufacturer's guidelines, speed-vac concentrated to ~60 μl, and acidified with 5 μl of formic acid. Samples can be stored at −20 °C until use.

3.3 MuDPIT Analysis

1. Samples are analyzed on an automated 12-step, 20-h MuDPIT analysis using an inline HPLC interfaced with an LTQ linear ion trap mass spectrometer (Thermo, San Jose, CA; see Note 3).
2. 100-μm fused silica capillary columns are pulled to a fine tip using a P-2000 laser puller. The column is then packed using a pressure cell with Magic C-18 RP resin then Luna SCX resin.
3. Peptides are loaded manually with a pressure cell onto the column (see Note 4).
4. The initial 80-min step of the MuDPIT was a 70-min gradient from 0 to 80 % buffer B followed by a 10-min hold at 80 % buffer B. The next 11 steps were 110 min long each beginning with a 5-min hold of 100 % buffer A. This was followed by a 2-min hold of *x* % buffer C/D, 3-min hold of 100 % buffer A, 10-min gradient of 0–10 % buffer B, and a final 90-min gradient of 10–45 % buffer B. For the percent of buffer C/D, *x* denotes the increase salt concentration for each step. Step 2 used 10 % buffer C and increases by 10 % for each subsequent step to 100 % at step 11. Step 12 uses 100 % buffer D.

3.4 Database Searching and Protein Sequencing

Obtained tandem mass spectra are analyzed using SEQUEST software algorithm to map spectra onto a set of mouse protein sequences from Swiss-Prot databases (8). The SEQUEST output is validated using STATQUEST software (9). STATQUEST determines the *p*-value cutoff to corrected peptide identification. Typical correct identification is set to 95 % likelihood to reduce false positives (see Note 5). Identified proteins reaching the cutoff are parsed into SQL-based database management system. The use of a database system can allow for efficient use of data, by allowing for queries and integration of protein parameters, such as the physical properties of molecular weight and isoelectric point and experimental data such as number of identified peptides and recorded spectra.

3.5 Bioinformatics

Proteins identified by two or more high-confidence peptides across technical runs are included in the data analysis. Proteins were assigned to the subfraction (cytosol, microsome, or mitochondria) where it contained the highest spectral count. This would denote its enrichment in that fraction. Proteins were assigned to a mixed category when the highest spectral count was present in at least two subfractions.

3.5.1 Hierarchical Clustering

Hierarchical clustering is a method of determining the similarity and dissimilarity of proteins within a particular dataset. Cluster 3.0 is a program that produces the cluster analysis of proteomic and genomic data (10). Datasets are converted to tab-delimited text file and divided by column by their subcellular fraction and by row by the proteins identified. Quantitative protein expression data must be used to determine the similarity or dissimilarity of the fractions. Protein abundance of each protein is semi-quantified using spectral counts and protein absence in a fraction is denoted by a very low nonzero value (e.g., 0.01) for cluster analysis. Using the TreeView software (11), heat maps can be generated to produce graphical representation of the cluster analysis (Fig. 1).

3.5.2 Protein Annotation

Protein annotation is the process of applying the biological and molecular properties to the identified protein. The web resource ExPASy (http://ca.expasy.org) provides access to different annotation tools such as the Mouse Genome Database (http://www.informatics.jax.org/) and Gene Ontology (GO (12)). The GO database provides protein characterization according to their biological process, molecular function, and cellular component. Protein clustering of datasets by GO terms can be accomplished using GoMiner (http://discover.nci.nih.gov/gominer/index.jsp) or Perl-based software GOClust (http://compbio.di.unito.it/tools/GOClust/index.html). Either program annotates protein accessions with GO terms and clusters proteins based on matching GO terms. The program can identify which GO terms are enriched in each fraction. These comparisons can also be translated across disease states if desired rather than across subfractions. These types

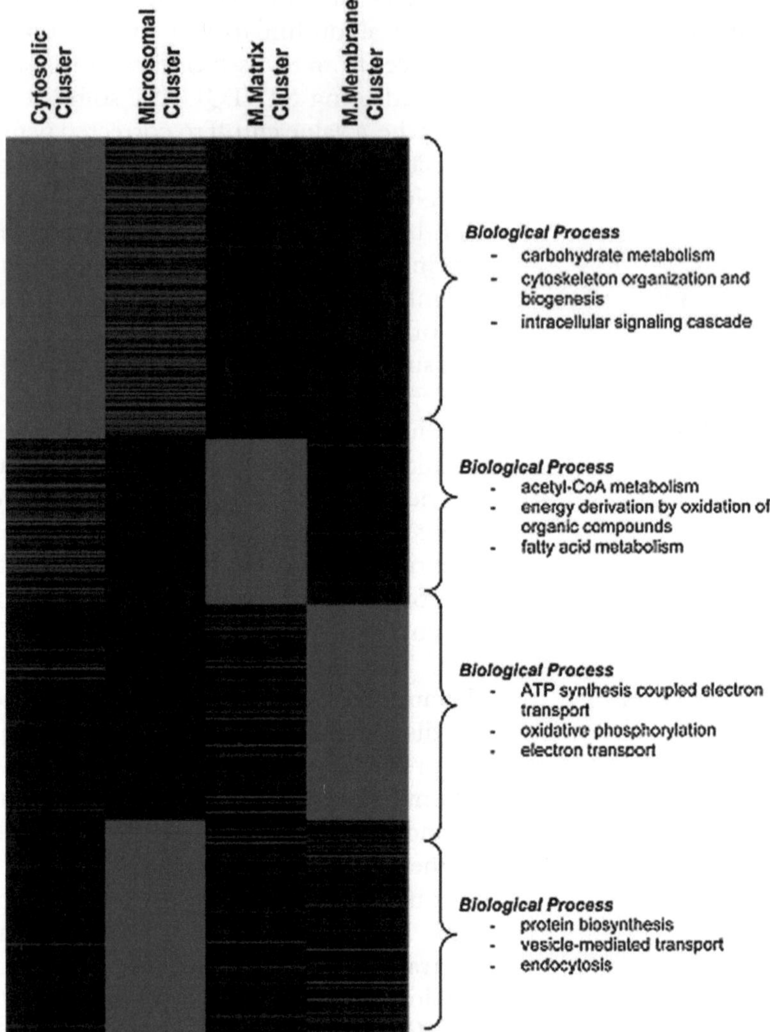

Fig. 1 Hierarchical clustering of the heart proteome divided by the compartments obtained by ventricular subfractionation. Semiquantitative measures of spectral counts provide the metric for cluster analysis. To the right shows each compartment's enriched GO terms. Enriched GO terms reflect the biological processes associated with the compartments that were intended to be subfractionated (Reprinted (adapted) with permission from Bousette et al. (6). Copyright 2009 American Chemical Society)

of analyses provide insight on possible affect biological processes that are affected by the disease studied. GO Term analysis provides a qualitative verifier of the subcellular enrichment of each fraction. For example, cytosolic proteins should show significant enrichment of GO terms such as cytosol and carbohydrate metabolism. Microsomal proteins should show GO term enrichment in endoplasmic reticulum and protein biosynthesis. Mitochondrial proteins should show GO term enrichment in oxidoreductase activity and mitochondrial matrix (Fig. 1).

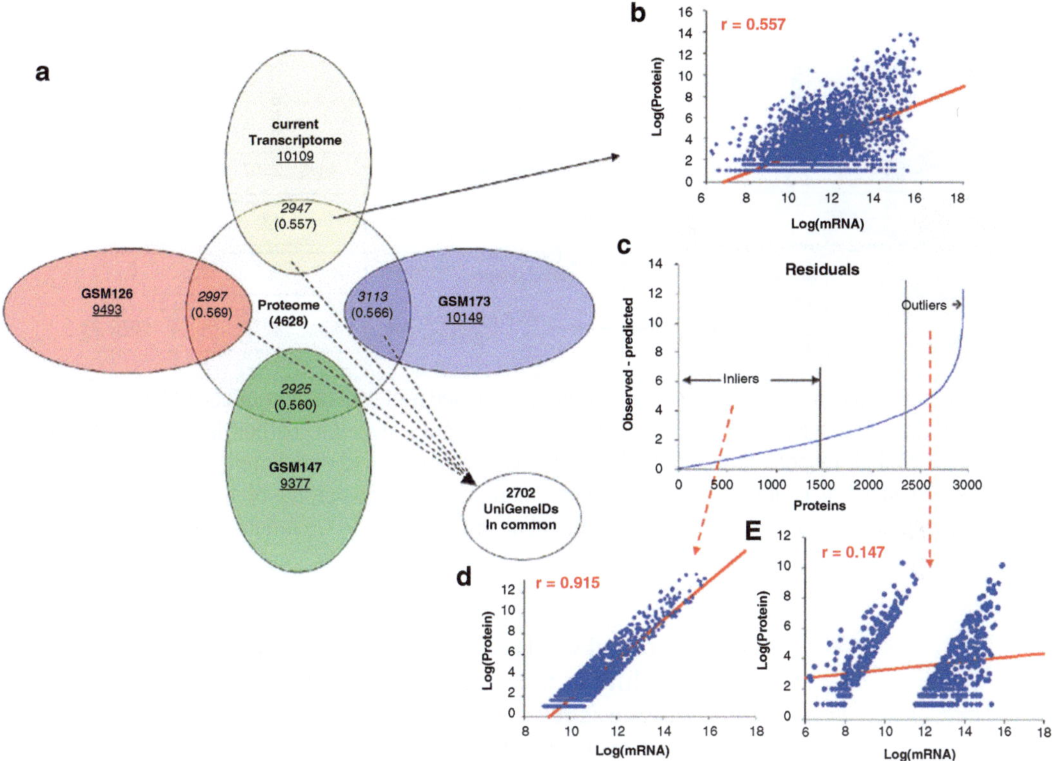

Fig. 2 Proteome and transcriptome comparisons. (**a**) Venn diagram representing common accessions between obtained proteome and four different microarrays. (**b**) Correlation analysis with current transcriptome and its respective proteome. (**c**) Residual plot showing identified outliers and inliers following regression analysis and (**d**) correlation analysis of the separated inliers and (**e**) outliers (Reprinted (adapted) with permission from Bousette et al. (6). Copyright 2009 American Chemical Society)

3.5.3 Microarray Analysis

Complementing the protein identification data together with a comparison of its mRNA expression data can provide insight on the role of posttranscription and posttranslational modifications that together play a role in protein expression. Microarray global mRNA analysis provides a robust tool to correlate the transcriptome with the proteome (Fig. 2). It is possible to identify proteins via its mRNA that may be absent in its proteome. This may be due in part to posttranscriptional and posttranslational factors. Use of the robust multichip analysis (RMA) method with Affymetrix full genome arrays provides a source of the entire murine genome. Significant probe intensities from biological replicates are identified by UniGene accessions and be directly compared to proteome data. Redundant probe averages are used for intensity values (see Note 6).

3.6 Quantitative Approaches

With spectral counts providing a semiquantitative measure of protein expression, other proteomic approaches should be considered if differential expression is to be examined between samples

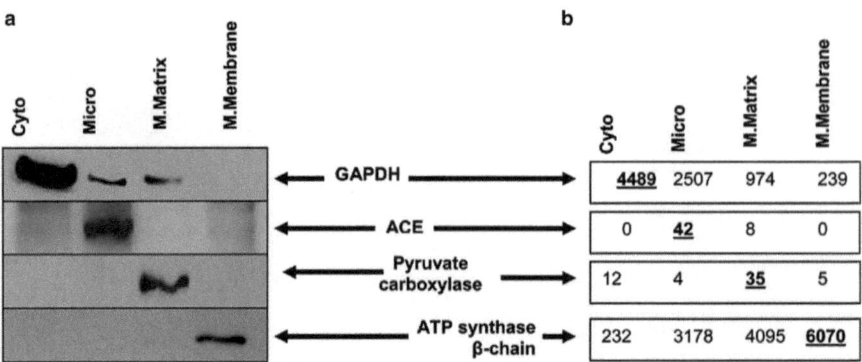

Fig. 3 Western blot validation of identified compartment-enriched proteins. (**a**) Immunoblots for glyceraldehyde 3-phosphate dehydrogenase (GAPDH), acetylcholinesterase (ACE), pyruvate carboxylase, and ATP synthase (β-chain) and (**b**) their respective spectral counts. Proteins with identified enrichment via proteomic analyses determined via spectral count show similar enrichment via immunoblotting of the subcellular fractions (Reprinted (adapted) with permission from Bousette et al. (6). Copyright 2009 American Chemical Society)

(such as normal vs. diseased). Quantitative proteomics can be relative or absolute in their quantification. Relative quantification usually employs a labeling of the peptide with a reporting ion that is fragmented at different m/z ratios. In isobaric tag for relative and absolute quantitation (iTRAQ (13)), different samples are labeled with similar molecular tags, and the reporter ions are fragmented differentially. In this analysis, the samples are pooled together, and the reporter ions provide a relative quantification of the peptides or proteins between the pooled samples.

Absolute quantitation can be done using selected reaction monitoring (SRM (14)). With a candidate protein, one must determine an ideal proteotypic peptide that is detectable by a triple quadrupole mass spectrometer. The three stages of the mass spectrometers have specialized roles. Q_1 selects for the precursor ion of the peptide of interest; Q_2 is the collision cell in which the precursor ion is fragmented into the fragment ion, which is selected in Q_3 for detection. SRM analysis spikes the sample to be analyzed with a heavy-isotope-labeled peptide standard of a known concentration to provide reference to the experimental peptide to be analyzed.

3.7 Verification

Traditional biochemical strategies complement large-scale proteomic studies that have inherent risks of false positives. Though this is minimized via stringent filtering criteria, Western blots (Fig. 3) will validate bioinformatic analyses. mRNA analysis by RT-PCR can also be used if suitable antibodies are unavailable.

3.8 Conclusion

MuDPIT analysis using tandem mass spectrometry with subfractionation of tissue will provide better detection of proteins.

Subfractionation of cardiac tissue can further be divided to other compartments such as membrane and secreted proteomes. Expanding proteomic analysis would provide insight on future therapeutic strategies regarding heart diseases. With focus on comparing healthy and diseased tissues, complementing discovery proteomics with a targeted approach will offer more insight and quantitation of protein expression profiles. Use of subfractionation in mitochondrial isolation can facilitate functional proteomics on this organelle. Stressing mitochondria can produce a different proteome (4) that may enhance healthy and pathological comparisons. In characterizing the murine cardiac proteome, we can produce datasets differential expressed proteins that may be candidates for biomarkers of modeled heart diseases. With an emphasis on translational medicine, if these candidates are reproducible in human cohorts and detectable in less invasive fluids (e.g., serum or urine) would produce an ideal biomarker for heart disease.

4 Notes

1. If a different rotor is used to ultracentrifuge microsomes, equivalent run times and speeds must be used to ensure similar pellet efficiency (k-factor) of the ultracentrifugation. Conversion calculators are available online for Beckman-Coulter rotors (https://www.beckmancoulter.com/).
2. The aim of acetone precipitation is to isolate the protein from incompatible detergents (such as TX-100) used to permeabilize and solubilize membranes in the sample. Acetone precipitation will result in a reduced yield that may cause loss of sample, a concern if the initial sample yield is low. The use of cleavable surfactants, such as PPS Silent® Surfactant which is cleaved in acidic conditions, is compatible with proteomic analysis and can replace TX-100.
3. Different numbers of MuDPIT steps can be used to optimize workflow. If there are time considerations, fewer steps to run samples can be used resulting in shorter run times at the cost of reduced sensitivity. Using a 6-step MuDPIT in triplicate would provide comparable coverage while shortening duty cycles. In practice, triplicates of biological replicates reduce the error from random sampling and providing sufficient protein detection.
4. Automated sample loading using the inline HPLC can be done. Under these conditions, ensure the volume loaded is no more than 75 % of the attached sample loop's volume to avoid sample pump contamination.

5. Increasing the cutoff to a more stringent value (e.g., 99 %) will decrease the FDR, but increases the probability for false-negative errors.
6. Previously published datasets can be accessible through GEO datasets (http://www.ncbi.nlm.nih.gov/gds). If comparable sets exist, one should determine correlation with genomes of similar tissues and use unrelated tissues to determine if correlations are not platform dependent.

Acknowledgments

Work in our laboratory was supported by grants from the Heart and Stroke Foundation (HSF) of Ontario (#T-6281; AOG), Canadian Institutes of Health Research (CIHR) (A.O.G.; MOP-84267), and Canadian Foundation for Innovation to A.O.G. A.O.G. is a Canada Research Chair and A.E. is an Ontario Research Chair in Biomarkers in Disease Management.

References

1. Gramolini AO, Kislinger T, Alikhani-Koopaei R et al (2008) Comparative proteomics profiling of a phospholamban mutant mouse model of dilated cardiomyopathy reveals progressive intracellular stress responses. Mol Cell Proteomics 7:519–533
2. McGregor E, Dunn MJ (2006) Proteomics of the heart. Circ Res 3:309–321
3. Gramolini AO, Kislinger T, Liu P, et al (2007) Analyzing the cardiac muscle proteome by liquid chromatography-mass spectrometry-based expression proteomics. Meth Mol Biol (Clifton, NJ) 357:15–31
4. Zhang J, Liem DA, Mueller M et al (2008) Altered proteome biology of cardiac mitochondria under stress conditions. J Proteome Res 7:2204–2214
5. Barallobre-Barreiro J, Didangelos A et al (2012) Proteomics analysis of cardiac extracellular matrix remodeling in a porcine model of ischemia/reperfusion injury. Circulation 125:789–802
6. Bousette N, Kislinger T, Fong V et al (2009) Large-scale characterization and analysis of the murine cardiac proteome. J Proteome Res 8:1887–1901
7. Washburn MP, Wolters D, Yates JR (2001) Large-scale analysis of the yeast proteome by multidimensional protein identification technology. Nat Biotechnol 19:242–247
8. Eng JK, McCormack AL, Yates JR (1994) An approach to correlate tandem mass-spectral data of peptides with amino-acid-sequences in a protein database. J Am Soc Mass Spectrom 5:976–989
9. Kislinger T, Emili A (2003) Going global: protein expression profiling using shotgun mass spectrometry. Curr Opin Mol Ther 5:285–293
10. Eisen MB, Spellman PT, Brown PO et al (1998) Cluster analysis and display of genome-wide expression patterns. Proc Natl Acad Sci USA 95:14863–14868
11. Saldanha AJ (2004) Java Treeview-extensible visualization of microarray data. Bioinformatics 20:3246–3248
12. Ashburner M, Ball CA, Blake JA et al (2000) Gene ontology: tool for the unification of biology. Nat Genet 25:25–29
13. Zieske LR (2006) A perspective on the use of iTRAQ (TM) reagent technology for protein complex and profiling studies. J Exp Bot 57:1501–1508
14. Elschenbroich S, Kislinger T (2010) Targeted proteomics by selected reaction monitoring mass spectrometry: applications to systems biology and biomarker discovery. Mol Biosyst 7:292–303

Chapter 2

Determining Protein Concentrations of the Human Ventricular Proteome

Arjen Scholten and Albert J.R. Heck

Abstract

Proteomics is mostly used for measurements of relative differences in protein concentrations. Although such analyses are meaningful for comparing differences between two and more conditions, they do not directly provide details on the absolute protein concentrations within a system. Now, proteomics is heading more towards absolute quantitative strategies with results being expressed in copies/cell or ng/mg tissue. In the cardiac context, such quantitative information is crucial for (1) evaluating the feasibility of selecting a certain protein as potential novel drug target, (2) the expected concentration excreted into the circulation when selecting a biomarker, and (3) to build a model of cardiac function at the molecular level. At the same time, by mass spectrometry-based proteomics, a wealth of spectral information is gathered that can be used to evaluate protein levels of a select set of novel disease-altered proteins using, for instance, single reaction monitoring. Here we describe how to build a quantitative map of the human left ventricular proteome using a simple yet effective mass spectrometry-based spectral count method.

Key words Absolute quantitation, Spectral count, Human left ventricle

1 Introduction

With the introduction of stable isotope labels, the large-scale comparison of relative differences in protein abundance has become routine (1). Lately, more and more studies use shotgun mass spectrometry data to evaluate absolute protein concentrations within a biological system in copies/cell or ng/mg (2–7). Further validation has shown that these methods correlate very well with other quantitative measures of absolute abundance, such as single reaction monitoring (6, 8), but also GFP Western blot (9) and mRNA-based readouts (5).

When performing cardiac proteomics in the quest for novel biomarkers, novel therapeutic targets or disease signatures, a quantitative library of the cardiac proteome is a useful resource to serve as reference set. To start this effort, we have recently mapped the complement of the human left ventricular proteome (2) using

Fernando Vivanco (ed.), *Heart Proteomics: Methods and Protocols*, Methods in Molecular Biology, vol. 1005, DOI 10.1007/978-1-62703-386-2_2,

the aforementioned mass spectrometry-based spectral count technology. Like the plasma proteome, the cardiac proteome has a challenging dynamic range. This means that a small set of very abundant proteins obscure more low abundant proteins. To increase depth, complementary sample preparation strategies can be used, for instance, using different proteases (10, 11), analysis platforms (e.g., gel-based protein separation, ion exchange-based peptide separation), and the use of multiple ion fragmentation techniques (12). Here we provide a detailed methodological protocol to build a quantitative proteome map of the human left ventricle based on a multifaceted proteomics approach. This methodology is easily transferable to other tissues and samples.

2 Materials

2.1 Tissue Lysis

1. Protease inhibitor cocktail (Complete mini, Roche Diagnostics).
2. Phosphatase inhibitor cocktail (type 1 and 2, Sigma).
3. Phosphate buffered saline (PBS): Prepare 10(×) stock with 1.37 M NaCl, 27 mM KCl, 100 mM Na_2HPO_4, and 18 mM KH_2PO_4, and adjust to pH 7.4 with HCl if necessary. Prepare a working solution by dilution of one part with nine parts Milli-Q water.
4. Lysis buffer: PBS buffer supplemented with 0.1 % v/v Tween 20, protease inhibitor cocktail (one tablet per 15 ml buffer), and 150 μl of phosphatase inhibitor cocktail II (Sigma) (see Notes 1 and 2).

2.2 SDS-PAGE Protein Separation and In-Gel Trypsin Digestion

1. 4–15 % and 10–20 % SDS-PAGE gradient gels (Bio-Rad), standard running buffer, SDS-PAGE loading buffer, and GelCode Coomassie staining solution (Thermo Scientific).
2. Dithiothreitol (DTT, 6.5 mM in 50 mM ABC) and iodoacetamide (54 mM in 50 mM ABC) solution are prepared fresh.
3. Trypsin (Roche Diagnostics), 0.1 μg/μl in 50 mM acetic acid. Stored in single-use aliquots of 10 μl at −80 °C.
4. 50 mM ammonium bicarbonate (ABC) in Milli-Q water and analysis grade acetonitrile.

2.3 In-Solution Digestion with Different Proteases

1. 50 mM ABC in Milli-Q water.
2. 8 M urea in 50 mM ABC.
3. Trypsin, chymotrypsin, and Lys-C (Roche Diagnostics) and Lys-N (from Grifola Frondosa, Seikagaku Corp.) are all dissolved at 0.1 μg/μl in 50 mM acetic acid and stored in single-use aliquots (10 μl) at −80 °C.
4. DTT and iodoacetamide solutions as described above.

2.4 Strong Cation Exchange Chromatography

1. Two Zorbax Bio-SCX Series II columns [0.8 mm (i.d.) × 50 mm, 3.5 μm material, Agilent Technologies].
2. A suitable HPLC, e.g., consisting of FAMOS autosampler (Dionex), Shimadzu LC-9A binary pump, and an SPD-6A UV detector.
3. Strong cation exchange (SCX) solvent A: 20 % acetonitrile, 0.05 % formic acid, pH 3.0 in Milli-Q water.
4. SCX solvent B: 500 mM KCl in 20 % acetonitrile and 0.05 % formic acid, pH 3.0 in Milli-Q water.

2.5 LC-MS/MS Analysis

1. Suitable high-resolution mass spectrometer, e.g., LTQ Orbitrap XL equipped with an electrospray ion source and the option to use electron transfer dissociation (ETD) as an additional fragmentation technique. The mass spectrometer is coupled online to a nanoflow HPLC system, e.g., Agilent 1200 series (see Note 3).
2. Trapping column, 20 mm, 100 μm i.d. packed with Aqua C18 reversed phase material (5 μm, Phenomenex).
3. Separation column, 400 mm, 50 μm i.d. packed with Reprosil-Pur C18-AQ reversed phase material (3 μm, Dr. Maisch).
4. Distally coated fused-silica emitter [360 μm (i.d.); 20 μm (i.d.); tip inner diameter, 10 μm, New Objective].
5. HPLC solvent A: 0.1 M acetic acid in Milli-Q water.
6. HPLC solvent B: 0.1 M acetic acid in 80 % acetonitrile in Milli-Q water.

2.6 Data Analysis

1. BioWorks (Thermo Electron) is used to extract mass spectrometric raw data into a searchable format.
2. Proteome Discoverer (Thermo Electron) is used to extract ETD generated data from the raw data files (see Note 4).
3. MASCOT (Matrix Science) is used as search engine for protein identification.
4. Scaffold (Proteome Software) is used to reduce protein redundancy and export FDR filtered datasets (see Note 5).
5. Microsoft Excel is used for combining datasets and calculating the protein abundance.

3 Methods

3.1 Data Gathering

A deep proteome analysis of the human left ventricle requires a multifaceted approach as the dynamic range of the cardiac proteome is challenging due to the presence of a few very abundant proteins. Here we describe how to use complementary chromatographic, mass spectrometric, and sample preparation

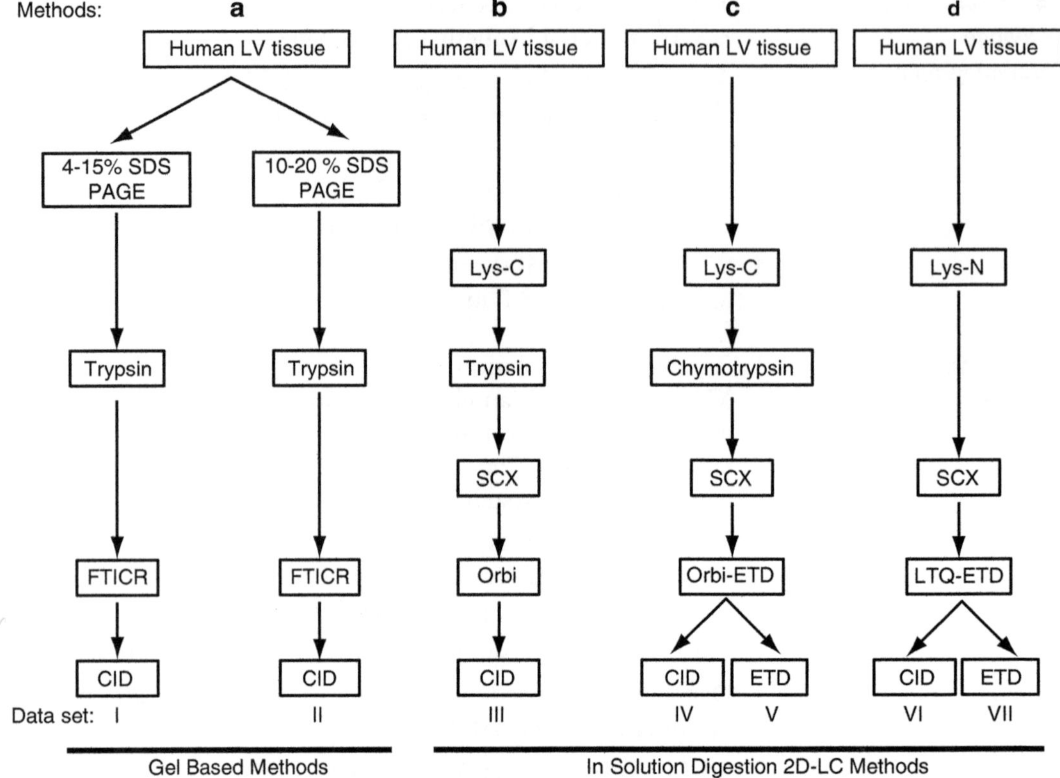

Fig. 1 Multifaceted proteomics approach. The use of four main approaches (Methods A–D) leads to the acquisition of seven LC-MS/MS datasets (I–VII). Method A involves SDS-PAGE separation of the protein extract on two different gradient gels, followed by in-gel trypsin digestion with trypsin and analysis on an LTQ-FTICR mass spectrometer using CID fragmentation. Methods B, C, and D use different digestion regimes with Lys-C and trypsin (B), Lys-C and chymotrypsin (C), and Lys-N (D) followed by two-dimensional peptide separation using SCX in the first and reversed phase in the second dimension. Peptides of Methods C and D are analyzed using both CID and ETD fragmentation (Adapted from Aye et al. (2))

strategies to accomplish this. Both gel-based protein separation and peptide-based strong cation exchange protocols are used to increase depth of coverage. Furthermore, we utilize different digestion protocols involving four different proteases (trypsin, Lys-N, and chymotrypsin) as well as two different peptide fragmentation techniques [collision-induced dissociation (CID) and electron transfer dissociation (ETD)]. The outline of our approach which involves four different methods (A–D) is depicted in Fig. 1.

3.2 Data Mining

To combine all data gathered in the different workflows into the four major (large) datasets and ultimately in one large dataset requires several steps that are outlined below. In addition, for a human cardiac catalogue to be meaningful, a sense of protein concentration would be a valuable addition. Here we describe in detail a simple spectral count-based approach to achieve this. Such information is invaluable when interpreting the relevance/ expression levels of for instance all members of the protein class of kinases within the human cardiac proteome. In such a way, novel insights can be gained towards the next kinase to target for functional verification within the cardiac context.

3.3 Tissue Lysis

1. Prechill a steel mortar and pestle in liquid nitrogen.
2. Human left ventricular tissue of a male donor with no prior cardiovascular disease (~1 cm^3) is taken from −80 °C and further frozen in liquid nitrogen.
3. Grind the tissue in the mortar and transfer the pulverized tissue with a cold spoon into an Eppendorf tube.
4. Add 1 ml of ice-cold lysis buffer and leave to lyse at room temperature for 5 min and another 10 min on ice.
5. Centrifuge the lysate at 20,000×g and 4 °C. Transfer the supernatant to a cold falcon tube. Extract the insoluble pallet twice more using steps 4 and 5 of Subheading 3.1.
6. Measure the protein concentration of the (combined) supernatant(s).

3.4 In-Gel Protein Separation and Digestion (Method A)

1. Run 50 μg of the heart lysate on two different SDS-PAGE gradient gels (4–15 and 10–20 %, Bio-Rad). Fix and stain with GelCode Coomassie Blue staining. Wash in Milli-Q water.
2. Slice the gel in ~70 gel pieces using a Mickle gel slicer (see Note 6).
3. Gel slices are washed (ABC and acetonitrile), reduced (6.5 mM DTT, 30 min, 56 °C), washed (ABC and acetonitrile), alkylated (54 mM iodoacetamide in the dark, 30 min), and washed (ABC, acetonitrile).
4. Incubate the gel slices with trypsin (0.1 μg per slice) overnight at 37 °C and collect the supernatant and two washes with 5 % formic acid solution.
5. Dry the samples in vacuo.

3.5 In-Solution Digestion (Methods B, C, and D)

1. Dilute 200 μg tissue lysate into 200 μl ABC supplemented with 96 mg urea (8 M final concentration).
2. Reduce (2 mM DTT, 15 min 56 °C) and alkylate (4 mM iodoacetamide, 30 min. room temperature, dark) all cysteine residues.
3. Method B: Digest with Lys-C (4 μg) for 4 h at 37 °C and then dilute the entire solution fourfold (2 M urea). Digest with trypsin (4 μg) overnight at 37 °C. Desalt the sample over C18 material and dry in vacuo. Reconstitute in SCX solvent A.

4. Method C: Perform lysate dilution, reduction, alkylation and Lys-C digestion and dilution to 2 M urea as described in Subheading 3.3, steps 1–3. Then digest the Lys-C digest further with chymotrypsin (4 μg) overnight followed by C18 desalting, in vacuo drying, and reconstitution into SCX solvent A.
5. Method D: Repeat steps 1 and 2 of Subheading 3.3. Add Lys-N in a 1:85 w/w ratio (so for 200 μg total lysate, this is 2.35 μg Lys-N). Digest overnight at 37 °C. Desalt, dry in vacuo, and reconstitute in SCX solvent A.

3.6 Strong Cation Exchange Fractionation (Methods B, C, and D)

1. These instructions assume the use of Zorbax Bio-SCX Series II columns [0.8 mm (i.d.) × 50 mm (l), 3.5 μm]. SCX is performed using two in-line coupled Zorbax Bio-SCX Series II columns, FAMOS autosampler, Shimadzu LC-9A binary pump, and SPD-6A UV detector. SCX is performed at pH = 3 (see Note 7).
2. Inject the equivalent of 200 μg protein lysate by FAMOS autosampler.
3. Load the injected sample to the column with SCX solvent A at a flow rate of 100 μl/min for 5 min.
4. Elute with a 1 %/min linear gradient of SCX solvent B at a flow rate of 50 μl/min for 45 min. Then equilibrate the column with SCX solvent A for 10 min.
5. A total of 50 SCX fractions (1 min each, 50 μl elution volume) are collected by using a suitable fraction collector.
6. Dry all fractions in vacuo and store at −30 °C until LC-MS/MS analysis.

3.7 LC-MS/MS Analysis

1. These instructions assume the use of a nanoflow liquid chromatography setup, directly coupled to, for instance, a LTQ Orbitrap XL mass spectrometer equipped with an electrospray ion source for MS analysis and an ETD source for alternative fragmentation.
2. Resuspend the dried samples from step 6 of Subheading 3.4 in 40 μl 10 % formic acid.
3. Inject 10 μl of the resuspended peptide mixtures individually onto an Agilent 1200 series LC system, equipped with a 20 mm Aqua C18 trapping column [100 μm (i.d.), packed in-house] and a 400 nm Reprosil-Pur C18-AQ analytical column [50 μm (i.d.), packed in-house].
4. Trapping is performed at 5 μl/min for 10 min in HPLC solvent A, and elution is achieved with a gradient of 10–35 % HPLC solvent B in 45 min in a total analysis time of 60 min. The flow rate is passively split to ~100 nl/min for peptide

separation. Nanospray is achieved using a distally coated fused-silica emitter.

5. The LTQ Orbitrap XL mass spectrometer is operated in a data-dependent mode, automatically switching between MS and MS/MS. Full-scan MS spectra (350–1,500 m/z) are acquired in the FT Orbitrap with a resolution of $R = 60{,}000$ at 350 m/z after accumulation to a target value of 500,000 ions in the linear trap. Parent ions were isolated for a more accurate measurement by performing a single-ion monitoring scan and fragmented by CID and ETD in data-dependent mode (two most intense ions, minimum intensity of 500). Ions were fragmented using CID with normalized collision energy of 35 and an activation time of 30 ms. ETD fragmentation was performed with supplemental activation. Fluoranthene was used as reagent anion, and ion/ion reaction in the linear ion trap was taking place for 100 ms.

3.8 Data Analysis and Combination of Raw Files

1. Process obtained .RAW files from each LC-MS/MS run with BioWorks. Combine the data of the individual LC-MS/MS runs into a single MASCOT generic file (.mgf) per method (see Note 8).
2. Perform MASCOT searching on the .mgf file against an appropriate database using the following criteria: Carbamidomethylation on cysteine residues as a fixed modification and methionine oxidation, serine/threonine/tyrosine phosphorylation, and N-terminal acetylation (proteins) as variable modifications. Allow two missed cleavages and a peptide mass tolerance of 10 ppm and an MS/MS fragment mass tolerance of 0.9 Da.
3. Filter the resulting data to a false discovery rate (FDR) of <1 % per method dataset to determine the cutoff threshold MASCOT score using standard procedures (see Note 9).
4. Load each.dat file into the Scaffold software package (Proteome Software) and apply the determined MASCOT score cutoff.
5. Export the required protein data (unique peptides, assigned spectra, sequence coverage, etc.) from Scaffold to Excel and combine all obtained datasets using the "vlookup" commands to build a comprehensive protein-oriented table (see Note 10) (Table 1). Further calculations can also be made in Microsoft Excel, e.g., total number of unique peptides and total number of observed spectra.

3.9 Quantitative Mapping of the Left Ventricular Proteome

1. The spectral counts of each complete gel lane or SCX separation (combine CID and ETD data when applicable; see Fig. 1) are turned into absolute quantitative information (ng/mg) with the F_{abb} factor. Calculate the F_{abb} of protein i by dividing

Table 1
Combining multiple datasets

UniProt	Protein description	MW (kDa)	Sum uniq. peps.	Number of identified spectra (per method)					Sum spectral counts	Average conc. (ng/mg)
				A-I	A-II	B	C	D		
P02144	Myoglobin	17,166	127	503	282	453	330	786	2,354	51,154.2
P68871	Hemoglobin subunit beta	15,980	93	335	164	338	380	139	1,356	26,334.6
P69905	Hemoglobin subunit alpha	15,239	69	161	115	323	380	143	1,122	23,677.1
P06732	Creatine kinase M-type	43,083	178	521	267	684	1,101	277	2,850	20,808.7
P05413	Fatty acid-binding protein, heart	14,840	87	205	142	270	370	9	996	19,862.4
P99999	Cytochrome c	11,731	63	82	113	106	12	180	493	16,856.1
P10916	Myosin regulatory light chain 2, ventricular/cardiac muscle isoform	18,771	92	173	156	154	201	109	793	14,428.1
P04406	Glyceraldehyde-3-phosphate dehydrogenase	36,035	120	263	58	504	519	159	1,503	12,342.3
P02768	Serum albumin	69,348	280	713	201	1,129	496	330	2,869	11,812.3
P07195	l-Lactate dehydrogenase B chain	36,620	113	341	59	299	295	178	1,172	9,698.1
P68400	Casein kinase II subunit alpha	45,126.5	6	2	0	3	0	1	6	34.6

Q13557	Calcium/calmodulin-dependent protein kinase type II delta chain	56,352.7	25	32	1	24	0	2	59	228.1
P17612	cAMP-dependent protein kinase catalytic subunit alpha	40,573.3	19	19	0	5	0	4	28	173.2
Q2M3C7	A-kinase anchor protein SPHKAP	186,439	2	0	0	3	0	0	3	3.2
Q8N3K9	Cardiomyopathy-associated protein 5	449,187	2	0	0	4	0	0	4	1.8
P54289	Voltage-dependent calcium channel subunit alpha-2/delta-	123,169	2	0	0	3	0	0	3	4.9

Protein-centric data presentation of the ten most abundant proteins found in the left ventricle sample (based on seven datasets, data aggregated for Methods C and D). Each dataset is individually searched (MASCOT) and filtered for FDR (RockerBox and Scaffold). Subsequently, each dataset is combined in excel using "vlookup" commands. Within this Excel table, the F_{abb} values and subsequently the average concentration in ng/mg can be calculated using the instructions of Subheading 3.7, steps 1–3. Also depicted are the examples within Fig. 2 in the lower abundance levels. MW = molecular weight, SUM Uniq. Peps. = the sum of the unique peptides observed over all runs, A-I = 4–15 % SDS-PAGE, A-II = 10–20 % SDS-PAGE (*see* Fig. 1), SUM spectral count = the sum of all observed spectra identified for that particular protein, Conc. = concentration

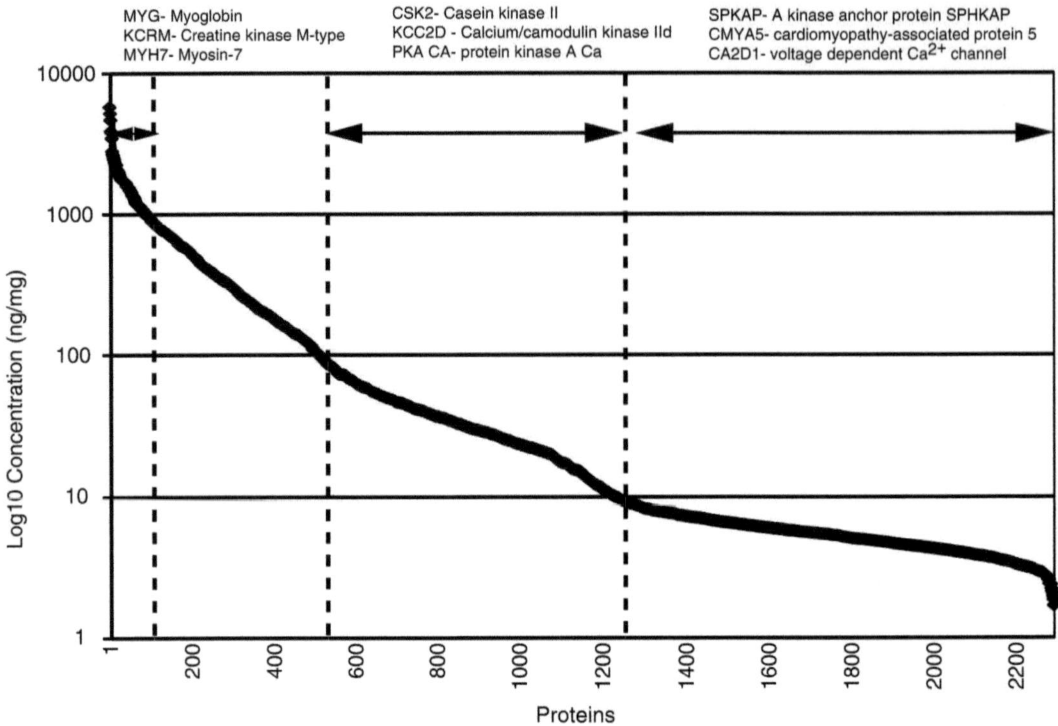

Fig. 2 Protein abundance map of the human left ventricular proteome (Adapted from Aye et al. (2)). The concentration calculation of the left ventricular proteome covers 2,277 proteins (at least two unique peptides) and four orders of magnitude. Depicted are representatives of three different abundance classes (top100, 500–1,250, and 1,250–2,277) to highlight that the first signaling proteins (kinases were chosen here) are at least 100-fold less abundant than the most abundant cardiac proteins. Examples of kinase signaling modulators and ion channels are residing mainly in the lowest part of the abundance map with copy numbers at least 1,000-fold below the most abundant proteins

the number of assigned spectra of protein i (SC_i) by its molecular weight (MW_i): $F_{abb,i} = SC_i/MW_i$.

2. Recalculate the F_{abb} values into absolute numbers by taking the proportion of the protein's F_{abb} towards the sum of all F_{abb} values in a particular dataset, $\sum(SC_k/MW_k)$. This proportion is then multiplied by a constant number, C, to indicate the total input per analysis [in this particular case, the equivalent of 25 μg total protein was injected (see Note 11)].

$$F_{abb,i} = \frac{SC_i / MW_i}{\sum (SC_k / MW_k)} \times C$$

3. Build a quantitative map by ranking the proteins based on their tissue/cellular abundance expressing the $F_{abb,i}$ in ng/mg (Fig. 2).

3.10 Conclusion

Here we describe a simple method to absolutely quantify the human left ventricular proteome. The basis for the quantitation relies on the use of spectral counts and the F_{abb}, a correction factor that normalizes the spectral counts of a protein to its molecular weight. This makes it a very simple yet effective approach as the only software tool required to calculate the protein concentrations is a spreadsheet software package. Although simple by nature, the F_{abb} factor correlates very well with other much more complex absolutely quantitative methods, such as APEX. The only prerequisite to make the F_{abb}-based absolute quantitation proteome-wide is to achieve a very in-depth analysis, i.e., aim for at least 100,000 high-quality spectra. As shown by us and others earlier, the use of multiple proteases, separation platforms, and peptide fragmentation techniques considerably adds protein coverage and peptide diversity (2, 11–18). This is however a requirement if we want to conquer the immense dynamic range challenge of the cardiac proteome to gain access to the interesting low abundant cardiac proteome with all its diversity in disease-relevant signaling proteins (see Note 12).

4 Notes

1. PBS buffer with 0.1 % Tween 20 can be prepared for multiple uses and stored at 4 °C. Buffer should be cold when used. Full lysis buffer should be prepared fresh only very prior to use to retain full efficiency of the protease and phosphatase inhibitors.
2. The described lysis buffer could be replaced by stronger buffers. For example, addition of 8 M urea to the lysis buffer increases lysis efficiency of more insoluble proteins. Sonication of the sample can also increase protein yield.
3. Alternatively, when one has access to an Orbitrap Velos, a decision tree using three fragmentation (CID, ETD, and HCD) methods could be used (12).
4. The Proteome Discoverer software package (v2.0 and up) is now also suitable to combine data extraction of different fragmentation regimes, which makes the use of BioWorks no longer required.
5. Proteome Discoverer (v2 and up) also allows data organization and multiple dataset comparisons. In addition, it has a built-in protein grouping algorithm to further reduce protein identification redundancy.
6. A Mickle gel slicer can cut small and consistently sized gel pieces of approximately 1 mm. Keep the gel wet with Milli-Q water to prevent sticking of gel pieces to the razor blade that is used to slice.

7. At pH = 3, all acidic residues (Asp and Glu) are uncharged, allowing all peptides to bind to the column.
8. Several software tools can be used here: MGFcombiner (MSQuant (19), RockerBox (20)), or Proteome Discoverer (Thermo). In addition, when using MSQuant, all individual files can be recalibrated for mass accuracy prior to combining the files, allowing a smaller mass cutoff when searching a more ideal false discovery rate filtering options (19).
9. Alternatively use RockerBox here to do the filtering automatically on your resulting MASCOT .dat file (20).
10. Combining datasets is not arbitrary as typical SCX-LC-MS/MS experiments tend to be extremely large and demand a lot of computing power. We use individual Scaffold outputs and combine them in Excel. Combining in Scaffold is an option but poses high demands in computing power. Another option is to combine different analysis files using Proteome Discoverer (v2.0 and up). Advantage of the latter is that the built-in protein grouping algorithm can be used to reduce protein redundancy automatically. Here computational power is also required. Combining 7 SCX-LC-MS/MS analyses typically generates file sizes easily exceeding 10 GB and requires a 32- or 64-core machine to be opened and manipulated. Combining at .dat file level and using protein grouping and annotation in MASCOT (v2.3 and up) is also an option. Browsing these data in the web-based MASCOT environment is however cumbersome when .dat file sizes approach 1 GB and up.
11. To ensure comparable results between different runs, experiments, and analysis platforms, load equal amounts of total protein onto the LC-MS/MS system. In this case, 50 μg protein was loaded per SDS-PAGE gel lane. Therefore, to use $C = 25$ μg, load 50 % of the peptide mixture obtained from each gel band onto the LC-MS/MS system (Method A). SCX separation is performed using the peptide mixture of the equivalent of 200 μg protein. Therefore, to retain quantitative information from each SCX fraction, 12.5 % should be injected.
12. Further curation of the dataset can be performed when assuming that tissue is always contaminated with blood, which means plasma and red blood cell proteins are also identified (hemoglobins, serum albumin, *see* Table 1). Before annotating a protein as present in the myocardium, careful cross-correlation with plasma and other circulating proteomes is advised. For this, published datasets, if they provide spectral counts or unique spectral counts, can also be quantitatively ranked using the F_{abb} factor, before comparison with the obtained proteome.

Acknowledgments

The authors would like to acknowledge the Netherlands Proteomics Centre embedded in the Netherlands Genomics Initiative for funding. This work was in part supported by the PRIME-XS project, Grant Agreement Number 262067, funded by the European Union Seventh Framework Program.

References

1. Bantscheff M, Schirle M, Sweetman G, Rick J, Kuster B (2007) Quantitative mass spectrometry in proteomics: a critical review. Anal Bioanal Chem 389:1017–1031
2. Aye TT, Scholten A, Taouatas N, Varro A, Van Veen TA, Vos MA, Heck AJ (2010) Proteome-wide protein concentrations in the human heart. Mol Biosyst 6:1917–1927
3. Huttlin EL, Jedrychowski MP, Elias JE, Goswami T, Rad R, Beausoleil SA, Villen J, Haas W, Sowa ME, Gygi SP (2010) A tissue-specific atlas of mouse protein phosphorylation and expression. Cell 143:1174–1189
4. Lee KK, Sardiu ME, Swanson SK, Gilmore JM, Torok M, Grant PA, Florens L, Workman JL, Washburn MP (2011) Combinatorial depletion analysis to assemble the network architecture of the SAGA and ADA chromatin remodeling complexes. Mol Syst Biol 7:503
5. Lu P, Vogel C, Wang R, Yao X, Marcotte EM (2007) Absolute protein expression profiling estimates the relative contributions of transcriptional and translational regulation. Nat Biotechnol 25:117–124
6. Malmstrom J, Beck M, Schmidt A, Lange V, Deutsch EW, Aebersold R (2009) Proteome-wide cellular protein concentrations of the human pathogen *Leptospira interrogans*. Nature 460:762–765
7. Schmidt A, Beck M, Malmstrom J, Lam H, Claassen M, Campbell D, Aebersold R (2011) Absolute quantification of microbial proteomes at different states by directed mass spectrometry. Mol Syst Biol 7:510
8. Picotti P, Bodenmiller B, Mueller LN, Domon B, Aebersold R (2009) Full dynamic range proteome analysis of *S. cerevisiae* by targeted proteomics. Cell 138:795–806
9. Ghaemmaghami S, Huh WK, Bower K, Howson RW, Belle A, Dephoure N, O'Shea EK, Weissman JS (2003) Global analysis of protein expression in yeast. Nature 425:737–741
10. Raijmakers R, Neerincx P, Mohammed S, Heck AJ (2010) Cleavage specificities of the brother and sister proteases Lys-C and Lys-N. Chem Commun (Camb) 46:8827–8829
11. Swaney DL, Wenger CD, Coon JJ (2010) Value of using multiple proteases for large-scale mass spectrometry-based proteomics. J Proteome Res 9:1323–1329
12. Frese CK, Altelaar AF, Hennrich ML, Nolting D, Zeller M, Griep-Raming J, Heck AJ, Mohammed S (2011) Improved peptide identification by targeted fragmentation using CID, HCD and ETD on an LTQ-Orbitrap Velos. J Proteome Res 10:2377–2388
13. Mi H, Thomas P (2009) PANTHER pathway: an ontology-based pathway database coupled with data analysis tools. Methods Mol Biol 563:123–140
14. Gauci S, Helbig AO, Slijper M, Krijgsveld J, Heck AJ, Mohammed S (2009) Lys-N and trypsin cover complementary parts of the phosphoproteome in a refined SCX-based approach. Anal Chem 81:4493–4501
15. Mohammed S, Lorenzen K, Kerkhoven R, van Breukelen B, Vannini A, Cramer P, Heck AJ (2008) Multiplexed proteomics mapping of yeast RNA polymerase II and III allows near-complete sequence coverage and reveals several novel phosphorylation sites. Anal Chem 80:3584–3592
16. MacCoss MJ, McDonald WH, Saraf A, Sadygov R, Clark JM, Tasto JJ, Gould KL, Wolters D, Washburn M, Weiss A, Clark JI, Yates JR 3rd (2002) Shotgun identification of protein modifications from protein complexes and lens tissue. Proc Natl Acad Sci USA 99:7900–7905
17. Schlosser A, Vanselow JT, Kramer A (2005) Mapping of phosphorylation sites by a multi-protease approach with specific phosphopeptide enrichment and NanoLC-MS/MS analysis. Anal Chem 77:5243–5250
18. Wu SL, Kim J, Hancock WS, Karger B (2005) Extended range proteomic analysis (ERPA): a new and sensitive LC-MS platform for high sequence coverage of complex proteins with extensive post-translational

modifications-comprehensive analysis of beta-casein and epidermal growth factor receptor (EGFR). J Proteome Res 4:1155–1170

19. Mortensen P, Gouw JW, Olsen JV, Ong SE, Rigbolt KT, Bunkenborg J, Cox J, Foster LJ, Heck AJ, Blagoev B, Andersen JS, Mann M (2010) MSQuant, an open source platform for mass spectrometry-based quantitative proteomics. J Proteome Res 9:393–403
20. van den Toorn HW, Munoz J, Mohammed S, Raijmakers R, Heck AJ, van Breukelen B (2011) RockerBox: analysis and filtering of massive proteomics search results. J Proteome Res 10:1420–1424

Chapter 3

Multidimensional Protein Identification Technology for Direct-Tissue Proteomics of Heart

Dario Di Silvestre, Francesca Brambilla, and Pier Luigi Mauri

Abstract

Multidimensional protein identification technology (MudPIT) is an invaluable approach to identify proteins at large-scale level. Here, we describe a procedure of investigation to functional characterize the proteomic profile of complex samples such as those from cardiac tissues. In particular, we focus on the main steps concerning sample preparation, MudPIT analysis, tandem mass spectra processing, and biomarker discovery using label-free approaches. Finally, we report a data-derived systems biology approach to identify groups of proteins of over-, under-, and normal expression.

Key words Proteomics, MudPIT, Profiling, Biomarker discovery, Systems biology, Cytoscape, Heart

1 Introduction

As category of *-omics* molecules, proteins play a central role in all biological systems. In fact, they represent the greatest portion of structural molecules and, either directly or indirectly, the effectors of most molecular function occurring into cell, organ, or organism at any given time. Due to a range of posttranslational modifications and protein–protein interactions, complexity of proteome exceeds that of genome leading to our current inability to globally interrogate it using a single methodology (1). However, the increasing need to analyze complex protein samples is driving the development of sophisticated analytical tools providing separation and analysis as far as possible to complete all components, reducing costs, and improving quality of measurements and the throughput (2).

The classic proteomic approach is based on two-dimensional polyacrylamide gel electrophoresis (2D-PAGE) (3), where the protein spots of interest are isolated and identified by mass spectrometry (MS) (4). This procedure is time-consuming, and even if resolving power of 2D-PAGE is excellent, it is considered to be lacking in proteome coverage, particularly for hydrophobic proteins (5) and for proteins having extreme isoelectric point (pI<4 or >9)

Fernando Vivanco (ed.), *Heart Proteomics: Methods and Protocols*, Methods in Molecular Biology, vol. 1005,
DOI 10.1007/978-1-62703-386-2_3, © Springer Science+Business Media New York 2013

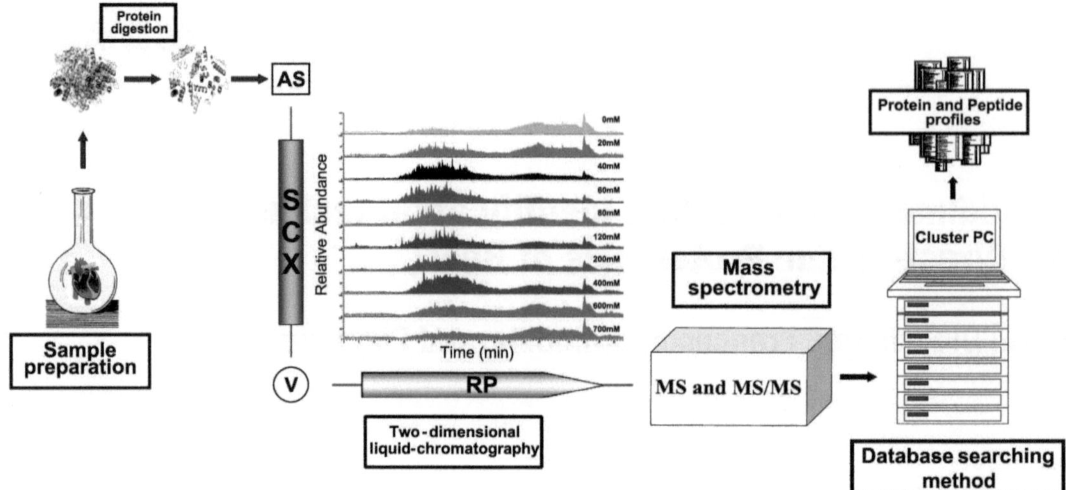

Fig. 1 MudPIT workflow. MudPIT is a fully automated high-throughput methodology for separating and identifying components of complex protein and peptide mixtures. It is composed by four main steps concerning sample preparation, liquid chromatography separation, mass spectrometry analysis and data processing. After enzymatic digestion of a protein mixture, MudPIT separates peptides by liquid chromatography combining a strong cation-exchange (SCX) with reversed-phase (RP) column. This allows greater separation of peptides that are directly interfaced with the ion source of a mass spectrometer. Eluted peptides are analyzed by detecting MS and MS/MS spectra. Finally, using a database searching method, MS/MS are interpreted by specific algorithms, such as SEQUEST, to obtain protein and peptide lists (*AS* auto sampler, *V* valve)

or molecular weight (MW < 10 kDa or >200 kDa). Many of these limitations are solved using a gel-free proteomic approach based on combination of two-dimensional liquid chromatography and electrospray ionization tandem mass spectrometry (2DC-MS/MS), or multidimensional protein identification technology, MudPIT) (6). It is a fully automated methodology for shotgun proteomics which involves simultaneously the following: (a) the generation of peptides by the enzymatic digestion of a complex protein mixture, (b) their separation by means of two in-line capillary columns, and (c) their sequencing by tandem mass spectrometry (MS/MS) analysis (Fig. 1). Finally, by means of a database searching method based on appropriate algorithms, such as the SEQUEST (7), the experimental tandem mass spectra are correlated to peptide sequences by comparison with the theoretical mass spectra calculated in silico from a protein sequence database.

Using MudPIT approach, hundreds of proteins per sample are characterized in wide p*I* and MW range, and hydrophobic proteins are identified, also (8). This big amount of experimental data is usually employed to characterize biomarkers using both label and label-free quantitative approaches. In particular, label-free strategy is less expensive and time-consuming, and different groups reported reproducible and useful data (9, 10). In addition, the availability

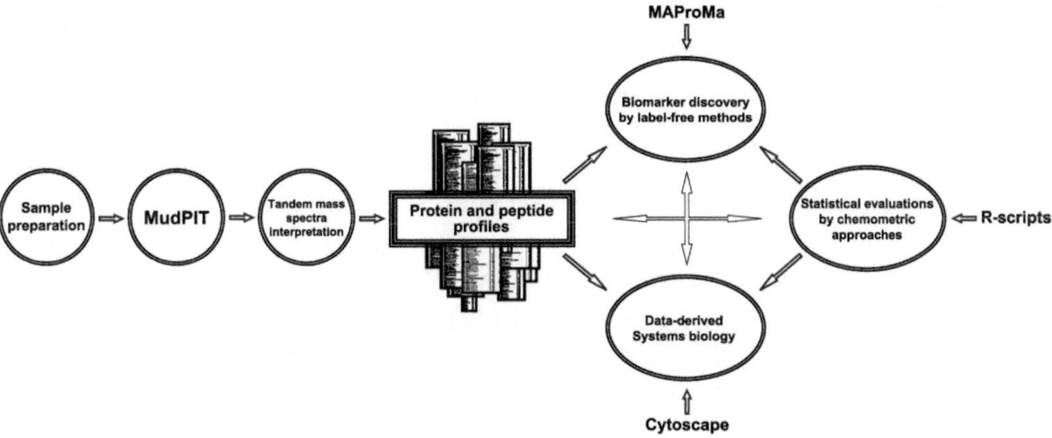

Fig. 2 Methods described in the chapter

of large network datasets offers the opportunity to investigate proteomics data using "data-derived systems biology" approaches by specific bioinformatics tools (11).

This chapter contributes an unbiased proteomic investigation procedure based on MudPIT approach and how the used key methods (Fig. 2) are currently applied in the context of heart proteomics. In addition to sample preparation and MudPIT analysis to identify proteomic profiles, it focuses on the main steps concerning tandem mass spectra processing and biomarker discovery using label-free quantitative approaches (12–14). Finally, it is reported a procedure to evaluate MudPIT data taking into consideration the functional relationship among proteins and specifically the protein–protein interactions.

2 Materials

2.1 Sample Preparation

Prepare all solutions using ultrapure water and analytical grade reagents. Prepare and store all reagents at room temperature (unless indicated otherwise). Conscientiously follow all waste disposal regulations when disposing waste materials.

1. Protein extraction buffer: 0.1 M NH_4HCO_3, pH 7.8. Add about 50 ml water to a 100 ml graduated cylinder. Weight 0.79 g NH_4HCO_3 and transfer to the cylinder. Add water to a volume of 90 ml. Mix and adjust pH value if necessary. Make up to 100 ml with water. Store at 4 °C.
2. RapiGest™ SF (Waters Co, Milford, MA, USA).
3. SPN™—Protein Assay kit (G-Biosciences, St. Louis, MO, USA).

4. Sequencing grade modified trypsin (Promega, Madison, WI, USA).
5. PepClean™ C-18 Spin Columns (PIERCE, Rockford, IL, USA).
6. Homogenizer (LabGEN 125, Cole-Palmer Instrument Company, Vernon Hills, IL, USA).
7. Formic acid, eluent additive for LC-MS, ~98 % (Fluka, Sigma-Aldrich Inc., St. Louis, MO, USA).
8. Trifluoroacetic acid (TFA), eluent additive for LC-MS (Fluka, Sigma-Aldrich Inc., St. Louis, MO, USA).
9. Acetonitrile (CH_3CN), HPLC gradient grade (Panreac Quimica S.A.U, Barcelona, Spain).

2.2 MudPIT Analysis

1. Biobasic SCX column, 0.320 i.d. × 100 mm, 5 μm (Thermo Electron Corporation, Bellefonte, PA, USA).
2. Biobasic C-18 column (0.180 i.d. × 100 mm, 5 μm (Thermo Electron Corporation, Bellefonte, PA, USA)).
3. Peptide trap (Zorbax 300 SB-C18, 0.3 i.d. × 5 mm, 5 μm, Agilent Technologies, Palo Alto, CA).
4. Eluent A: 0.1 % formic acid in water.
5. Eluent B: 0.1 % formic acid in acetonitrile.
6. Nano-LC electrospray ionization source (ESI) (Thermo Fisher, San Jose, CA, USA).
7. LTQ or LTQ Orbitrap (Thermo Fisher, San Jose, CA, USA).

2.3 MS/MS Data Processing

1. BioWorks 3.3.1 software (University of Washington, licensed to Thermo Fisher Scientific) equipped with SEQUEST algorithm.
2. Nonredundant (nr) protein sequence database (ftp://ftp.ncbi.nlm.nih.gov/blast/db/FASTA/).
3. Cluster PC (see Note 1).

2.4 Qualitative and Semiquantitative Evaluation of Protein Lists Obtained by MudPIT

1. MAProMa software (12).
2. R-scripts equipped with *XlsReadWrite*, *clue*, *clValid* library (http://www.r-project.org/).
3. RapidMiner software (http://rapid-i.com).
4. JMP software (http://www.jmp.com).

2.5 Data-Derived Systems Biology

1. Cytoscape platform (http://www.cytoscape.org/).
2. BioNetBuilder plugin (http://err.bio.nyu.edu/cytoscape/bionetbuilder/).

3 Methods

3.1 Sample Preparation

Carry out all procedures at room temperature unless otherwise specified.

1. Suspend from 10 to 15 mg (see Note 2) of heart frozen tissue in 250 μl of 0.1 M NH_4HCO_3, pH 7.8. Add 1 μl of caprylic acid to avoid foaming during homogenization of tissue. Homogenize tissue with a handheld homogenizer keeping sample cold (in ice).
2. Centrifuge sample at 5 g for 2 min and determine protein concentration of the supernatant using SPN Protein Assay kit (see Note 3).
3. Mix sample in order to resuspend the pellet. Take a volume of resuspended sample containing about 30 μg of proteins; add 0.1 M NH_4HCO_3, pH 7.8, RapiGest 2 %, and CH_3CN to reach a final volume of 80 μl having 10 % CH_3CN and 0.2 % RapiGest (see Note 4).
4. Add sequencing grade trypsin (see Note 5) in a ratio 1:50 enzyme–substrate and incubating at 37 °C o/n. Add the second aliquot of trypsin in a ratio 1:100 enzyme–substrate and incubating at 37 °C for 4 h (see Note 6).
5. After incubation, evaporate CH_3CN in a vacuum system (60 °C for about 5 min) and then adjust the pH to 2.0 adding TFA to stop the reaction and precipitate RapiGest (see Note 7).
6. Incubate the mixture for 45 min at 37 °C and then centrifuge at 13,000 rpm (or 11,3×g) for 10 min collecting the supernatant containing peptides.
7. Desalt and concentrate the supernatant using PepClean C-18 Spin Columns (see Note 8). Add 20 μl of H_2O, 0.1 % formic acid to PepClean C-18 eluate and concentrate it in a vacuum system (60° for about 10 min); repeat three times.
8. Finally, reconstitute sample in H_2O, 0.1 % formic acid in order to reach a protein concentration of about 1 μg/μl before the MudPIT analysis.

3.2 MudPIT Analysis

MudPIT analysis is performed using a ProteomeX-2 system configuration (Thermo Fisher Scientific, San Jose, CA, USA) implemented on a LTQ or LTQ Orbitrap mass spectrometer. It is advised to analyze sequentially samples to quantitatively compare (see Note 9).

1. Load digested peptide mixture (5 μl) onto a capillary strong cation exchange (SCX) column and elute using ten steps of increasing ammonium chloride concentration (0, 20, 40, 60, 80, 120, 200, 400, 600, 700 mM) (see Note 10). Fractions are

captured in turn onto two peptide traps for concentration and desalting prior to final separation by reversed-phase (RP) C-18 column (see Note 11). Elute peptides through an ACN gradient (Eluent A and Eluent B). Set the gradient profile to 5 % Eluent B for 5 min, 5–40 % Eluent B in 45 min, 40–80 % Eluent B in 10 min, 80–95 % Eluent B in 5 min, and 95 % Eluent B for 10 min; the flow rate is 100 μl/min split in order to achieve a final flow rate of 1.5 μl/min (see Note 12).

2. Eluted peptides are analyzed using a linear ion trap (LTQ) mass spectrometer equipped with a nano-LC electrospray ionization source. Nanospray is achieved using a uncoated fused-silica emitter (New Objective, Cambridge, MA, USA) (360 μm o.d./50 μm i.d./30 μm tip i.d.) held to 1.5 kV, and the heated capillary is held at 185 °C.
3. Acquire full mass spectra in positive mode and over a 400–2,000 m/z range, followed by five MS/MS events sequentially generated in a data-dependent manner on the first, second, third, fourth, and fifth most-intense ions selected from the full MS spectrum, using dynamic exclusion for MS/MS analysis. In particular, acquire MS/MS scans setting a normalized collision energy of 35 % on the precursor ion and, when a peptide ion was analyzed twice, applying an exclusion duration of 0.5 min (see Note 13).

3.3 MS/MS Data Processing

1. Into Cluster PC equipped with BioWorks 3.3.1 software save and store the .raw files containing the experimental tandem mass spectra (MS/MS) produced by MudPIT approach (see Note 14).
2. Download from NCBI Website the nonredundant protein sequence database (nr) (see Note 15). Using Database Utilities tool of BioWorks 3.3.1 software, retrieve the protein sequence annotated by a string indicating the organism's name under investigation.
3. By means of the SEQUEST algorithm, contained on BioWorks 3.3.1, correlate the experimental tandem mass spectra (MS/MS) to theoretical spectra calculated in silico from protein sequence database of reference. In particular, set "No enzyme" search parameter (see Note 16) and parent and fragment ion tolerance of 2 and 1 amu, respectively.
4. By means of Multiconsensus Result tool of BioWorks 3.3.1, combine the .srf files (see Note 17) obtained processing the .raw mass spectra against the protein sequence database.
5. For a reliable peptide identification, use the following filtering criteria: retain spectra/peptide matches only if they have a minimum Xcorr of 1.5 for +1, 2.0 for +2, and 2.5 for +3

change state, set $\Delta cn \geq 0.1$, consensus score value ≥ 10, and the threshold of peptide probability to $\leq 10^{-3}$ (see Note 18).

6. Export and save, in excel format, the lists containing the sampling statistics associated to proteins (see Note 19) and peptides identified. Alternatively, may export and save the lists in XML format.
7. To assess the false positives (FP) identifications and false discovery rates (FDR), repeat the procedure of MS/MS data processing from steps 1 to 5 using a database composed of reversed amino acid sequences (see Note 20).

3.4 Qualitative and Semiquantitative Evaluation of Protein Lists Obtained by MudPIT

1. Using the Autoformat tool of the MAProMa software, arrange the raw protein lists to maintain, for each protein identified, Reference, Accession number, SEQUEST-based SCORE (SCORE), Spectral count (SpC), p*I*, and MW (see Note 21).
2. By means of Multiconsensus Comparison tool of MAProMa, select and compare the protein lists, arranged previously, to obtain a $n \times p$ data matrix (n represents the number of proteins and p the number of samples) containing SpC and SCORE values for each protein identified (see Note 22).
3. To evaluate the analytical reproducibility of MudPIT approach, plot the SpC (or SCORE) values for couple of replicate samples (technical and biological) and assess the linear relationship evaluating correlation coefficient (R-squared value) and slope of the resulting graph. Alternatively, to evaluate reproducibility of MudPIT considering at the same time, all analyzed samples process $n \times p$ data matrix (with SpC or SCORE values) by unsupervised learning methods, such as principal component analysis (PCA) and hierarchical clustering (HC) (see Note 23).
4. If observe a low reproducibility of MudPIT data, normalize them using Total Signal normalization procedure (see Note 24). Normalized SpC values (*n*SpC) are obtained by dividing SpC value (*p*SpC) of each protein for the value obtained by summing SpC values of all proteins belonging to the same sample (ΣSpC). Use the same procedure to normalize SCORE values.
   - *n*SCORE = *p*SCORE/ΣSCORE.
5. Perform pairwise comparisons of samples applying DAve and DCI algorithm of MAProMa software to evaluate the differentially represented proteins (6, 8, 9, 11–14). Specifically:
   - $(X-Y)/(X+Y)/0.5$ represent DAve (differential average)
   - $(X+Y)\times(X-Y)/2$ represent DCI (differential confidence index)

   where X is the normalized SCORE of the protein in the first sample while Y represents the normalized SCORE of the same protein but in the second sample. To increase differential reliability of proteins, set threshold value of acceptability of ≥400 (or ≤−400) for DCI and 0.4 (or ≤−0.4) for DAve (see Note 25).

6. To improve the knowledge of discovery process, perform pair-wise comparison of samples applying the statistical G-test according to Sokal and Rohlf (15):

$$G = 2f_1 \ln\left(\frac{f_1}{\hat{f}_1}\right) + 2f_2 \ln\left(\frac{f_2}{\hat{f}_2}\right)$$

where f_1 and f_2 are the normalized SpC for the protein in sample 1 and sample 2, respectively; $\hat{f}_1$ and $\hat{f}_2$ are expected spectral counts for the protein in sample 1 and sample 2, respectively. Assume the protein is equally expressed, thus $\hat{f}_1 = \hat{f}_2 = (f_1 + f_2)/2$.

Use the *G*-value to assess whether the protein is differentially expressed according to the chi-square distribution table with one degree of freedom. Proteins with *G* larger than 3.841 are differentially expressed with $P < 0.05$ (see Note 26).

3.5 Data-Derived Systems Biology

1. By means of Cytoscape platform, install BioNetBuilder plugin to retrieve protein–protein interaction networks of the species of interest.
2. After installation, start BioNetBuilder and type in the text field "Species search string" the name of the species and press the Search button (see Note 27). Then select the matching TAXID and species for your network.
3. Set node label priorities (see Note 28).
4. If need, refine search of protein–protein interactions selecting nodes by GO annotations.
5. Select the source databases for retrieving protein–protein interactions and for each one set the interaction types to include into network (see Note 29).
6. Select node attributes to attach to nodes in the network (see Note 30).
7. Name your network and export it and its attributes using XGMML format. Press the "Finish" button and BioNetBuilder will then retrieve nodes, edges, and attributes from the servers.
8. By select functions of Cytoscape, select in the network nodes identified experimentally by MudPIT (see Note 31). Save the resulting subnetwork considering the nodes identified experimentally and excluding the isolated nodes.
9. By VizMapper tool of Cytoscape, label identified nodes and biomarkers by different colors and sizes to highlight subnetworks which change consistently between investigated conditions, thus to identify groups of interacting proteins of over-, under-, and normal expression.

4 Notes

1. Processing of tandem mass spectra produced by MudPIT approach may be time-consuming. Data processing by software versions for Cluster PCs is very useful to reduce it.
2. When prepare several samples to be analyzed and compared, if possible use the same amount of tissue per sample. It helps the extraction of a similar amount of total proteins, thus a better quantitative evaluation of analyzed samples.
3. Be sure to determine the protein concentration on clear supernatant without removing it from pellet. Spin column method for protein estimation is compatible with many laboratory agents, such as detergents including SDS up to 2 %. It requires from 0.5 to 10 μg of proteins per assay. We find that 1–2 μl of supernatant are optimal for kit. No protein standards are needed, and both spectrophotometer and microplate reader are suitable.
4. RapiGest is a detergent, compatible with the liquid chromatography and mass spectrometry analyses, to enhance in-solution digestion of proteins without inhibiting enzyme activity unlike other denaturants. To prepare it, reconstitute the lyophilized powder in 50 μl of 0.1 M NH_4HCO_3, pH 7.8 to reach the concentration w/v of 2 %. The solution is stable for 1 week at 4 °C. The recommended final concentration in sample is 0.1 % (w/v), but hydrophobic proteins may require higher RapiGest concentration (0.2 %). Add 8 μl of 2 % RapiGest to sample containing about 30 μg of proteins and boil the mixture at 100 °C for 5 min. Cool the sample and add 8 μl of CH_3CN before adding proteolytic enzymes.
5. Trypsin is a protease that cleaves proteins on the carboxyl side of arginine (R) or lysine (K) residues, except when either is followed by proline. Due to its highly specific cleavage, it is routinely used in proteomics for peptide mapping and protein sequencing. In fact, the prediction of the set of peptides potentially produced by digestion reduces the time-consuming of data processing based on database search methods.
6. The addition of a further aliquot of trypsin improves the digestion efficiency.
7. RapiGest is degraded at low pH and simply removed from the sample by precipitation. Be sure that pH is lower than 2 because the RapiGest insoluble residue is not always visible.
8. Each PepClean C-18 Spin Columns can bind up to 30 μg of total peptide from 10 to 150 μl sample volume. Mix three parts sample to one part of sample buffer (2 % TFA in 20 % CH_3CN). Prepare the column by adding 200 μl of activation solution (50 % CH_3CN) and centrifuge (1.5 g for 1 min). Repeat this step twice.

Then, load sample on column and centrifuge (1.5 g for 1 min). To ensure complete binding, recover flow-through and reload sample on column. After centrifugation, add 200 μl wash solution (0.5 % TFA in 5 % CH_3CN) and centrifuge (1.5 g for 1 min). Repeat this step three times. Elute sample by adding 20 μl of elution buffer (70 % CH_3CN) twice.

9. Samples analyzed sequentially are investigated at the same or near capacity of MS instrument. In this way, it is reduced the experimental bias due to the instrument performance drift and thus the false identification of differentially expressed proteins.
10. Using ten salt steps of increasing ammonium chloride (NH_4Cl) concentration in the first dimension (0, 20, 40, 60, 80, 120, 200, 400, 600, 700 mM), the most part of peptides elutes in the first eight steps, while steps of ammonium chloride to 600 and 700 mM completely wash the column. A powerful and highly resolving separation method better fractionate peptides prior to entering the mass spectrometer. This aspect improves the MS acquisition of data obtaining the best representation of the proteins in the mixture in a large dynamic range.
11. In the ProteomeX-2 system, the 10 port valve is equipped with two C-18 traps that alternatively capture the SCX fractions. Peptides are desalted on the traps and then eluted using the CH_3CN gradient on the C-18 column.
12. In the ProteomeX-2 system, the optimal flow rate after the SCX column is 12 μl/min, 6–10 μl/min after C-18 traps, and 1.5 μl/min after C-18 column. The final low flow rate assures the ionization efficiency on the nano-LC electrospray ionization source.
13. Acquire five MS/MS spectra for each full MS spectra working in a low-resolution mode with a linear ion trap (LTQ). Acquire four MS/MS spectra for each full MS spectra working in high-resolution ($R = 60{,}000$) mode with a LTQ Orbitrap. In both methods, set the dynamic exclusion mode in order to not acquire redundant MS/MS spectra for the same m/z.
14. Spectra acquired by MudPIT are stored in a number of .raw files corresponding to the salt steps used in the separation method applied.
15. NCBI nonredundant (nr) database is the biggest protein databases used for proteomic experiments (16). It compiles all protein sequences available from "GenBank" translations, Protein Data Bank (PDB), UniProtKB/Swiss-Prot, PIR, and PRF. Even if it possesses a high level of redundancy, it is widely used for studies involving non-model organisms because it has a better representation of these species. It is available in .fasta format, and every sequence processed is associated to a "GenInfo" (gi) numbers.

16. Even if protein mixture is digested with trypsin, "No enzyme" setting allows also searching of peptide sequences obtained by unspecific cleavages. Even if on the one hand this is time-consuming, on the other hand it allows the identification of a bigger number of peptide sequences.
17. For each .raw file is produced a .srf one. Files in .srf format contain the results of .raw mass spectra processing.
18. Xcorr is not a probabilistic score, thus validation by statistical approaches are useful to ensure a correct identification of peptide sequences. BioWorks 3.3.1 is equipped to provide a probabilistic evaluation similarly to that obtained by other widely used software, such as Peptide and Protein Prophet (17).
19. Accession, Reference, Protein probability, SEQUEST-based SCORE (SCORE), Peptide Hits (Spectral count (SpC)), molecular weight (MW), and isoelectric point (p*I*) are associated to identified proteins. Alternatively, in place of p*I* may export the Protein Coverage. Before save and export the list of proteins, be sure you did not apply the filter for selecting Distinct Peptides. In this way, it is considered the total number of MS/MS spectra (Spectral count) taken on peptides from a given protein. This value is linearly correlated with the protein abundance and allows the semiquantitative comparison of proteins in different lists (18).
20. To evaluate FP and FDR, several strategies based on decoy database (19) are utilized. FDR is defined as the number of peptides from decoy sequences (*N*d) divided by the number of peptides from target sequences (*N*t) (FDR = *N*d/*N*t), while the estimated number of true positives (TPs) is then *N*t − *N*d. Usually, if FDR exceed the 5 %, it is advised to use more stringent criteria of filtering.
21. By means of 2D MAP tool of MAProMa software, protein lists may be presented in the usual form for biologists (maps). In fact, it automatically plots in a virtual 2D map the MW vs. p*I* for each protein identified, assigning it a color according to a range of a sampling statistics (SCORE or SpC) derived by SEQUEST data handling. This representation permits to have a rapid overview of the proteins identified and also to highlight immediately proteins with extreme molecular weight and/or p*I*.
22. Statistical evaluation of SCORE and SpC allows quantification of proteome differences by label-free quantitative approaches. In the course of our studies, by analyzing various samples, such as microorganisms, human cells, and tissues, we observed a direct relationship between the SEQUEST-based SCORE values and relative abundance of identified proteins (12, 13). In the same way, other authors showed that it is possible to obtain a relative quantification of proteins comparing SpC values (18).

23. Several software, both free of change (i.e., RapidMiner) and commercial (i.e., JMP), allow easily processing of $n \times p$ data matrixes. We propose the use of *in-house* R-script, based on *XlsReadWrite*, *clue*, *clValid* library. It processes the $n \times p$ data matrix obtained by MAProMa and automatically shows the PCA and HC tree. In this way, it is possible to evaluate grouping of biological and technical replicates, thus the MudPIT reproducibility. Specifically, HC is performed by Euclidean metric and *Ward*'s method.
24. In addition to Total Signal strategy, other valid procedure of normalization, concerning the proteomic approaches based on liquid chromatography coupled to mass spectrometry, has been evaluated in other studies (20, 21). In particular, log preprocessing (by ln), Z normalization, Maximum Signal, or Row Sigma is available.
25. Conventionally, signs (+/−) indicate if proteins are upregulated in the first or in the second sample, respectively. A value of DAve ≥ 0.4 (or ≤ −0.4) corresponds to SCORE ratio ≥1.5. Coupled to a threshold value ≥400 (or ≤ −400) for DCI, it allows, with a good reliability, to identify differentially expressed proteins. However, DAve and DCI threshold values may be decreased when calculated considering mean SCORE values derived from replicate analyses (DAve ≥ 0.2 or < −0.2 and DCI ≥ 200 and ≤ −200). On the contrary, when there is an available single analysis per sample/condition, a better reliability of the differentially expressed proteins may be assured increasing the threshold values (DAve ≥ 0.8 or < −0.8 and DCI ≥ 800 and ≤ −800).
26. In general, statistical validation is desired if one wants to use spectral counting for protein quantitation. In particular, it is required when there are available few replicate analyses. In this scenario, G-test is a likelihood ratio test for goodness of fit (18). If three or more replicate analyses per sample are available, in addition to DAve, DCI, and G-test, evaluation of the differentially expressed proteins may be performed by feature selection methods, such as the Golub index, the Student *t*-test, a strategy based on the weighting used in a forward-support vector machine (SVM-F) model, and SVM recursive feature elimination (20).
27. The name of the species can be incomplete. BioNetBuilder automatically lists the species (and the corresponding TAXID) containing into name the string inserted in the text field "Species search string."
28. In this step, choose how label nodes in the network. To automatically select nodes using the protein lists identified by MudPIT, set as first label priority the same identifier (ID)

reported in the protein sequence database used for tandem mass spectra processing (i.e., GenInfo (GI) numbers).

29. To obtain a network containing as far as possible complete set of interaction, select all source databases (BIND, Biogrid, DIP, Intact, Interlogger, KEGG, MINT, MPPI, Prolinks, and GO) available by BioNetBuilder. By default, all types of interaction are included in the network. To reduce the network dimension, select the interactions of interest (protein–protein, protein–DNA, protein–RNA, enzymatic compounds, genetic interaction, etc.). In addition, may filter the interactions setting a confidence value (*P*-value threshold).
30. The most useful ones are the URL attributes since they allow a direct connection for a node to a Web page with information about that node in a public database like NCBI, PIR, Prolinks, and the Yeast Resource Center.
31. For each sample or biological condition analyzed by MudPIT, put in a text file the identifiers (GI number) of the proteins identified experimentally. In the same way, create different text files containing proteins resulted up- and down-regulated by applying label-free quantification approaches. Use these files to automatically select the nodes in the network.

Acknowledgments

This work was supported by CARIPLO Foundation (2008.2504, 2007.5312 and project - Proteomic platform, Operational Network for Biomedicine Excellence in Lombardy). The authors thank Marta G. Bitonti for MAProMA software.

References

1. Ly L, Wasinger VC (2011) Protein and peptide fractionation, enrichment and depletion: tools for the complex proteome. Proteomics 11:513–534
2. Nilsson T, Mann M, Aebersold R et al (2010) Mass spectrometry in high-throughput proteomics: ready for the big time. Nat Methods 7:681
3. Horgan GW (2007) Sample size and replication in 2D gel electrophoresis studies. J Proteome Res 6:2884–2887
4. Yates JR, Ruse CI, Nakorchevsky A (2009) Proteomics by mass spectrometry: Approaches, advances, and applications. Annu Rev Biomed Eng 11:49–79
5. Braun RJ, Kinkl N, Beer M et al (2007) Two-dimensional electrophoresis of membrane proteins. Anal Bioanal Chem 389:1033–1045
6. Mauri PL, Scigelova M (2009) Multidimensional protein identification technology for clinical proteomic analysis. Clin Chem Lab Med 47:636–646
7. Eng JK, McCormack AL, Yates JR (1994) An approach to correlate tandem mass spectral data of peptides with amino acid sequences in a protein database. J Am Soc Mass Spectrom 5:976–989
8. Comunian C, Rusconi F, De Palma A et al (2011) A comparative MudPIT analysis identifies different expression profiles in heart compartments. Proteomics 11:2320–2328
9. Mauri P, Deho' G (2008) A proteomic approach to the analysis of RNA degradosome composition in *Escherichia coli*. Methods Enzymol 447:99–117

10. Park SK, Venable JD, Xu T et al (2008) A quantitative analysis software tool for mass spectrometry-based proteomics. Nat Methods 5:319–322
11. Di Silvestre, D., Daminelli, S., Brunetti, P. et al. (2010) Bioinformatics tools for mass spectrometry-based proteomics analysis. *Reviews in Pharmaceutical & Biomedical Analysis* Bentham eBooks, pp. 30–51
12. Mauri P, Scarpa A, Nascimbeni AC et al (2005) Identification of proteins released by pancreatic cancer cells by multidimensional protein identification technology: a strategy for identification of novel cancer markers. FASEB 19:1125–1127
13. Regonesi ME, Del Favero M, Basilico F et al (2006) Analysis of the Escherichia coli RNA degradosome composition by a proteomic approach. Biochimie 88:151–161
14. Simioniuc A, Campan M, Lionetti V et al (2011) Placental stem cells pre-treated with a hyaluronan mixed ester of butyric and retinoic acid to cure infarcted pig hearts: a multimodal study. Cardiovasc Res 90:546–556
15. Sokal RR, Rohlf FJ (1994) Biometry: the principles and practice of statistics in biological research, 3rd edn. Freeman, New York
16. Griss J, Côté RG, Gerner C et al (2011) Published and Perished? The Influence of the Searched Protein Database on the Long-Term Storage of Proteomics Data. Mol Cell Proteomics 10:M111.008490
17. Nesvizhskii AI, Keller A, Kolker E et al (2003) A statistical model for identifying proteins by tandem mass spectrometry. Anal Chem 75:4646–4658
18. Zhang B, VerBerkmoes NC, Michael A et al (2006) Detecting Differential and Correlated Protein Expression in Label-Free Shotgun Proteomics. J Proteome Res 5:2909–2918
19. Wang G, Wu WW, Zhang Z et al (2009) Decoy Methods for Assessing False Positives and False Discovery Rates in Shotgun Proteomics. Anal Chem 81:146–159
20. Carvalho PC, Hewel J, Barbosa VC et al (2008) Identifying differences in protein expression levels by spectral counting and feature selection. Genet Mol Res 7:342–356
21. Carvalho PC, Fischer JS, Chen EI et al (2008) PatternLab for proteomics: a tool for differential shotgun proteomics. BMC Bioinformatics 9:316–329

Chapter 4

Global Protein Quantification of Mouse Heart Tissue Based on the SILAC Mouse

Anne Konzer, Aaron Ruhs, Thomas Braun, and Marcus Krüger

Abstract

Metabolic labeling of living organisms with stable isotopes has become a powerful tool for global protein quantitation. The SILAC (stable isotope labeling with amino acids in cell culture) approach is based on the incorporation of nonradioactive-labeled isotopic forms of amino acids into cellular proteins. The effective SILAC labeling of immortalized cells and single-cell organisms (e.g., yeast and bacteria) was recently extended to more complex organisms, including worms, flies, and even rodents. The administration of a $^{13}C_6$-lysine (heavy) containing diet for one mouse generation leads to a complete exchange of the natural (light) isotope $^{12}C_6$-lysine. SILAC-labeled organisms are mainly used as a heavy "spike-in" standard into nonlabeled counterparts, and the combination with high-performance mass spectrometers allows for global proteomic screening. Here we used the fully labeled SILAC mice to identify proteins based on SILAC pairs from isolated cardiomyocytes, and we analyzed β-parvin-deficient hearts. Our approach confirmed the absence β-parvin and revealed simultaneously a clear up regulation of α-parvin in heart tissue. In this protocol, we describe the generation of a SILAC mouse colony and show two approaches to perform a proteome-wide analysis of heart tissue. Thus, the SILAC mouse spike-in approach is a readily available procedure and allows for a straightforward systematic analysis of disease models and knockout mice.

Key words SILAC, SILAC mouse, Metabolic labeling, Cardiomyocytes proteome, β-Parvin

1 Introduction

Over the past decades, cardiovascular diseases have increased tremendously and are the biggest cause of death worldwide (1). In recent years, substantial progress in the exploration of this multifactorial disease has been made by the use of systematic screening techniques, including microarrays and deep sequencing. Although genomic approaches are well developed and important, they mostly disregard regulations at the posttranscriptional and posttranslational level. Thus, the knowledge of protein expressions and protein modifications is essential to functionally analyze complex cellular systems more in depth.

Since a couple of year's mass spectrometry-based proteomics has become a suitable method for the systematic identifications of

Fernando Vivanco (ed.), *Heart Proteomics: Methods and Protocols*, Methods in Molecular Biology, vol. 1005, DOI 10.1007/978-1-62703-386-2_4,

proteins and it is now possible to detect more than 10-k proteins from mammalian cells in relatively short measurement times(2). However, MS-based proteomics of more specialized tissue, such as skeletal muscle tissue and body fluids, results clearly in lower proteome coverage compared to undifferentiated cell culture cells. This can be explained by the large dynamic range of protein expressions in some tissues, which is larger than five orders of magnitude. For example, muscle tissue consists mainly of proteins building up the contractile apparatus. Proteins with lower expression levels, including transcription factors, are therefore hard to detect by mass spectrometry. One way to reduce the sample complexity is the biochemical pre-fractionation of cellular compartments to achieve deeper proteome coverage. However, one has to take into account that each subcellular fractionation step increases the number of samples and required considerable measurement time.

Another bottleneck in proteomics is the accurate and reliable protein quantitation of complex samples. Besides applicable label-free approaches, the use of stable isotope labeling for protein quantitation has emerged to the most popular method in typical shotgun proteomics experiments. Isotope labeling incorporates a heavy isotope (^{13}C, ^{15}N, ^{2}H) of a small molecule or amino acid into the peptide, either by chemical reactions or metabolic labeling. After combining the light and heavy population, peptide pairs with a defined mass difference can be detected, and the comparison of peak intensities after mass spectrometric analysis allows finally accurate peptide quantification (3).

Although the stable isotope labeling of amino acids in cell culture (SILAC) method was initially developed for cell culture systems (4) recently, a wide variety of model organisms were labeled with SILAC amino acids, including bacteria (5), yeast (6), worms (7), flies (8), and rodents (9, 10). The SILAC labeling of the mouse model is based on a lysine-free diet supplemented with the $^{13}C_6$-lysine isotope (Lys6). The administration of the Lys6 diet over one mouse generation results in fully labeled animals, and SILAC mouse tissue is used commonly as a “spike-in” SILAC approach to introduce a labeled standard in the control and experimental condition. Similar to the SILAC mouse, rodents can also be labeled with the heavy isotope of nitrogen (^{15}N). The SILAM (stable isotope labeling in mammals) method is also used as a “spike-in” approach, and the ^{15}N method can also be extended to label prototrophic organisms, including bacteria and plants. A general overview of the SILAC “spike-in” approach and the calculation of ratios between light samples are outlined in Fig. 1.

In this protocol, we describe briefly the generation of the SILAC mouse, and we analyze the proteome of isolated mouse cardiomyocytes. We then compare normal heart tissue and heart tissue lacking the expression of β-parvin.

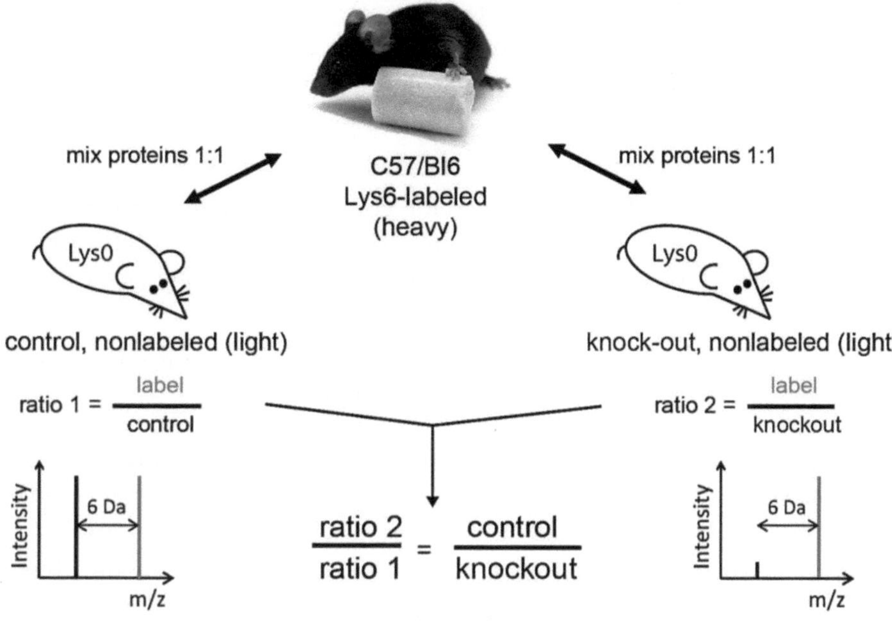

Fig. 1 The SILAC mouse as an internal standard for protein quantitation. Lys-6-labeled mouse tissue is "spiked in" in nonlabeled (light) samples of a control (*left site*) and a knockout mouse (*right site*). After mass spectrometric measurements, peptides are identified as SILAC pairs and the intensities of the light (Lys0) and heavy (Lys6) peak leads to relative protein quantitation (r1 = H/L and r2 = H/L). Calculation of direct ratios between the light samples is achieved by dividing r2/r1 which results in a direct comparison of nonlabeled animals. In this example, the selected peptide is clearly downregulated in the knockout animal

2 Materials

For all solutions and solvents, use ultrapure water (Milli-Q water) and HPLC-grade reagents.

2.1 SILAC Mouse

1. $^{13}C_6$-lysine (Lys6) mouse diet was purchased from Silantes. The SILAC diet contains 1 g of the $^{13}C_6$-lysine isotope. Alternatively, it is also possible to use other lysine-free diets supplemented with stable isotopes to generate a SILAC mouse diet (see Note 1).
2. Standard mouse strain for example C57BL/6 mice or mouse strain of choice.

2.2 Heart Tissue/Cardiomyocytes Isolation and Protein Extraction

1. Anesthesia with sodium pentobarbital in 0.9 % NaCl solution (50 mg pentobarbital/kg mouse weight).
2. Ice-cold 1× PBS.
3. For cardiac perfusion, use 50 ml syringe and a 22-gage (G) needle.
4. Store isolated tissue in liquid nitrogen.

5. Perfusion buffer (25 ml): 3.5 mg Liberase DH (Roche) and 3.75 mg trypsin (Sigma) in calcium-free PBS (see Note 2).
6. Myocyte stopping buffer: 12.5 μM $CaCl_2$ and 10 % dialyzed FCS in perfusion buffer (see Note 3).
7. Store isolated tissue/cells in liquid nitrogen.
8. SDS lysis buffer for complete protein extraction: 4 % SDS, 0.1 M dithiothreitol (DTT) in 0.1 M Tris/HCl pH 7.6.
9. For tissue homogenization: ULTRA-TURRAX (IKA Works) and Sonicator.
10. For protein concentration measurements: DC protein assay (Bio-Rad) or alternate methods.

2.3 1D SDS-PAGE and in Gel Digestion

1. For protein separation: NuPAGE 4–12 % Bis-Tris gels, 4× LDS sample buffer, MOPS running buffer, and antioxidants (Invitrogen).
2. For protein visualization: colloidal blue staining (Invitrogen).
3. Absolute ethanol (Sigma).
4. Ammonium bicarbonate (ABC) digestion buffer: 50 mM ammonium bicarbonate in water. Prepare solution fresh.
5. Reduction buffer: 100 mM DTT stock solution. Store aliquots at −20 °C.
6. Alkylation buffer: 550-mM Iodoacetamide (Sigma) stock solution in 50-mM ABC-buffer. Aliquots can be stored at −20 °C. Iodoacetamide is light sensitive and should be stored in the dark.
7. Endopeptidase LysC: 0.5-μg/μl LysC dissolved in ABC-buffer.

3 Methods

Metabolic labeling with stable isotopes of living animals has been successfully used to measure the flux of small molecules and other metabolites (11). With the development of high-performance mass spectrometers, it is now possible to measure proteome-wide incorporation rates of stable amino acids in cell culture systems as well as in living animals (12, 13).

However, for relative protein quantitation of two experimental conditions, a full incorporation of the labeled amino acid is necessary. An incomplete labeled protein in the heavy labeled population will be detected as a light signal and will contribute to the unlabelled signal, which results lastly to quantification errors. In a typical cell culture experiment, a complete labeling with heavy amino acids is achieved after five cell doublings (14). In the case of living animals, one has to take the efficient recycling of unlabelled amino acids into newly synthesized proteins into account. In practice, a complete labeling of a mouse is not achievable by extending the feeding period for several months. Instead one has to breed at least one mouse generation (F1) on the Lys6 diet to obtain a change for >96 % of the proteins to the heavy state. The SILAC mouse is used as an internal protein standard which will be spiked in into the light control and experimental condition, respectively. Direct comparison of the light conditions can be achieved by dividing the heavy/light (control) by the heavy/light (experiment) measurements (Fig. 1). The advantage of the "ratio-of-ratio" approach is that one takes changes of the protein expression levels due to the different diet or mouse strains into account. The versatility of the in vivo SILAC approach in mice has been demonstrated by several other studies (15–18).

Here, we describe the SILAC labeling of a mouse, and we show an example for a proteome analysis based on SILAC pair identifications from isolated cardiomyocytes. Moreover, we use the in vivo SILAC approach to analyze the proteome of β-parvin-deficient heart tissue. The described protocol can be adapted to other protein isolation methods and mouse models.

3.1 Mouse Labeling and SILAC Incorporation Rates

The most critical part of SILAC labeling is the presence of a mouse diet containing only the labeled amino acid. Here we use Lys6 for labeling because lysine is an essential amino acid and is not known to be converted to other amino acids (see Note 4).

1. Start with female mice (filial generation 0, F0) from the mouse strain of interest. The most commonly used strain for the generation of transgenic and knockout animal models is the inbred strain C57BL/6.
2. An initial feeding period of the F0 generation (~4–6 weeks) is recommended to adopt animals to the diet and to pre-label animals with Lys6 (Fig. 2a, b).
3. After mating and breeding of the F1 generation for ~2 month organs, including heart tissue, typically display a median incorporation rate of ~96 %. If the incorporation of Lys6 is not sufficient, continue with breeding to the F2 generation (Fig. 2d) (see Note 5).
4. A rough estimate of the food consumption during the labeling period up to the F1 generation is shown in Fig. 2a. Clearly, the food consumption is dependent on the size, age, and mouse strain. Here we estimated 3–5 g food per day per C57/Bl6 mouse.
5. An example for Lys6 incorporation in cardiomyocytes after 2 weeks of SILAC labeling is shown in Fig. 2d. Selected MS spectra indicate proteins with slow and fast incorporation rates.

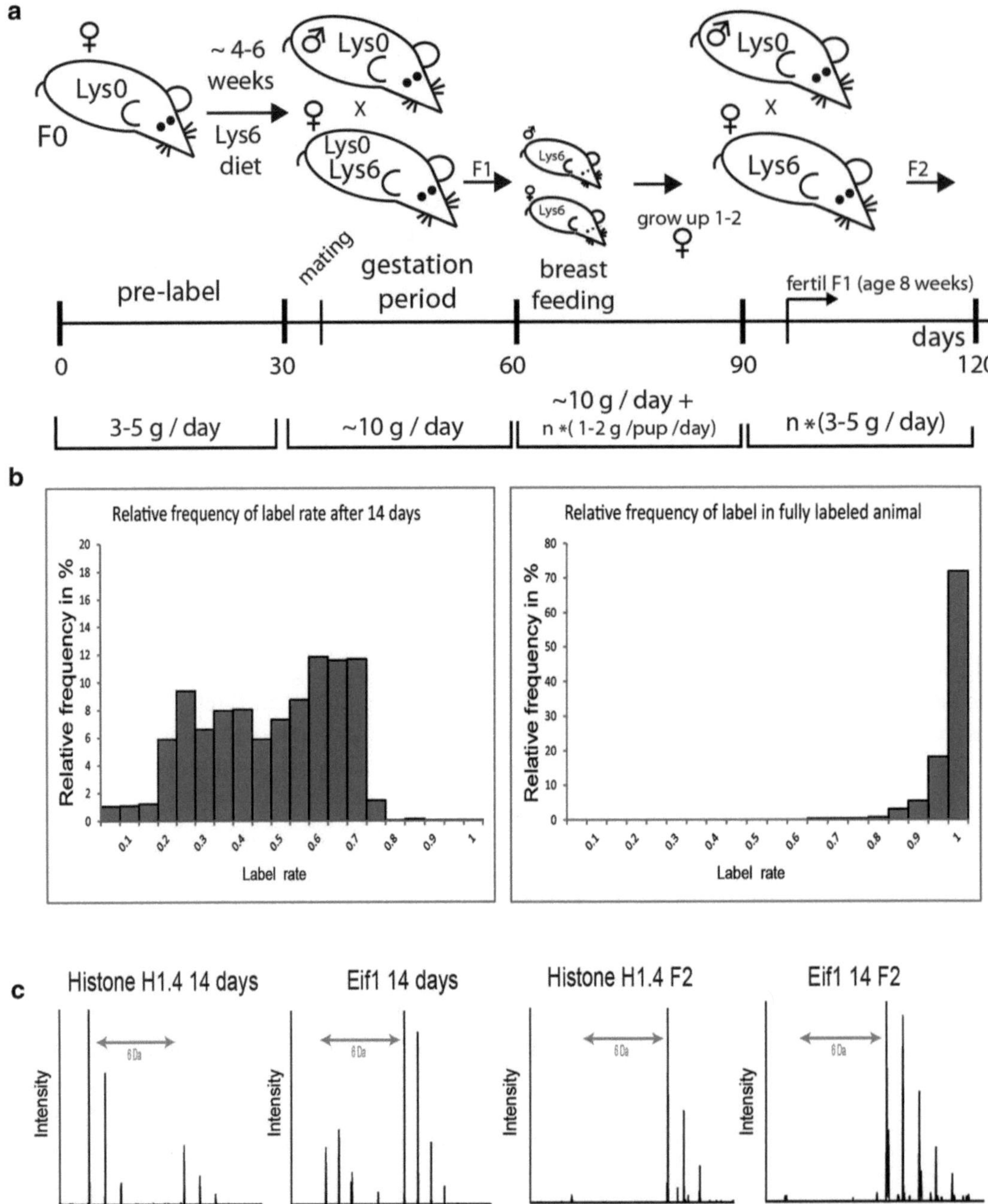

Fig. 2 SILAC labeling and incorporation rates. (**a**) Estimated time frame for the generation of a F1 SILAC mouse colony. (**b**) The left panel shows the Lys6 incorporation rate of cardiomyocytes after 2-week labeling time. The widespread distribution is due to different protein turnover rates. The left panel shows the label efficiency of adult cardiomyocytes of the F1 generation. The incorporation rate can be calculated by % labeling = (SILAC ratio × 100)/(SILAC ratio + 1). (**c**) Selected MS spectra of peptides with slow (histone) and fast (eukaryotic translation initiation factor 1) Lys6 incorporation rates. Both proteins are completely labeled in the F1 generation

3.2 Isolation of Cardiomyocytes and Protein Extraction

The heart consists not only of cardiomyocytes but several other cell types such as fibroblasts, endothelial cells, and neuronal cells that contribute to the architecture and function of the heart. To focus our analysis more on contractile cells, we isolated cardiomyocytes based on a collagenase digestions step as described in ref. 19:

1. Anesthetize a C57 BL/6 mouse with sodium pentobarbital (50 μg/g mouse weight) in 0.9 % NaCl solution by intraperitoneal injection. Thrombocytes aggregation can be blocked by heparin injection (100 U/ml in PBS).
2. Open the thorax and transfer the heart together with the lung and thymus in ice-cold, calcium-free PBS solution. Remove fat tissue and thymus to isolate the aortic arch (see Note 6).
3. Cannulate the heart by sliding a cannula into the aorta up to the aortic valve and fix it with a thread. Remove the lung and remaining tissue from the heart.
4. Connect the heart with a perfusion system and perform a retrograde perfusion of calcium-free PBS solution at 37 °C. Add 25 ml perfusion buffer to the system to dissociate cardiomyocytes. After ~5 min, the heart appears swollen and pale (see Note 7).
5. After the enzymatic tissue disaggregation, remove the heart from the perfusion system by cutting below the atria. Transfer the ventricles into 2.5 ml perfusion buffer and shearing the heart into smaller pieces with forceps and by gentle pipetting. Add 2.5 ml myocyte stopping buffer to the cell suspension to stop enzymatic activity.
6. Transfer the cell suspension into a 15 ml centrifuge tube and sediment the cell debris by gravity for 2–3 min. The supernatant, which contains the intact cardiomyocytes, has to be transferred into a new 15 ml centrifuge tube.
7. Resuspend the pellet with 2.5 ml myocyte stopping buffer, centrifuge at 300 rpm for 1 min, and combine the supernatant with the cardiomyocytes from step 6.
8. Filtrate (filter pore size 100 μm) the combined supernatants to remove remaining cell debris, centrifuge at 800 rpm for 1 min, wash the cells with 2 ml PBS, centrifuge at 800 rpm for 1 min, and freeze the cells in liquid nitrogen.
9. Add 200 μl SDS lysis buffer to 40 μg cardiomyocytes for cell lysis and protein denaturation.
10. For complete cell lysis, a glass douncer (Kontes Glass Company) has to be used prior to sample heating at 95 °C for 5 min.
11. Sonicate samples to reduce viscosity.
12. Samples have to be clarified by centrifugation at 16,000 × g for 5 min.
13. Estimate protein concentration using Bradford or alternate methods.

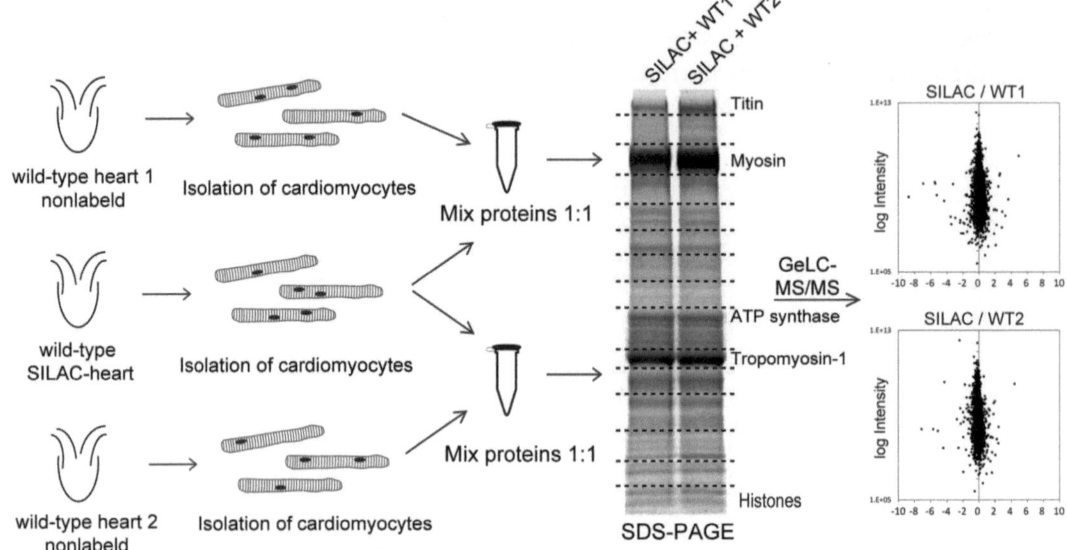

Fig. 3 Proteome analysis of cardiomyocytes based on SILAC pairs. The workflow illustrates the isolation of cardiomyocytes derived from hearts of two nonlabeled wild-type animals and a SILAC mouse. After cell homogenization, nonlabeled samples were mixed with the SILAC mouse sample and separated by 1D SDS-PAGE. Each lane was cut into 15 pieces prior to GeLC-MS/MS. Histograms contain SILAC ratios of both samples plotted against the log peak intensities, and most of the quantified proteins show a ratio close to 1:1

3.3 Labeled Cardiomyocytes (Heart Tissue) as Internal Standard for Protein Quantification and Proteome Analysis by GeLC-MS

After mixing equal protein amounts of the heavy sample to the respective light samples (Fig. 3), we separated in this protocol the proteins by a 1D SDS-PAGE (20) and performed a protein digestion with the endopeptidase LysC. In addition, alternate gel-free separation methods such as isoelectric focussing, reversed phase chromatography, and size exclusion chromatography are useful to perform a more in-depth analysis of the cardiac proteome:

1. Load maximal ~40 μg protein lysate (20 μg heavy cardiomyocytes extract + 20 μg light cardiomyocytes extract) mixed with loading buffer on a NuPAGE 4–12 % Bis-Tris gel with MOPS running buffer and visualize proteins by colloidal blue staining (Invitrogen) or other MS-compatible protein dyes.
2. Slice gel lanes into 10–15 pieces and cut each gel piece into smaller pieces of ~1 mm^2. Transfer the samples in separate 1.5 ml Eppendorf tubes. Heavily stained bands correspond mainly to very abundant proteins (e.g. myosin heavy chain) and should therefore cut out to their own slices (Fig. 3).
3. Wash gel pieces with 100 μl 1:1 ABC-buffer/EtOH for 20 min.
4. Repeat step 3. After each washing step, discard the supernatant. The colloidal dye should be washed out completely.
5. Dehydrate gel pieces by adding 100 μl abs. EtOH for 10 min, discard the supernatant and dry the pieces using vacuum centrifugation for 5 min. Samples can be stored at 4 °C after this step.

6. Incubate the gel pieces with 100 µl of 10 mM DTT in ABC-buffer at 56 °C for 45 min to reduce disulfide bonds of the proteins. Discard the supernatant.
7. Alkylate proteins by adding 100 µl 55 mM iodoacetamide (IAA) in ABC-buffer for 30 min at RT in the dark.
8. Wash gel pieces with 100 µl ABC-buffer for 15 min.
9. Dehydrate gel pieces with 100 µl abs. Ethanol for 15 min.
10. Wash gel pieces with 100 µl ABC-buffer for 15 min.
11. Dehydrate gel pieces twice with 100 µl 100 % EtOH for 15 min and dry the pieces using vacuum centrifugation for 5 min.
12. Add ~40 µl of 12 ng/µl LysC in ABC-buffer and let the gel pieces swell at 4 °C for 15 min. The volume of the protease depends on the size of the gel pieces. The gel pieces should absorb the complete digestion solution until saturation.
13. Add further ~100 µl ABC-buffer to cover the gel pieces and incubate the samples overnight at 37 °C.
14. After overnight digestion, transfer the supernatant into a new Eppendorf tube. The supernatant already contains digested peptides.
15. To recover digested peptides from the gel, sequentially extract gel pieces twice with 100 µl of 30 % Acetonitrile/3 % TFA for 10 min, twice with 70 % Acetonitrile for 10 min, and with 100 % Acetonitrile for 10 min. All extracts from one tube are pooled together, respectively.
16. Evaporate the Acetonitrile using a vacuum centrifuge and clean peptides, for example, with stop-and-go extraction tips (stage tips) (21).
17. Measure samples by mass spectrometry (see Note 8).

3.4 Proteomic Analysis of β-Parvin-Deficient Hearts

As an example for a quantitative proteome analysis of a knockout model, we describe the analysis of β-parvin-deficient hearts. β-parvin-deficient animals are viable and do not show an obvious heart phenotype:

1. Isolate heart tissue (or cardiomyocytes) and extract proteins with SDS based lysis buffer as described in Subheading 3.2 (see Note 9).
2. According to the protein concentration, mix the heavy sample with the light wild-type and the light knockout (β-parvin −/−) sample 1:1, respectively. Fractionate proteins with 1D SDS-PAGE and visualize proteins with colloidal blue staining.
3. Equally sized proteins bands were excised from both gel bands and processed for in gel digest with LysC as described in Subheading 3.3.

3.5 Data Analysis

1. Mass spectrometric data analysis can be performed by several software tools. Here we used the freely available MaxQuant (22).
2. After processing the RAW data for protein identification and quantitation, a mass spectrometric heat map provides an overview of detected SILAC pairs (Fig. 4a). The 3D isotopic cluster shows an example for the mass and the elution profile of a typical SILAC pair.
3. For the calculation of direct ratios, use normalized SILAC ratios. Proteins identified with very high or low SILAC ratios have to be carefully evaluated since keratins or other contaminants usually have no heavy counterpart. It is useful to plot the log-transformed ratios (light WT/light −/−) against the summarized peak Intensities of the light and heavy proteins to monitor the global ratio distribution. Most of the ratios should be localized around a 1:1 ratio. Alternatively, SILAC ratios can be grouped into bins and plotted against their relative frequency.
4. As expected, β-parvin was measured with a low ratio indicating a clear downregulation (~16-fold) in β-parvin −/− hearts (Fig. 4b) (see Note 10). In addition, we detected an over twofold increase of α-parvin, which indicates a potential compensation for the β-parvin function. Selected SILAC peptide pairs of α- and β-parvin are shown in Fig. 4c (see Note 11).

4 Notes

1. A detailed protocol for in-house food preparation is described in ref. 23. Most importantly, the diet should not contain the nonlabeled amino acid of the respective SILAC amino acid.
2. Use ultrapure collagenase and trypsin to disaggregate heart tissue since crude enzymes will lead to damage of cell surface proteins and lastly to cell death.
3. Preferentially use only dialyzed serum with a cutoff of 10 kDa to avoid contamination with light amino acids.
4. The most commonly used SILAC amino acids are lysine and arginine. However any other amino acid can be used for labeling. The advantage of the lysine and arginine combination is that after proteolytic digestion with trypsin, all tryptic peptides have at least on label in their peptide sequence. Since we are using only lysine for in vivo labeling, one should use the endopeptidase LysC, which cleaves only after lysine. In case of lysine labeling and trypsin digestion, all arginine-containing peptides are not quantifiable. However, LysC digest leads to longer peptides and results in a reduced number of mass spectrometric accessible peptides. Further developments on instrumentation and software will most likely improve the detection of

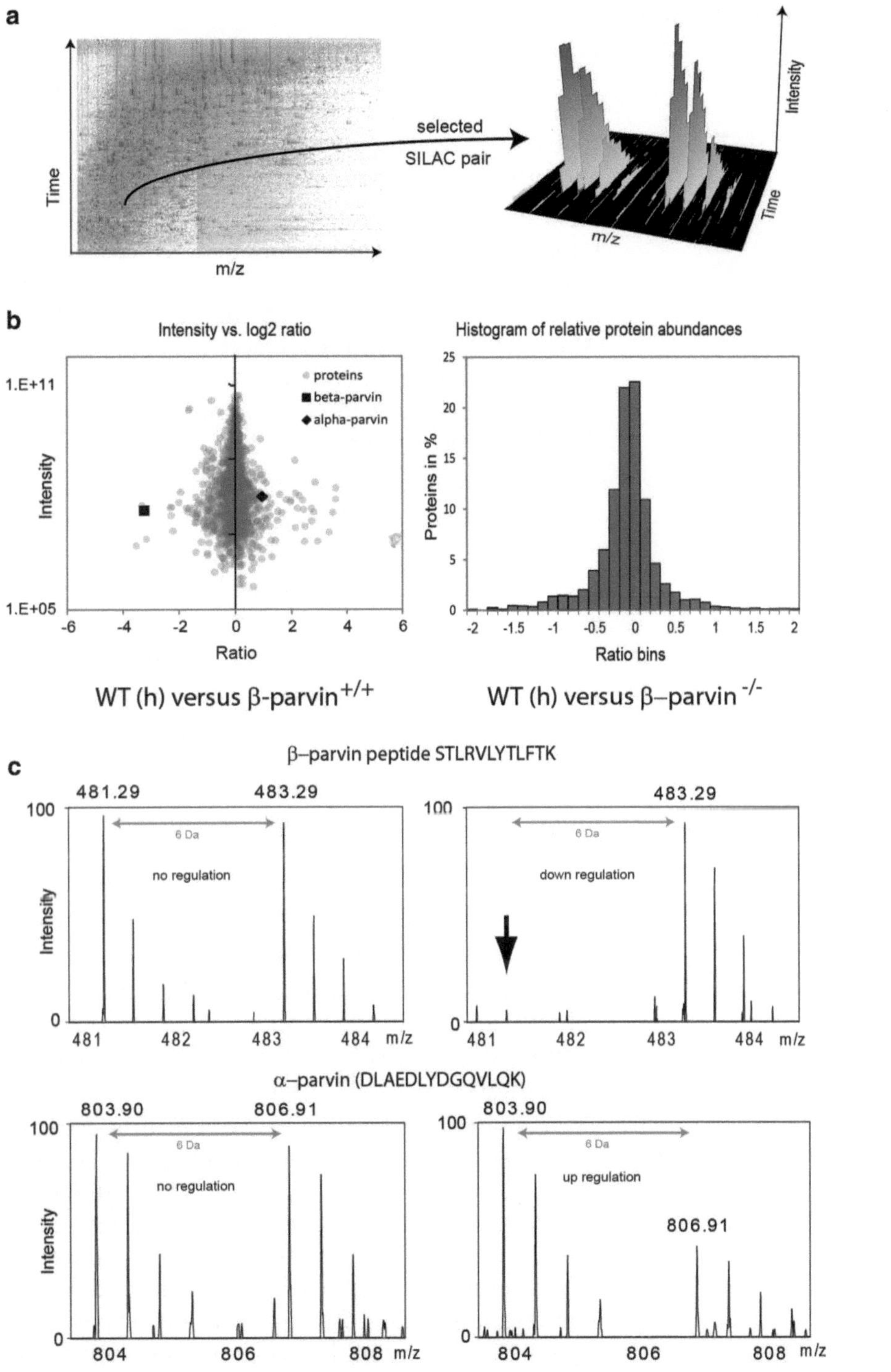

Fig. 4 Analysis of β-parvin-deficient hearts. (**a**) Mass spectrometric heat map indicates density of detected peptides (*dark areas* indicates high intensity whereas *light* areas reflect lower peptide intensities). A three-dimensional SILAC pair with a 1:1 ratio is shown on the right site. (**b**) Ratio distribution between β-parvin knockout and wild-type animals. The frequency histogram on the right site confirms the correct normalization of the dataset. The β-parvin ratio is more than 15-fold downregulated, and the assignment of a β-parvin SILAC ratio is due to the detection of background peaks at the same mass range. (**c**) The selected peptide spectra (left WT right KO) of β- and α-Parvin confirm the observed ratios

longer peptides. Although we do not tested arginine labeling in mice, it was shown for other organisms like yeast and flies that a substantial proportion of arginine is converted to proline and other amino acids (24).

5. Lys6 incorporation rates can be checked by taking of blood samples after different labeling periods. However, serum proteins show a rather fast Lys6 incorporation, and they do not necessary represent the Lys6 incorporation of other tissues. Alternatively, cutting a small piece (~2 mm) from the tip of the mouse tail will represent the incorporation rate of a solid tissue. We anticipate a food consumption of approx. 1 kg Lys6 diet for the first generation.

6. Avoid cutting the aorta to short which can lead to a problematic cannulation or a damaged aortic valve. The length of the aorta should be about 2–3 mm.

7. Cannulating and connecting the heart to the perfusion system should be done very quickly to prevent incomplete perfusion, structural damage, and subsequently protein degradation.

8. In this study, we used liquid chromatography combined with a high-performance tandem mass spectrometer for protein identification (LC-MS/MS). In general, the setup consists of an online coupled nano-HPLC system to separate peptides via reversed phase chromatography, electrospray ionization (ESI), and peptide identification/quantification with a high-performance mass spectrometer. Preferentially, a fast mass spectrometer with high resolution and high mass accuracy should be used (e.g., LTQ Orbitrap Velos).

9. To wash out blood, we recommend a cardiac perfusion with cold PBS. A detailed description of the standard operation procedure can be downloaded from www.research.buffalo.edu/laf/resources/sop/2A12.pdf.

10. The peptide quantification by MaxQuant is carried out in two subsequent steps. First, all unambiguous SILAC pairs are quantified. In the second run, MaxQuant attempts to quantify SILAC pairs having seemingly one of the peaks missing. Usually, this occurs if the regulation is very strong or the lower peak intensity is in range of the background noise. However, identification of the correct peak is mostly possible due to the high mass accuracy of the instrument. In rare cases, a random peak overlaps to the required m/z range and is therefore wrongly selected for quantification. To avoid this feature, the second search can be omitted be deactivating the "requantify" option in MaxQuant.

11. For single measurements statistical tools like t-test and ANOVA are not available. To get a rough estimate of which proteins are significantly regulated, one can use the intensity-dependent p

value (significance B) between heavy/light ratios calculated by MaxQuant. Alternatively, one can use the inverse of the normal cumulative distribution of Excel (NORM.INV). The parameters of the $\log_2$ ratio distribution together with the required probabilities (e.g., 0.05 for lower and 0.95 for upper cutoff) reflect possible cutoff values. In case for a typical spike-in SILAC experiment, we expect usually a standard deviation of 15–20 %, and therefore, a fold change greater than 1.5 should be a measurable outlier. However, the analysis of multicellular tissues from different animals results in a slightly higher variation (±30 %), and a fold change greater than 2 should be considered as a conservative threshold for outliers. Nevertheless we strongly recommend performing at least two repeats of all measurements. This enables access to many powerful statistical tools, similar to large-scale genomic data.

Acknowledgments

This work was supported by the Max Planck Society, the Excellence Initiative "Cardiopulmonary System," and the University of Giessen-Marburg Lung Center (UGMLC). In addition, we like to thank Silantes for the development of the Lys6 diet and Ingo Thievessen for providing the β-parvin knockout mice.

References

1. Mathers CD, Loncar D (2006) Projections of global mortality and burden of disease from 2002 to 2030. PLoS Med 3:e442
2. Nagaraj N et al (2011) Deep proteome and transcriptome mapping of a human cancer cell line. Mol Syst Biol 7:548
3. Ong SE, Mann M (2005) Mass spectrometry-based proteomics turns quantitative. Nat Chem Biol 1:252–262
4. Ong SE et al (2002) Stable isotope labeling by amino acids in cell culture, SILAC, as a simple and accurate approach to expression proteomics. Mol Cell Proteomics 1:376–386
5. Soufi B et al (2010) Stable isotope labeling by amino acids in cell culture (SILAC) applied to quantitative proteomics of *Bacillus subtilis*. J Proteome Res 9:3638–3646
6. Gruhler A et al (2005) Quantitative phosphoproteomics applied to the yeast pheromone signaling pathway. Mol Cell Proteomics 4:310–327
7. Larance M et al (2011) Stable-isotope labeling with amino acids in nematodes. Nat Methods 8:849–851
8. Sury MD, Chen JX, Selbach M (2010) The SILAC fly allows for accurate protein quantification in vivo. Mol Cell Proteomics 9:2173–2183
9. McClatchy DB, Yates JR, III (2008) Stable Isotope Labeling of Mammals (SILAM). CSH Protoc pdb prot4940
10. Kruger M et al (2008) SILAC mouse for quantitative proteomics uncovers kindlin-3 as an essential factor for red blood cell function. Cell 134:353–364
11. Waterlow JC (1995) Whole-body protein turnover in humans—past, present, and future. Annu Rev Nutr 15:57–92
12. Schwanhausser B et al (2009) Global analysis of cellular protein translation by pulsed SILAC. Proteomics 9:205–209
13. Doherty MK et al (2005) Proteome dynamics in complex organisms: using stable isotopes to monitor individual protein turnover rates. Proteomics 5:522–533
14. Ong SE, Mann M (2006) A practical recipe for stable isotope labeling by amino acids in cell culture (SILAC). Nat Protoc 1:2650–2660

15. Walther DM, Mann M (2011) Accurate quantification of more than 4000 mouse tissue proteins reveals minimal proteome changes during aging. Mol Cell Proteomics 10(M110): 004523
16. Scholten A et al (2011) In-depth quantitative cardiac proteomics combining electron transfer dissociation and the metalloendopeptidase Lys-N with the SILAC mouse. Mol Cell Proteomics 10(O111):008474
17. Jordan SD et al (2011) Obesity-induced overexpression of miRNA-143 inhibits insulin-stimulated AKT activation and impairs glucose metabolism. Nat Cell Biol 13:434–446
18. Drexler HC et al (2011) On marathons and sprints: an integrated quantitative proteomics and transcriptomics analysis of differences between slow and fast muscle fibers. Mol Cell Proteomics 11:M111.010801
19. O'Connell TD, Rodrigo MC, Simpson PC (2007) Isolation and culture of adult mouse cardiac myocytes. Methods Mol Biol 357: 271–296
20. Shevchenko A et al (2006) In-gel digestion for mass spectrometric characterization of proteins and proteomes. Nat Protoc 1:2856–2860
21. Rappsilber J, Ishihama Y, Mann M (2003) Stop and go extraction tips for matrix-assisted laser desorption/ionization, nanoelectrospray, and LC/MS sample pretreatment in proteomics. Anal Chem 75:663–670
22. Cox J et al (2009) A practical guide to the MaxQuant computational platform for SILAC-based quantitative proteomics. Nat Protoc 4:698–705
23. Zanivan S, Krueger M, Mann M (2012) In vivo quantitative proteomics: the SILAC mouse. Methods Mol Biol 757:435–450
24. Bicho CC et al (2010) A genetic engineering solution to the "arginine conversion problem" in stable isotope labeling by amino acids in cell culture (SILAC). Mol Cell Proteomics 9:1567–1577

Chapter 5

Global Proteomic Profiling and Enrichment Maps of Dilated Cardiomyopathy

Ruth Isserlin, Daniele Merico, and Andrew Emili

Abstract

Global expression profiling is a promising technique to help uncover perturbations associated with heart disease. With the large amount of expression data generated, we demonstrate how a list of differentially expressed proteins or genes can be translated into a list of perturbed pathways and functions using conventional overrepresentation analysis such as gene set enrichment analysis, DAVID, and BiNGO. As outputs from ORA can be daunting in themselves, we further demonstrate how to visualize enrichment results as Enrichment Maps for easy interpretation and analysis.

Key words Mass spectrometry, Protein expression, Proteomics, Pathway analysis, Cytoscape, Enrichment analysis, Enrichment Map, Gene set enrichment analysis (GSEA)

1 Introduction

Cardiovascular disease (CVD) and heart failure (HF) specifically is a leading cause of mortality. HF represents an increasing share of CVD-related mortality as a result of an aging population and successes in treating heart attacks. Despite recent progress, the molecular mechanisms of cardiomyopathies that result in decreased cardiac contractility and ultimately overt HF remain unclear. Often it is difficult to distinguish the resulting compensatory response from the initial cause, but in order to devise better diagnosis and treatments of HF, it is vital that we uncover all the processes and pathways involved in the progression of the disease.

Global proteomic profiling is a promising approach for characterizing protein perturbations associated with cardiovascular disorders or any disease. Using exhaustive shotgun tandem mass spectrometry (MS/MS)-based sequencing to examine quantitative changes in protein expression patterns, one can measure protein relative abundance, represented as spectral counts, for thousands of proteins across multiple time points and cellular locations.

Fernando Vivanco (ed.), *Heart Proteomics: Methods and Protocols*, Methods in Molecular Biology, vol. 1005,
DOI 10.1007/978-1-62703-386-2_5, © Springer Science+Business Media New York 2013

Translating these profiles into meaningful biological insights can subsequently be achieved through overrepresentation analysis (ORA) which aims at detecting statistically significant differential expression of functionally related proteins or "gene sets." A "gene set" is a collection of gene products that share some biological attribute or feature, for example, all the components annotated to a common pathway or all the gene products with a specific protein domain.

Numerous tools are available to perform ORA (reviewed by Huang et al. (1)), most requiring different types of data and/or annotation input all generating slightly different outputs, but the typical end result is usually lists of overrepresented gene sets (ideally with an associated statistic). Some popular tools, like BiNGO (2) and David (3), are used on subsets of the total protein list selected a priori, such as all significantly overexpressed proteins (chosen using some differential score threshold), to summarize the strongest signals generated from an expression profiling experiment. Alternatively, other tools, such as GSEA (4), analyze the ranking of all the gene products according to a differential statistic, and then the annotated gene sets are tested to see if members lie more toward the top or bottom of the ranking than expected by chance alone. This latter approach capitalizes on all the expression signals available as opposed to just those with values on the extreme tails of the distribution.

Regardless of the strategy used, the list of overrepresented gene sets generated by either of these methods can often be as long and daunting to interpret as the initial list of differentially expressed proteins. To aid in the interpretation of these results, we have developed Enrichment Map (5), a Cytoscape plugin, to intuitively group, visualize, and compare enrichment results across different disease time points or experimental conditions. This has been applied successfully to reveal dysregulated known apoptotic regulators in a time-course study of heart failure in a transgenic mouse model of dilated cardiomyopathy (6). Below we highlight our preferred implementation of ORA and EM analysis for interpreting gene product (mRNA or protein) expression profiles as might be generally generated from studies of other cardiovascular conditions.

2 Materials

Material preparation can be divided into two sections. First all software required for the analysis needs to be installed followed by collation and formatting of experimental results into required file formats.

2.1 Software Installation

Install GSEA, Cytoscape, and Enrichment Map Cytoscape plugin:

1. GSEA installation: Prior to installing GSEA, verify that your computer has Java 6 installed (see Note 1). In your web browser, navigate to the GSEA home page (http://www.broadinstitute.org/gsea/). Click on *Downloads* in the page header (see Note 2). From the "javaGSEA Desktop Application" right click on "*Launch with 1 Gb memory*" (see Note 3). Click on "*Save Taget as…*" and save shortcut to your desktop or your folder of choice so you can launch GSEA for your analysis without having to navigate to it through your web browser.
2. Cytoscape installation: In your web browser, navigate to the Cytoscape home page (http://www.cytoscape.org/). Click on *Download* in the page header (see Note 2). Select your platform-specific installer and download the latest version. Double click on the downloaded executable and follow instructions in order to install the application.
3. Enrichment Map Cytoscape plugin installation: Enrichment Map is still actively being developed but is only released periodically. Often there are new features in the beta version that have yet to be incorporated into the stable release version. In order to use these new features, you must install the beta version of the plugin (see Note 4). Choose one of the following options:
   (a) Stable release installation: Launch Cytoscape. (On Windows, click on Start/Programs/Cytoscape/Cytoscape, on Mac OsX click on Applications/Cytoscape.) In Cytoscape, click on Plugins/Manage Plugins. Under "Available to Install," expand "Analysis" by clicking on the "+." Click on the highest version of Enrichment Map (e.g., EnrichmentMap v.1.2). Click on "*Install*" (see Note 5).
   (b) Beta release installation: Navigate to the Enrichment Map home page (http://www.baderlab.org/Software/EnrichmentMap). Download the zipped development version. Unzip the downloaded archive and copy the .jar file to your Cytoscape plugin directory. (On Windows with a default Cytoscape installation, this directory can be found at "C:\Program Files\Cytoscape_v#.#.#\plugins," and on Max OSX with a default Cytoscape installation, this directory can be found at "/Applications/Cytoscape_v#.#.#/plugins") (see Note 6).

2.2 Required Files and Formats

No matter which ORA method one chooses, there are three essential files that need to be created or obtained in order to complete the analysis: an expression dataset or gene product ranking (e.g., differential expression score), a gene set definition file, and an enrichment results file.

2.2.1 Sample Data

Sample protein abundance data was generated using our previously published data from our PLN-R9C dilated cardiomyopathy mouse time-course proteomic profiling study (7). Protein samples were collected from ventricular tissues from two distinct strains of mice, one a transgenic expressing a mutated phospholamban protein leading to DCM similar to the human condition, and the other a healthy control. Three time points were collected representing early-stage (8 weeks), mid-stage (16 weeks), and end-stage (24 weeks). The sample data can be downloaded from http://www.baderlab.org/Data/ProteomicsEM and consists of:

1. R9C early-stage disease files:
   (a) Protein expression file.
   (b) Ranked protein expression file.
   (c) GSEA enrichment results using ranked protein expression file.
   (d) BiNGO enrichment results using all significantly (p-value < 0.05) overexpressed proteins.
   (e) DAVID enrichment results using all significantly (p-value < 0.05) overexpressed proteins.
2. R9C mid-stage disease files:
   (a) Protein expression file.
   (b) Ranked protein expression file.
   (c) GSEA enrichment results using ranked protein expression file.
   (d) BiNGO enrichment results using all significantly (p-value < 0.05) overexpressed proteins.
   (e) DAVID enrichment results using all significantly (p-value < 0.05) overexpressed proteins.

2.2.2 Protein Expression

A file listing the expression values or significance measure for each protein across multiple experiments.

1. Raw protein expression (*.txt): Represented as a matrix where each row is a different protein identified by some identifier (gene symbol, Entrez gene ID, or UniProt accession are common choices) and description followed by any number of columns where each column is a different experimental condition, disease stage, phenotype, or replicate. The value in each cell represents the magnitude of expression of the given protein in the specified experiment (see Note 7). The file should contain a header row with unique names for each column.

   <Gene/protein ID> (tab) <description> (tab) <condition 1> (tab) <condition 2> …
2. Scored differential protein expression (*.rnk): Represented as a matrix where each row is a different protein identified by some

identifier (gene symbol, Entrez gene ID, or UniProt accession are common choices) followed by one column representing the comparison between two or more experimental conditions. Different from the raw protein expression, the value in each cell is a rank or static indicating the relative differential expression the protein has between the two states or conditions (i.e., disease and control) (see Note 8).

<Gene/protein ID>(tab)<condition 1 rank/statistic>

2.2.3 Gene Set Definitions

A gene set is a collection of genes that share some attribute or function. Gene set definitions are required to be in the following file format (.gmt) (see Note 9):

<Gene set Name>(tab)<description>(tab)<gene 1>(tab)<gene 2>...

Gene sets can be created from functional categories, pathways, structural similarities, or any experimental data or trend. There are many different sources that release gene set definition files, but the challenge is collating multiple resources while maintaining consistent identifier usage. A good source for collated gene sets including gene ontology annotations, disease phenotypes, pathways, transcription factors, and microRNAs (with a choice of identifier including Entrez gene, gene symbol or UniProt accession) can be downloaded from http://download.baderlab.org/EM_Genesets/ (5).

2.2.4 Enrichment Results

In its simplest form, enrichment results consist of a list of gene sets found to be enriched in the protein expression set. Enrichment Map recognizes multiple formats from popular enrichment tools including GSEA (see Subheadings 3.1 and 3.2 for methods to generate), DAVID (see Subheading 3.3 for method to generate), and BiNGO (see Subheading 3.4 for method to generate).

In addition to these formats, Enrichment Map also accepts a generic file format so no matter which enrichment program you use as long as its results are formatted in the following format they can be used:

<Gene set Name>(tab)<Description>(tab)<*p*-value>(tab)<corrected *p*-value (see Note 10)>(tab)<Phenotype (see Note 11)>

3 Methods

3.1 Generating Enrichment Results from Raw Protein Expression Files Using GSEA

1. Double click on GSEA icon you created when you installed it.
2. Click on "*Load data*" in left panel.
3. Click on "*Browse for files...*" in newly opened "Load data" pane.

4. Navigate to directory where you stored your expression and gene set file. Select raw protein expression (.txt) file and gene set (.gmt) file. Click on "*Open.*"
5. Wait until confirmation box appears indicating that all files loaded successfully. Click on "*Ok.*"
6. Click on "*Run GSEA*" in left panel.
7. Click on the arrow next to the "Expression dataset" text box. Select the expression set you wish to run the analysis on.
8. Click on "..." next to the text box of "Gene Set Database."
9. Click on "Gene Matrix (local gmx/gmt)" tab.
10. Select gmt file you uploaded and click on "*Ok.*"
11. Click on "..." next to the text box of "Phenotype labels."
12. Under Options, click on "*Create an on-the-fly phenotype...*" button.
13. Under "Samples for class A (one per line)," enter the column names from the expression file that belong to class A (using the example file for early disease, this list should be Early_Disease_1 <new line> Early_Disease_2 <new line> Early_Disease_3 <new line> Early_Disease_4 <new line> Early_Disease_5 <new line> Early_Disease_6 <new line> Early_Disease_7 <new line> Early_Disease_8 <newline> Early_Disease_9 <newline> Early_Disease_10 <new line>).
14. Under "Samples for class B (one per line)," enter the column names from the expression file that belongs to class B (using the example file for early disease, this list should be Early_Control_1 <newline> Early_Control_2 <newline> Early_Control_3 <new line> Early_ Control_4 <new line> Early_ Control_5 <new line> Early_Control_6 <newline> Early_Control_7 <newline> Early_Control_8 <newline> Early_Control_9 <newline> Early_Control_10 <new line>).
15. Under "Enter a brief name for class A," enter "Disease."
16. Under "enter a brief name for class B," enter "Control."
17. Make sure the "Dataset" selected is correct (if not click on the arrow and update the Dataset).
18. Click on "*Apply to Dataset.*"
19. Click on "*OK*" (see Note 12).
20. Click on "*Close.*"
21. Under "Select one phenotype", "Disease_versus_Control" and "Control_versus_Disease" should appear. Select "Disease_versus_Control."
22. Click on "*OK.*"
23. Click on the down arrow next to the text box for "Collapse dataset to gene symbols." Select "false."

24. Click on the down arrow next to the text box for "Permutation type." Select "gene_set."
25. Click on "Show" next to "Basic fields."
26. Click in text box next to "Analysis name" and rename (example—early_stage_expression_gsea_enrichment_results).
27. Click on "..." next to "Save results in this folder" text box. Navigate to the folder where you wish to save the results (preferably the same directory where all the input files have been saved).
28. Click on "*Run*" in the bottom right corner.
29. Enrichment results file will be saved to folder specified and will be used in Subheading 3.5.

3.2 Generating Enrichment Results from Scored Differential Expression Files Using GSEA

1. Double click on GSEA icon you created when you installed it.
2. Click on "*Load data*" in left panel.
3. Click on "*Browse for files...*" in newly opened "Load data" pane.
4. Navigate to directory where you stored your expression and gene set file. Select scored expression (.rnk) file and gene set (.gmt) file. Click on "*Open*."
5. Wait until confirmation box appears indicating that all files loaded successfully. Click on "*Ok*."
6. Click on the Tools menu and select "*GSEA PreRanked*."
7. In "run GSEA on Preranked Gene list" Window Click on "..." next to text box of "Gene Set Database."
8. Click on "Gene Matrix (local gmx/gmt)" tab.
9. Select gmt file you uploaded and click on "*Ok*."
10. Click on the down arrow next to the text box for "Ranked List." Select the ranked list you wish to run the analysis on.
11. Click on the down arrow next to the text box for "Collapse dataset to gene symbols." Select "false."
12. Click on "*Show*" next to "Basic fields."
13. Click in text box next to "Analysis name" and rename (example—early_stage_ranked_gsea_enrichment_results).
14. Click on "..." next to "Save results in this folder" text box. Navigate to the folder where you wish to save the results (preferably the same directory where all the input files have been saved).
15. Click on "*Run*" in the bottom right corner.
16. Enrichment results file will be saved to folder specified and will be used in Subheading 3.5.

3.3 Generating Enrichment Results from Scored Differential Expression Files Using BiNGO

1. Open Cytoscape.
2. Click on Plugins/Start Bingo v# # #.
3. Enter the name "top up-regulated genes" in the text box marked "Cluster Name."
4. Select the box "Paste Genes From Text."
5. Define a set of genes you are interested in testing, i.e., all significantly upregulated genes. Select and copy all genes. Paste in large text box.
6. Change "Select ontology file" to "GO_full."
7. Change "Select organism/annotation" to "Homo sapiens" (or species relevant to your study).
8. Select the box "Check box for saving Data."
9. Click on "*Save BiNGO Data file in*:"
10. Navigate to desired folder and Click on "*Save*."
11. Click on "*Start BiNGO*."
12. Enrichment results file will be saved to folder specified and will be used in Subheading 3.6.

3.4 Generating Enrichment Results from Scored Differential Expression Files Using DAVID

1. Go to DAVID website—http://david.abcc.ncifcrf.gov/.
2. Define a set of genes you are interested in testing, i.e., all significantly upregulated genes. Select and copy all genes.
3. In "Upload" tab of DAVID interface Paste genes in text box marked "Step1: Enter Genelist."
4. Select "Official Gene Symbol" in "Step 2: Select Identifier" if using gene symbols.
5. Select "Gene list" in "Step 3: Select List Type."
6. Click on "*Submit list*."
7. Select species: "Homo sapiens" (or species relevant to your study).
8. Click on "*Functional Annotation Chart*."
9. Save file to desired directory. This is the file that can be used in Subheading 3.7.

3.5 Creating an Enrichment Map from GSEA Enrichment Results

(A) Manually loading all required files

1. Open Cytoscape.
2. Click on Plugins/Enrichment Map/Load Enrichment Map Results.
3. Click on "..." next to "GMT" in User Input/Gene Sets in left hand Enrichment Map frame in the Cytoscape Control Panel.
4. Navigate to the directory where you stored all the inputs and outputs for the analysis. Select your gene set file (.gmt).
5. Click on "*Open*."

6. Click on "…" next to "Expression" in User Input/Datasets/Dataset 1 in left hand Enrichment Map frame in the Cytoscape Control Panel.
7. Navigate to the directory where you stored all the inputs and outputs for the analysis. Select your expression or ranked file (.txt or .rnk).
8. Click on "*Open.*"
9. Click on "…" next to "Enrichments 1" in User Input/Datasets/Dataset 1 in left hand Enrichment Map frame in the Cytoscape Control Panel.
10. Navigate to the directory where you stored the GSEA results.
11. Double click on directory with the results you wish to view (e.g., early_stage_ranked_gsea_enrichment_results.GSsea Preranked.#############).
12. Select the results file named gsea_report_for_CLASSA.#############.xls.
13. Click on "*Open.*"
14. Click on "…" next to "Enrichments 2" in User Input/Datasets/Dataset 1 in left hand Enrichment Map frame in the Cytoscape Control Panel.
15. Navigate to the directory where you stored the GSEA results.
16. Double click on directory with the results you wish to view (e.g., early_stage_ranked_gsea_enrichment_results.GSsea Preranked.#############).
17. Select the results file named gsea_report_for_CLASSB.#############.xls.
18. Click on "*Open.*"
19. Tune parameters in "Parameters" section (see Note 13).
20. Click on "*Build.*"

(B) Using an .rpt file—an .rpt file is a configuration file created by GSEA indicating the names of all the input and output files used for a specific analysis

1. Open Cytoscape.
2. Click on Plugins/Enrichment Map/Load Enrichment Map Results.
3. Click on "…" next to "Expression" in User Input/Datasets/Dataset 1 in left hand Enrichment Map frame in the Cytoscape Control Panel.
4. Navigate to the directory where you stored the GSEA results.
5. Double click on directory with the results you wish to view (e.g., early_stage_ranked_gsea_enrichment_results.GSsea Preranked.#############).

6. Select the .rpt file from the results directory (e.g., early_stage_ranked_gsea_enrichment_results.GSsea Preranked.#############.rpt).
7. Click on "*Open.*"
8. All fields in Enrichment Map Panel should be populated (see Note 14).
9. Tune parameters in "Parameters" section (see Note 13).
10. Click on "*Build.*"

3.6 Creating an Enrichment Map from BiNGO Enrichment Results

1. Open Cytoscape.
2. Click on Plugins/Enrichment Map/Load Enrichment Map Results.
3. In "Analysis Type" select "DAVID/BiNGO."
4. Click on "..." next to "Expression" in User Input/Datasets/Dataset 1 in left hand Enrichment Map frame in the Cytoscape Control Panel.
5. Navigate to the directory where you stored all the inputs and outputs for the analysis. Select your expression (.txt).
6. Click on "*Open.*"
7. Click on "..." next to "Enrichments" in User Input/Datasets/Dataset 1 in left hand Enrichment Map frame in the Cytoscape Control Panel.
8. Navigate to the directory where you stored the BiNGO results.
9. Select the results file (e.g., earlydisease_upregulated_BiNGO_results.txt).
10. Click on "*Open.*"
11. Tune parameters in "Parameters" section (see Notes 13 and 15).
12. Click on "*Build.*"

3.7 Creating an Enrichment Map from DAVID Enrichment Results

1. Open Cytoscape.
2. Click on Plugins/Enrichment Map/Load Enrichment Map Results.
3. In "Analysis Type" select "DAVID/BiNGO."
4. Click on "..." next to "Expression" in User Input/Datasets/Dataset 1 in left hand Enrichment Map frame in the Cytoscape Control Panel.
5. Navigate to the directory where you stored all the inputs and outputs for the analysis. Select your expression (.txt).
6. Click on "*Open.*"
7. Click on "..." next to "Enrichments" in User Input/Datasets/Dataset 1 in left hand Enrichment Map frame in the Cytoscape Control Panel.

8. Navigate to the directory where you stored the DAVID results.
9. Select the results file (e.g., DAVID_earlystage_enrichment_results.txt).
10. Click on "*Open.*"
11. Tune parameters in "Parameters" section (see Note 13).
12. Click on "*Build.*"

3.8 Exploring the Enrichment Map

1. The results from Subheadings 3.5–3.7 will be an Enrichment Map where each node represents an individual enriched term and each edge connecting two nodes represents the degree of overlap between them calculated based on the coefficient specified in the input panel (Jaccard, Overlap, or Jaccard + Overlap).
2. Clusters of highly connected nodes represent a group of annotations that are very similar and do not indicate that the particular function is more significant or more important than a lone node. Different areas have been studied more and are represented in multiple datasets.
3. Clicking on an individual node in the network will create a heat map of the expression values for all proteins that are part of that gene set in the Data Panel "EM Geneset Expression viewer" tab.
4. Clicking on an individual edge in the network will create a heat map of the expression values for all proteins that are part of the overlap between the two nodes the edge connects in the Data Panel "EM Overlap Expression viewer" tab.
5. Clusters can be summarized using the Cytoscape plugin WordCloud (8).

4 Notes

1. GSEA can only run with Java 6 or higher. To verify your Java version, navigate to http://www.java.com/en/download/installed.jsp. Click on "*Verify Java Version*" to get your currently installed version. Once the program has assessed your Java version, the refreshed page will include your current version (e.g., "Your Java version: Version 6 Update 25"). As long as your version is 6 or higher, there is no need to update Java even if there is a newer version. If your Java version is less than 6, click on "*Download Java Now*" to update.
2. GSEA and Cytoscape are freely available tools, but in order to download it, you are required to register and agree to their licensing agreement. To register to GSEA, click on "*Click*

here"; enter your name, email address, organization, and country in the designated fields; and agree to the licensing agreement. For all subsequent downloads, you simply enter the email address registered in order to get to the download site. To register to Cytoscape, simply enter your name, organization, and email address; select "I accept terms of use:"; and click on "*Proceed to Download*" every time you visit their download site.

3. Ideally it is better to launch GSEA with 1 Gb memory, but older computers might not have the memory resources to do this. If GSEA does not launch with 1 Gb memory use, the "*Launch with 512 Mb memory*" instead.
4. The beta version has not been tested to the same extent as the stable version and therefore might not perform optimally. Some of the additional features found in the beta version can be indispensable to your analysis so is well worth using. Any bugs or suggestions can be reported to enrichmentmap-dev@googlegroups.com and are often promptly addressed.
5. If you are unable to find the Enrichment Map plugin listed under "Available to Install/Analysis," verify that you have not already installed by checking "Currently Installed/Analysis." If it is already installed but you wish to upgrade it to the latest version, click on "*close*" in the Manage Plugins Window. Select Plugins/Update Plugins. If there is a newer version, a Window will appear. Double click on "*Updatable Plugins*," double click on "*Analysis*," double click on "*EnrichmentMap v.#.#*," select most recent version, and click on "*Update Selected.*" Once it has finished installing, click on "*Close.*"
6. If the Enrichment Map plugin is already installed on your computer and was installed through the Cytoscape plugin manager, it is very important that the older version is erased as Cytoscape will by default preferentially use the version installed by the plugin manager. To verify which version of the Enrichment Map plugin you are using, go to "Plugins/Enrichment Map/About." The version is listed in the title of the about box. If the version does not reflect the one you installed close Cytoscape. On Windows, delete the older version from "C:\Documents and Settings\User\.cytoscape\#.#\plugins," and on Mac OSX, delete the older version from "~/.cytoscape/#.#/plugins."
7. Expression values can be represented as spectral counts, mass spectrometry intensities, or any form of measurement as long as the values in the matrix are measured consistently across all experiments for all proteins and are relatively reproducible across replicates. Different technologies will require varying normalization methods to account for noise and discrepancies generated by the technique.

8. The statistic or rank used for the analysis is very dependent on the experimental design. Commonly *p*-values are generated to highlight differentially expressed proteins between two conditions. Depending on the technique used to measure protein expression, there are different tests that can be used to generate these *p*-values (e.g., *t*-test, ks-test, or more complicated linear models). It is important to note that often the *p*-value has no indication toward which distribution the statistic is significant (i.e., condition A or condition B) and instead simply proves or disproves the null hypothesis (which is defined based on the test being used). For some tests, this can be calculated from the test statistic generated, for instance, the *t*-statistic generated by the *t*-test is positive when condition A is significant and negative when condition B is significant. In this case, using the *t*-statistic as the rank will produce the best results for a program like GSEA. Not all statistical tests produce signed statistics so an easy way to convert your list of differentially expressed proteins into a ranked list (from most upregulated to most downregulated) is to calculate $(1-p\text{-value}) \times$ sign of [log 2 (condition A/condition)].
9. Gene set collections are required when running GSEA and Enrichment Map. If using BiNGO or David, the gene sets are not required to be in a gmt file and are instead deduced from the output files generated.
10. Although the generic file format requires a corrected *p*-value if the method used does not generate one, then you can simply duplicate the values in the *p*-value column and place them into the corrected *p*-value column.
11. Phenotype values are used to identify if the gene set is enriched in up- or downregulation or in either of the two phenotypes being compared in the two-class analysis. Phenotype values can be 1 or (−1). Sets with the value of 1 will be red, and set with the value of (−1) will be blue.
12. If the columns you entered are not in the file or contain spelling mistakes, the phenotypes will not be successful. If an error message appears, verify that the names of the columns entered match exactly what is found in the expression file.
13. The parameters used will be dependent on the type of data used for the analysis. Generally you can start with more permissive values (*p*-value cutoff of 0.05 and FDR *Q*-value cutoff of 0.25). If there are excessive amounts of results, you can tune this setting from within the map generated.
14. If any of the text in the Enrichment Map Panel is red, then the application is unable to find that particular file in the place the .rpt file has specified it is. This often happens when the results directory is moved from one computer to another. In order to

fix this, wherever there is red text, click on "…" next to it and navigate to the correct file.

15. Unless you have created your own annotation files for BiNGO, the only files available are gene ontology annotation files. Due to the hierarchical structure of GO to avoid a hairball in the resulting Enrichment Map changing the edge weight calculation from the default "Overlap Coefficient cut-off" to "Jaccard Coefficient cut-off" drastically improves the results.

References

1. Huangda W, Sherman BT, Lempicki RA (2009) Bioinformatics enrichment tools: paths toward the comprehensive functional analysis of large gene lists. Nucleic Acids Res 37:1–13
2. Maere S, Heymans K, Kuiper M (2005) BiNGO: a Cytoscape plugin to assess overrepresentation of gene ontology categories in biological networks. Bioinformatics 21: 3448–3449
3. Dennis G Jr et al (2003) DAVID: database for annotation, visualization, and integrated discovery. Genome Biol 4:P3
4. Subramanian A et al (2005) Gene set enrichment analysis: a knowledge-based approach for interpreting genome-wide expression profiles. Proc Natl Acad Sci USA 102:15545–15550
5. Merico D et al (2010) Enrichment map: a network-based method for gene-set enrichment visualization and interpretation. PLoS One 5:e13984
6. Isserlin R et al (2010) Pathway analysis of dilated cardiomyopathy using global proteomic profiling and enrichment maps. Proteomics 10:1316–1327
7. Gramolini AO et al (2008) Comparative proteomics profiling of a phospholamban mutant mouse model of dilated cardiomyopathy reveals progressive intracellular stress responses. Mol Cell Proteomics 7:519–533
8. Oesper L et al (2011) WordCloud: a Cytoscape plugin to create a visual semantic summary of networks. Source Code Biol Med 6:7

Chapter 6

Characterization of the Human Myocardial Proteome in Dilated Cardiomyopathy by Label-Free Quantitative Shotgun Proteomics of Heart Biopsies

Elke Hammer, Katrin Darm, and Uwe Völker

Abstract

Proteomic profiling of heart tissue might help to discover the molecular events related to or even causing cardiovascular diseases in human. However, this material is rare and only available from biopsies taken for diagnostics, e.g., assessment of inflammatory events or virus persistence. Within this chapter, we describe a workflow for the quantitative proteome analysis of heart biopsies. Starting with 1–2 mg of tissue material, crude protein extracts were prepared, digested with LysC and trypsin, and then analyzed by LC-ESI-tandem mass spectrometry. Due to the low technical variance, the method can be used for label-free quantitation of disease-specific alterations in the human heart. Methods discussed include homogenization of biopsy tissue, sample preparation, proteolytic digestion, as well as data analysis for label free quantitation.

Key words Proteomics, Label-free quantitation, Human heart biopsies

1 Introduction

Dilated cardiomyopathy (DCM) is a chronic myocardial disease characterized by progressive depression of myocardial contractile function and by ventricular dilatation. It is one of the major causes of heart failure and the leading causes for heart transplantation (1). Quantitative proteomics of cardiac tissue for characterization of the disease status or prediction of progression is still a challenge (2, 3). Proteomic studies on human specimen have been reported early on relying on analysis of silver-stained 2-D gels (4, 5). However, since many proteins of the sarcomere are represented by multiple protein spots, the number of distinct proteins that are covered is much lower than reported for cell culture analyses. 2-DE can nicely resolve protein species with different p*I* or molecular mass but is limited in coverage of proteins of low abundance or with extreme p*I* and molecular weight as well as of highly hydrophobic proteins like membrane proteins, which play key roles in many cellular processes. A more comprehensive overview of disease-associated alterations in

Fernando Vivanco (ed.), *Heart Proteomics: Methods and Protocols*, Methods in Molecular Biology, vol. 1005,
DOI 10.1007/978-1-62703-386-2_6, © Springer Science+Business Media New York 2013

the cardiac proteome can likely be gained when mass spectrometry-centered approaches are applied (6, 7). Proteins in a sample are proteolytically digested and the resulting peptides are subjected to ionization and mass analysis following pre-fraction by HPLC. Signal intensities recorded can be used for relative quantitation, and fragment spectra are acquired in parallel in the same run providing information of the amino acid sequence of peptides underlying a subset of LC-MS features. The development of semi-automatic LC-MS workflows covering extraction, annotation, and quantitation of LC-MS features has brought tremendous progress for label-free quantitative analysis which is especially valuable for quantitative profiling of human tissues which are not accessible to metabolic labelling and which are often present in minute amounts severely compromising subsequent peptide or protein-labelling approaches such as iTRAQ (isobaric tags for relative and absolute quantitation) or ICPL (isotope-coded protein labelling) (8, 9).

2 Materials

2.1 Tissue Homogenization

1. Micro pistils.
2. Big and small forceps.
3. Liquid nitrogen in a Dewar vessel.
4. Microdismembrator S (Sartorius AG, Göttingen, Germany).
5. Clean Teflon beakers with wolfram beads (Sartorius AG).
6. Ice-cold acetone (−20 °C, 100 %).
7. Solution 8 M urea/2 M thiourea, UT.
8. 100 and 1,000 μl tips (cut and uncut).
9. Low polymer binding reaction tubes.
10. Table top centrifuge at 4 °C.

2.2 Protein Digestion LysC/Trypsin Method

1. 20 mM ammonium bicarbonate buffer, pH 7.8 (Sigma, A6141).
2. 25 mM dithiothreitol (DTT) in 20 mM ammonium bicarbonate buffer.
3. 100 mM iodoacetamide (Sigma, No. I6125, light sensitive) in 20 mM ammonium bicarbonate buffer.
4. 100 % acetic acid, purified (Sigma, No. 380121 or Roth No. HN65.1), used as a 5 % solution.
5. 100 ng/μl LysC (Sigma, Chemical Company, St. Louis, MO, USA, P3428) prepared in water.
6. 100 ng/μl trypsin modified, sequencing grade (Promega, Madison, WI) prepared in water.
7. Thermomixer (Eppendorf, Westbury, NY, USA).

2.3 Desalting of Peptide Extracts

1. All solutions are prepared with water HPLC gradient grade (e.g., Baker, No. 4218).
2. 5 % acetic acid (Sigma, No. 380121 or Roth No. HN65.1) to stop digestion.
3. 80 % acetonitrile (Baker, No. 9017) in 1 % acetic acid (1,200 μl acetonitrile 100 % + 300 μl 5 % acetic acid).
4. 50 % acetonitrile in 1 % acetic acid (750 μl acetonitrile 100 % + 450 μl water + 300 μl 5 % acetic acid).
5. 30 % acetonitrile in 1 % acetic acid (450 μl acetonitrile 100 % + 750 μl water + 300 μl 5 % acetic acid).
6. 1 % acetic acid (300 μl 5 % acetic acid + 1,200 μl water).
7. ZipTip® pipette tips (ZipTip u-C18, Millipore, Schwalbach, Germany).
8. Vacuum concentrator (Eppendorf, Westbury, NY, USA).

2.4 Chromatography-Mass Spectrometry

1. Buffer A: 2 % acetonitrile, 0.1 % acetic acid (solvents as described above).
2. Buffer B: 100 % acetonitrile, 0.1 % acetic acid.
3. HPLC, nanoACQUITY UPLC (Waters, Manchester, UK).
4. LC column nanoACQUITY UPLC column BEH130-C18 Symmetry, 100 μm × 100 mm.
5. High-resolution mass spectrometer, LTQ FTICR (linear ion trap coupled Fourier transform ion cyclotron resonance) mass spectrometer (Thermo Electron Corp., Bremen, Germany), or any other mass spectrometer providing high-resolution and high mass accuracy data.
6. Xcalibur interface software (Thermo Electron Corp., Waltham, MA, USA).

2.5 Mass Spectrometry Data Analysis

1. Sorcerer SEQUEST (SageN Research, Inc., Milpitas, CA, USA).
2. Rosetta Elucidator (Ceiba Solutions, Boston, MA, USA).

3 Methods

3.1 Tissue Homogenization and Protein Precipitation

1. Pool tissue of 3–4 biopsies of 0.5–1.5 mg each (see Note 1) and homogenize with a micro pistil in low polymer binding reaction tubes while cooling with liquid nitrogen. Transfer the powder to a precooled Teflon beaker containing a wolfram bole and add 500 μl of ice-cold acetone (see Note 2). Perform cell disruption in a Microdismembrator (or other bead mill) at 2,600 rpm for 2 min (see Note 3). Homogenized material is transferred to a tube with ice-cold acetone and kept at −20 °C for 1 h for protein precipitation.

2. Precipitate protein using a high-speed spin (16,000 × *g*) for 30 min at 4 °C using microcentrifuge.
3. Remove all the supernatant and air dry the protein pellet for approximately 5–10 min (see Note 4).
4. Resolve pellet in 50–100 μl 8 M urea/2 M thiourea (UT, see Note 5) by aspirating and vortexing.
5. Remove non-soluble material by a second centrifugation step (16,000 × *g*, 30 min, 4 °C) and transfer the protein-containing supernatant to a new reaction tube.
6. Determine protein concentration (see Note 6).

3.2 Protein Digestion

1. Protein extract volume corresponding to 2 μg protein is adjusted with 20 mM ammonium bicarbonate to a final volume of 10 μl (see Note 7).
2. Reduce the solution with 1 μl DTT (final concentration 2.5 mM) for 1 h at 60 °C and then cool the sample to room temperature.
3. Add 7 μl ammonium bicarbonate and alkylate the solution by adding 2 μl iodoacetamide (final concentration 10 mM) in the dark for 30 min at 37 °C.
4. Add LysC for a final enzyme: protein ratio of 1:100. Incubate at 37 °C for 3 h.
5. Add modified trypsin at an enzyme: protein ratio of 1:10, and incubate overnight (16–18 h) at 37 °C (see Note 8).
6. Stop digestion by adding acetic acid (final concentration 1 %) (see Note 9).

3.3 Desalting of Peptide Extracts for Column Loading

1. Activate a C18-ZipTip (volume: 10 μl) with 100 % acetonitrile (three times), followed by equilibration of the material by aspirating with 80, 50, and 30 % acetonitrile in 1 % acetic acid (five times each) and 1 % acetic acid (two times) (see Note 10).
2. Bind peptides on the ZipTip by aspirating and dispensing the sample in the same tube ten times.
3. Aspirate wash solution (10 μl of aqueous 1 % acetic acid) into tip and dispense to waste five times.
4. Elute the peptides with 50 % ACN, 1 % acetic acid and 80 % ACN in 1 % acetic acid subsequently (5 μl per extraction step).
5. Combine eluates in a HPLC microvial and evaporate in a vacuum concentrator until only 1 μl remains. Add 2 % acetonitrile in 0.1 % acetic acid (buffer A) to adjust the peptide concentration to roughly 100 ng/μl.

3.4 Reverse-Phase Separation of Tryptic Peptides and Tandem Mass Spectrometry

1. Peptide separation is achieved using a nonlinear 320 min gradient of 5–60 % ACN in 0.1 % acetic acid at a constant flow rate of 400 nl/min on a nanoACQUITY UPLC reverse-phase column (C18, 100 μm × 100 mm) operated on a nanoACQUITY UPLC system. The following acetonitrile gradient mixed from buffer A (2 % acetonitrile in 0.1 % acetic acid) and buffer B (0.1 % acetic acid in 100 % acetonitrile) should be delivered: 1 % B for 35 min (see Note 11), 1–5 % B over 1 min, 5–25% B over 209 min, 25–60 % B over 60 min, up to 99 % B within 1 min, 99 % B for 1 min, 1 % B over 1 min, and hold for an additional 10 min (see Note 12).
2. Keep column temperature at 35 °C throughout the whole run. Keep samples at 6 °C in the autosampler and no longer than 10 h before analyzed (see Note 13).
3. Mass spectrometer is set to acquire MS/MS data in a data-dependent mode for precursor ions with charge 2 or 3. After a survey scan in the FTICR analyzer (scan m/z 300–2,000 Da, resolution 60,000), MS/MS in the linear ion trap are triggered on the five most intensive ions exceeding a minimum threshold of 1,000. Target ions already selected for fragmentation should be dynamically excluded for 90 s to optimize peptide coverage (see Note 14).

3.5 Data Processing for Protein Identification

1. Peak lists can be generated by *ReadW* or *msn-extract* algorithms and subjected to search algorithms such as SEQUEST or MASCOT, which correlates the experimental tandem mass spectra with theoretical mass spectra from a protein sequence database in order to identify the peptide sequences detected.
2. Proteins are identified via an automated database search against a forward–reverse database which allows calculation of false-positive rates and annotation based on statistical evaluation of search results by, e.g., *Peptide and ProteinProphet* (10). Parent mass tolerance (MS) is set to 10 ppm and fragment mass tolerance to 1 Da. For reduced and alkylated samples, carbamidomethylation of cysteine is set as fixed modification, and methionine oxidation is considered as optional modification. For label-free quantitation experiments, missed cleavages allowed should be set to 0.
3. For label-free quantitation only high-quality peptides should be used. Therefore, annotation should be performed at a false-positive rate of <1 % for peptide identifications in combination with a probability peptide score of >0.8.

3.6 Label-Free Quantitation in the Software Package Elucidator®

Label-free quantitation relying on peptide intensities is based on the retention time alignment of chromatograms and identification of feature-peaks defined by their unique time and m/z boundaries. We employed the Rosetta Elucidator software package to quantify the detected MS signals (features) across all individual LC-MS/MS runs. This procedure enables the evaluation of the technical quality for each isotope group detected, which is finally summarized to peptide intensities (11). The approach also allows quantitation of low-abundant proteins, since MS/MS events in runs with higher intensity of a signal are assigned to low-intensity signals of other runs too. This is of particular importance in the analysis of muscle tissues since the dynamic range of the sample is strongly influenced by the highly abundant proteins of the contractile apparatus.

1. Import RAW files containing MS and MS/MS spectra into Elucidator system (Ceiba solutions).
2. Define analysis levels like treatment group and biological and technical replicates.
3. Set parameters for retention time alignment, noise filtering, and feature extraction in *Peak Teller* (see Note 15).
4. Filter features for high quality regarding peak time or m/z scores (>0.8) and overall quality $p<0.05$.
5. Annotate features by the peptide amino acid sequences and protein names obtained by SEQUEST or MASCOT search described in Subheading 3.7 (see Note 16).
6. Create a set of features representing peptide or peptide-like molecules ($z>1$) for normalization of intensities with a treatment group and across treatment groups.
7. Filter the data set for unique peptides before calculation of protein intensities and ratios (see Note 17).

3.7 Protein Annotation

Based on known or predicted molecular or biological properties extracted from publications, protein annotation or assignment of proteins to subcellular compartments can be performed. Several useful Web pages provide comprehensive information to which biological process, pathway, or sub-proteome a protein can be assigned. A short summary of all these information for each protein and useful links are provided by the ExPASy molecular biology server (http://ca.expasy.org/).

A comparison of protein classes covered by the analysis with those displaying alterations under the experimental conditions studied (enrichment analysis) allows the prediction of biological processes or molecular function particularly influenced. For this purpose, the protein analysis through evolutionary relationships classification software (PANTHER—http://www.pantherdb.org) can be used (see Note 18).

4 Notes

1. In order to reduce sampling error, which might induce high inter-sample proteomic variability between individual biopsies, analyses were performed with protein extracts from multiple biopsies (3–4) per patient.
2. Teflon beakers should be kept in liquid nitrogen as well as beads for precooling. For complete transfer of the homogenized tissue powder from the reaction tubes to the Teflon beaker, add 500 μl 100 % ice-cold acetone and use a cutted 1,000 μl pipette tip to aspirate the material and dispense the mixture to the bottom of the Teflon beaker. Be sure that beakers are closed tightly.
3. The acetone freezes in the liquid nitrogen but will thaw while the tissue powder will be homogenized completely in the dismembrator. Ensure that no liquid acetone leaks from the beakers which will lead to loss of material. If leaking of acetone is observed, extraction can be stopped already after approx. 1 min without loosing homogenization efficiency.
4. As an alternative to air drying of the pellet, a vacuum concentrator can be used for 2–5 min.
5. The volume of UT buffer used for dissolving the pellet depends on the size (50–100 μl). The use of UT buffer results in an irreversible denaturation of the proteins (including proteases) while keeping their primary structure. Furthermore, proteins are partially extracted from membranes thus making them available for further analysis.
6. Before estimating the protein concentration of the extract, protein samples should be frozen for a minimum of 2 h. Concentrations were determined using the Bradford protein assay (BioRad, No. 500-0006, München, Germany) with bovine serum albumin as a standard in a range of 1–12 μg/μl.
7. Protein concentration should be set to 1 μg/μl for all samples before starting the sample preparation procedure. Following the protocol provided, the denaturing UT solutions will be diluted to concentrations in which the proteolytic enzymes are still active (≤2 M). Diluting the sample with ammonium bicarbonate adjust the pH to >7.5, which provides optimal digestion conditions for the enzymes.
8. For heart protein samples, optimal digestion was obtained using a combination of LysC and trypsin, whereas incomplete digestion was observed with trypsin only (12). The increase of the enzyme to protein ratio up to 1:10 did not lead to unspecific cleavage events.
9. Addition of acetic acid decreases the pH, which leads to the reduction of trypsin activity. At the same time, setting the pH

to 2–3 ensures the proper binding of peptides to the RP material during the desalting step by forcing ions to form non-dissociated molecules ($COO^{-} \rightarrow COOH$).

10. During pipetting of solutions, development of air bubbles should be avoided until the peptide binding is finished. Air bubbles would affect the equilibration and binding capacity of the C18 material resulting in a loss of peptides.
11. The specific characteristics of Waters UPLC pre-columns cause a loss of separation quality when gradients longer than 3 h are used. Thus, we did not use a pre-column for sample loading, which therefore had to be carried at a flow rate of 400 nl/min. This results in a long loading time of, i.e., 10 min for 4 μl injection volume. In total, we run the system for 35 min at 2 % acetonitrile for loading and washing the sample. This time can be reduced significantly in nano-LC systems where long gradients can be run also with trap-column configurations (i.e., Easy-nLC, Thermo, Bremen, Germany).
12. Extending the run time from 100 to 320 min permits higher sample loading (500 ng instead of 200 ng) and resulted in an increased number of proteins from 106 to 485 (two peptides or more per protein).
13. Higher temperatures on column improve separation and use of a column oven ensures high reproducibility of retention which helps to reduce technical variance and improves label-free quantitation.
14. When peptides are ionized by electrospray, in more than 95 % charge of the resulting molecules will be two or higher, whereas low molecular weight contaminants are predominantly single-charged ions. Thus, exclusion of single-charged ions from fragmentation directs MS/MS fragmentation to peptides and avoids extensive selection of single-charged contaminants as precursors. Exclusion time should be adjusted according to peak width in the chromatogram and the quantitation method used. For intensity-based methods one high-quality spectrum recorded per peak might be sufficient (exclusion time can be set as 1.5-fold peak width). If spectral counting is considered for semiquantitative quantitation, acquisition of a peak height-dependent number of spectra is essential and exclusion time has to be much shorter, e.g., about 25 % of peak width.
15. Peak Teller is a complex algorithm that aligns chromatograms, thereby correcting for retention time shifts and identifying features (peaks representing peptides) across all MS runs. During alignment and feature extraction, a background subtraction is performed, before intensities are calculated (11). The corresponding parameters "feature noise filtering strength" and "noise filtering strength–intensity" can be optimized effectively by running the workflow until the step of

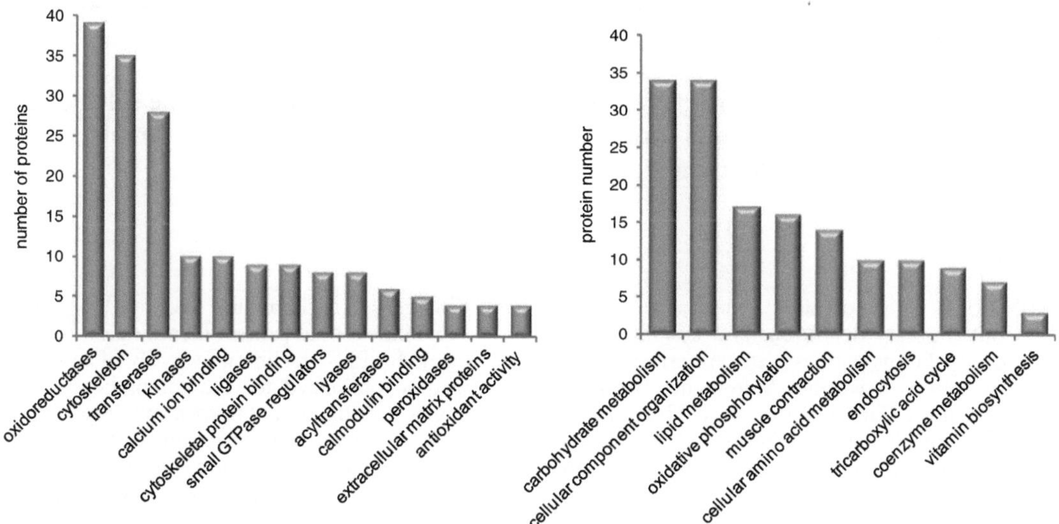

Fig. 1 Molecular proteins and biological processes mainly influenced by iDCM. Proteins displaying altered levels in myocardial biopsies of iDCM patients in comparison to those of subjects with normal left ventricular function (controls) were assigned to molecular functions and biological processes by PANTHER analysis (http://www.pantherdb.org). Only protein classes with enrichment of the function or process among the regulated proteins in comparison to the total number of proteins assigned to this class in NCBI ($p < 0.05$) were considered

"DTA extraction" for a small fraction of the data space, e.g., selecting a small time window of about 5 min and a mass range of 100 Da of a total chromatogram. Combining the highest number of DTA files with the lowest number of features ensures high processing speed of the workflow but limits loss of low-abundant peptide identifications.

16. Searches were performed with the setting "0 missed cleavages" to avoid influence of missed cleavage peptides on quantitation.
17. Data were filtered for protein-specific (unique) peptides, and only these peptides were considered for quantitation. The user has to be aware that this approach reduces the total number of peptides in the analysis. However, homologous peptides might display different ratio in comparison to protein-specific peptides and therefore compromise significance of protein ratios.
18. The lists of all identified proteins and those displaying different levels of proteins in human myocardial biopsies of iDCM patients in comparison to controls were subjected to protein annotation by PANTHER (http://www.pantherdb.org). Major changes in abundance were observed for mitochondrial (34.5 %) and cytoskeletal proteins (24.7 %), but also metabolic pathways were affected during iDCM (see Fig. 1). The large number of proteins involved in energy-providing pathways which displayed decreased intensity in iDCM patients points to a limited energy supply in the myocardium. Increase levels of cell structure proteins might indicate disturbances of cellular integrity.

Acknowledgments

The study was performed within the framework of the collaborative research project SFB/TR19 and supported by grants of the Deutsche Forschungsgemeinschaft.

References

1. Maron BJ, Towbin JA et al (2006) Contemporary definitions and classification of the cardiomyopathies: an American Heart Association Scientific Statement from the Council on Clinical Cardiology, Heart Failure and Transplantation Committee; Quality of Care and Outcomes Research and Functional Genomics and Translational Biology Interdisciplinary Working Groups; and Council on Epidemiology and Prevention. Circulation 113:1807–1816
2. McGregor E, Dunn MJ (2006) Proteomics of the heart: unraveling disease. Circ Res 98: 309–321
3. McDonough JL, Neverova I et al (2002) Proteomic analysis of human biopsy samples by single two-dimensional electrophoresis: Coomassie, silver, mass spectrometry, and Western blotting. Proteomics 2:978–987
4. Knecht M, Regitz-Zagrosek V et al (1994) Characterization of myocardial protein composition in dilated cardiomyopathy by two-dimensional gel electrophoresis. Eur Heart J 15:37–44
5. Corbett JM, Why HJ et al (1998) Cardiac protein abnormalities in dilated cardiomyopathy detected by two-dimensional polyacrylamide gel electrophoresis. Electrophoresis 19: 2031–2042
6. Kislinger T, Gramolini AO et al (2005) Multidimensional protein identification technology (MudPIT): technical overview of a profiling method optimized for the comprehensive proteomic investigation of normal and diseased heart tissue. J Am Soc Mass Spectrom 16:1207–1220
7. Kline KG, Wu CC (2009) MudPIT analysis: application to human heart tissue. Methods Mol Biol 528:281–293
8. Ross PL, Huang YN et al (2004) Multiplexed protein quantitation in *Saccharomyces cerevisiae* using amine-reactive isobaric tagging reagents. Mol Cell Proteomics 3:1154–1169
9. Flory MR, Griffin TJ et al (2002) Advances in quantitative proteomics using stable isotope tags. Trends Biotechnol 20:S23–S29
10. Nesvizhskii AI, Keller A et al (2003) A statistical model for identifying proteins by tandem mass spectrometry. Anal Chem 75:4646–4658
11. Neubert H, Bonnert TP et al (2008) Label-free detection of differential protein expression by LC/MALDI mass spectrometry. J Proteome Res 7:2270–2279
12. Hammer E, Goritzka M et al (2011) Characterization of the human myocardial proteome in inflammatory dilated cardiomyopathy by label-free quantitative shotgun proteomics of heart biopsies. J Proteome Res 10:2161–2171

Chapter 7

Systems Proteomics of Healthy and Diseased Chromatin

Haodong Chen, Emma Monte, Thomas M. Vondriska, and Sarah Franklin

Abstract

Differences in chromatin-associated proteins allow the same genome to participate in multiple cell types and to respond to an array of stimuli in any given cell. To understand the fundamental properties of chromatin and to reveal its cell- and/or stimulus-specific behaviors, quantitative proteomics is an essential technology. This chapter details the methods for fractionation and quantitative mass spectrometric analysis of chromatin from hearts or isolated adult myocytes, detailing some of the considerations for applications to understanding heart disease. The state-of-the-art methodology for data interpretation and integration through bioinformatics is reviewed.

Key words Chromatin, Cardiovascular disease, Proteomics, Mass spectrometry, Bioinformatics

1 Introduction

How the same genome encodes multiple proteomes in the hundreds of cell types in a given multicellular eukaryote is one of the great unanswered questions of biology. The packaging of DNA in the nucleus occurs on the basis of the following structural hierarchy: a segment (~147 bp) of DNA wraps around a protein complex containing two copies each of four core histones (H2A, H2B, H3, and H4), constituting a nucleosome; this octameric protein complex in turn forms higher-ordered structures of less well-defined architecture through interactions with linker histones (like H1) and other chromatin structural proteins. These higher-order chromatin regions determine the overall shape and presentation of each region of the genome.

During mitosis, the structure of the chromosomes is well established—as are the highly orchestrated structural events during cytokinesis. The 3D structure of the non-mitotic genome, however, is poorly understood. The packing task of the nucleus is daunting: in the case of a cardiac cell, for example, the two copies of the ~3 billion base pair genome must be collapsed into ~350–400 μM^3 of nuclear space; accommodation must also be made for RNA and

Fernando Vivanco (ed.), *Heart Proteomics: Methods and Protocols*, Methods in Molecular Biology, vol. 1005,
DOI 10.1007/978-1-62703-386-2_7,

protein (we have measured >1,000 proteins in the cardiac nucleus (1), including >300 bound to chromatin (1, 2)), while still maintaining conformational flexibility to enable large, rapid changes in gene expression. New insights into genomic structure provided by techniques like chromosomal conformation capture (3, 4) have demonstrated a nonrandom structure of the genome in an interphase nucleus and suggest that coordinated packaging of chromosomal territories is a critical task to enable global gene expression programs in eukaryotes. Not addressed in these studies are the proteins contributing to the chromatin backbone and how quantitative changes in these chromatin structural proteins influence global gene expression in disease.

One of the most important problems in biology is explaining how genome-wide changes in transcription are facilitated by structural remodeling of chromatin. Thirty-seven years after the original nucleosome hypothesis was proposed (5), the field is still coming to understand how chromatin dynamics are structurally regulated in vivo with high-resolution nucleosome mapping (6). The recent explosion of studies using chromatin immunoprecipitation (ChIP) and DNA sequencing to map the localization of select proteins across the genome (7–10) has added new detail to the picture of gene regulation. Extensive work has been done to show how various classes of proteins (such as deacetylases, acetyltransferases, and methyltransferases, among others) modulate chromatin accessibility and thereby gene expression (11–15); however, not revealed in these studies is how specific genes are targeted and how global remodeling is integrated across the genome. To initiate transcription, RNA polymerases themselves, as well as a cadre of chromatin-remodeling enzymes, reorganize and evict nucleosomes to free up first the promoter region, and subsequently downstream exons, for transcription (16–18).

Because adult cardiac myocytes do not readily divide, it is well established that the injured heart will grow surviving myocytes to maintain cardiac output. This hypertrophic process is a common antecedent to heart failure in a variety of conditions (19) and involves ordered reprogramming of gene expression to convert the cell to a more primitive phenotype (the so-called fetal gene program) (20, 21). To be transcribed, genes silenced during development must transition from a heterochromatic (tightly packed) to euchromatic (loosely packed) environment. This presents an intriguing scenario from the standpoint of chromatin regulation: cardiac transcriptional responses must somehow "find" the right genes to impinge upon through a combination of consensus motifs, associating proteins and—*this part is a hypothesis*—reception by a permissive global chromatin structure, poised for a given cellular phenotype.

Proteomics is a critical tool to answer questions of global chromatin regulation in a fully differentiated cell (the cardiomyocyte),

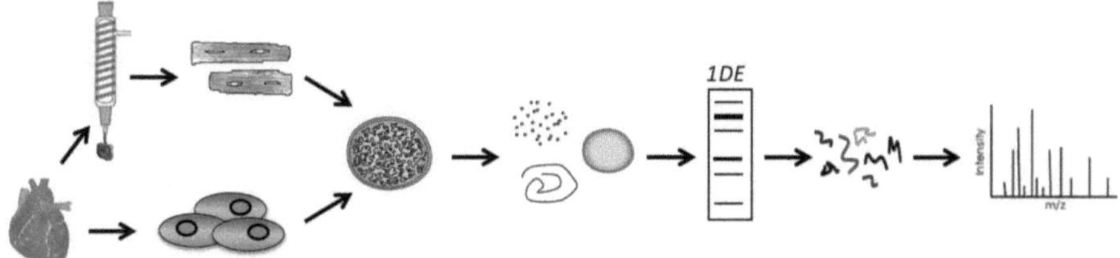

Fig. 1 Flow chart for fractionation and quantitative mass spectrometric analysis of chromatin. Whole mouse hearts are homogenized, or adult cardiac myocytes are lysed, and the lysate enriched for nuclei. Each nuclei sample can either be separated into nucleoplasm and chromatin fractions or into an acid-extracted fraction that enriches for histones. Proteins from each are separated by one-dimensional SDS-PAGE and the bands trypsin digested and run by LC/MS/MS

with a complex gene expression profile and highly specialized physiological phenotype. This protocol describes methodologies for label-free quantitative analysis of subproteomes within the nucleus and addresses unique challenges for analyzing heart tissue, including multiple cell types, abundant nonnuclear contractile proteins (Fig. 1), and bioinformatic and computational biology approaches (Fig. 2) for data interpretation.

2 Materials

Unless otherwise indicated, all solutions are made in deionized water, with the exception of the digestion and LC/MS/MS steps, for which all solutions should be HPLC grade and made in HPLC water unless specified otherwise.

2.1 Fractionation Components

1. Phosphate buffered saline (PBS).
2. Protease/phosphatase inhibitor mix: 0.1 mM phenylmethanesulfonylfluoride, protease inhibitor cocktail pellet (Roche, catalogue number 04 693 159 001), 0.2 mM sodium orthovanadate, 0.1 mM sodium fluoride, and 10 mM sodium butyrate. Rather than make a stock mix, these components should be added individually directly to the following buffers when they are made.
3. Whole lysate buffer: 20 mM Tris pH 7.4, 150 mM NaCl, 1 mM ethylenediaminetetraacetic acid (EDTA), 1 mM ethylene glycol tetraacetic acid (EGTA), 1 % SDS, 2.5 mM sodium pyrophosphate, 1 mM glycerophosphate with protease/phosphatase inhibitor mix. Store at –20 °C for up to 1 week.
4. 100 μM strainer (catalogue number 352369) (BD Falcon).
5. Langendorff system.
6. Heart perfusion buffer: 100 mM potassium glutamate, 10 mM aspartic acid, 25 mM KCl, 10 mM KH_2PO_4, 2 mM $MgSO_4$,

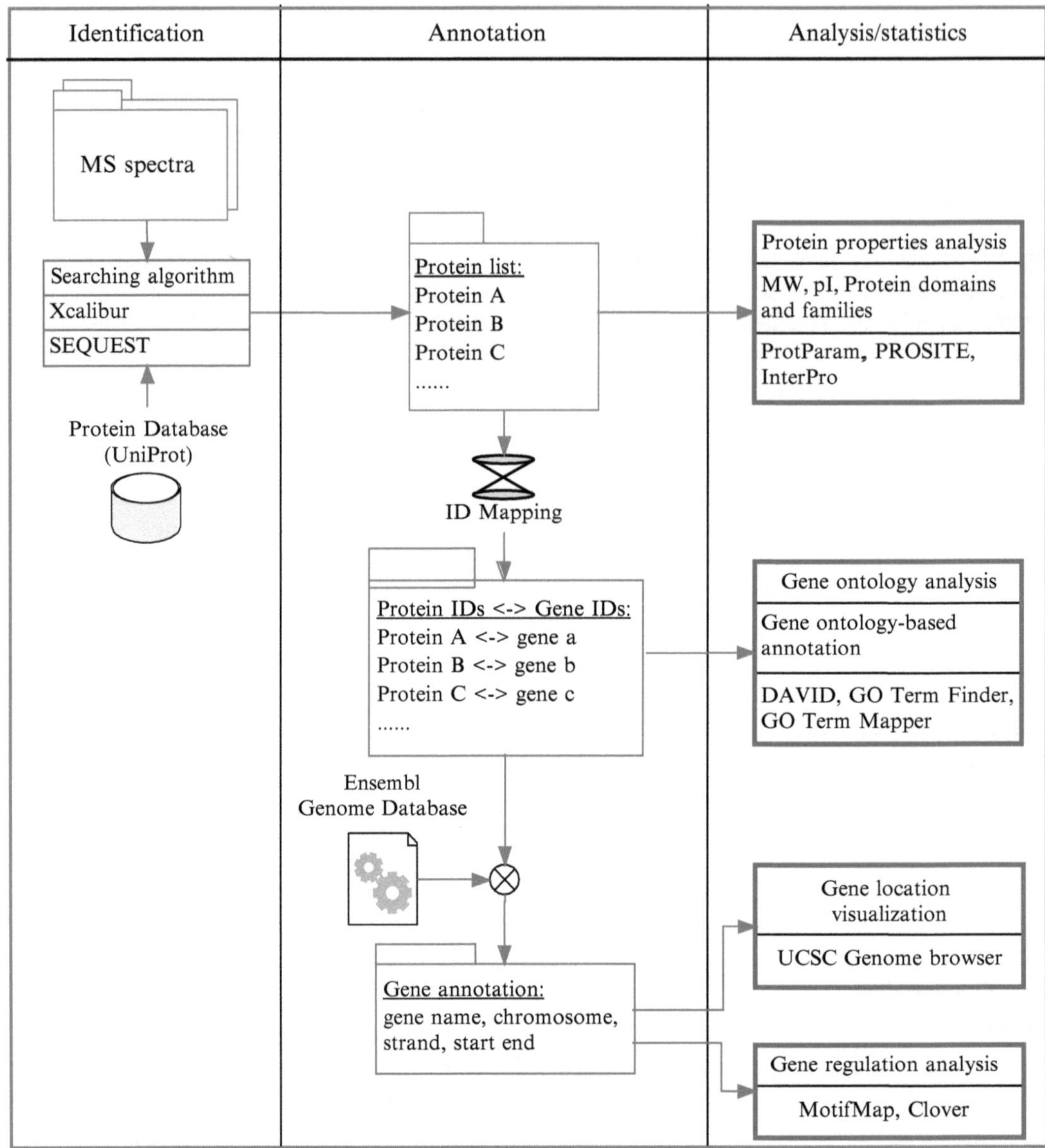

Fig. 2 Flow chart for bioinformatic analysis of proteomic data. Raw MS spectra are obtained from LC/MS/MS and the results searched against a protein database (UniProt) using the SEQUEST algorithm, which is embedded in Xcalibur. This list can be used directly to obtain global features of the proteome, such as physical-chemical properties and protein domains/families. After converting protein identifiers to gene identifiers, gene ontology-based annotation can be conducted. The location of genes within chromosomes can be extracted from Ensembl genome database and visualized by UCSC genome browser. Candidate regulatory elements, i.e., binding motifs of transcription factors, can be found using MotifMap database or Clover algorithm. Refer to Table 1 for the availability of these tools

20 mM taurine, 5 mM creatine, 0.5 mM EGTA, 5 mM 4-(2-hydroxyethyl)-1-piperazineethanesulfonic acid (HEPES), and 20 mM glucose. Adjust pH to 7.2 with KOH. Add glucose just before use. Warm the perfusion buffer to 37 °C before use.

7. Collagenase: Collagenase type II from Worthington is recommended. Dissolve the enzyme into perfusion buffer before use. We used a concentration of 1,000 mg/l but this should be determined empirically.
8. Lysis buffer: 0.15 % Nonidet P-40 (NP-40), 10 mM Tris pH 7.5, 15 mM NaCl, with protease/phosphatase inhibitor mix. Store at −20 °C for up to 1 week.
9. Sucrose buffer: 10 mM Tris pH 7.5, 15 mM NaCl, 24 % sucrose (w/v), with protease/phosphatase inhibitor mix. Make fresh.
10. Transfer pipette.
11. PBS/EDTA: 1 mM EDTA in PBS. Store at room temperature for months.
12. TRIS, SDS, EDTA buffer: 50 mM Tris pH 7.4, 10 mM EDTA, 1 % sodium dodecyl sulfate (SDS), with protease/phosphatase inhibitor mix. Store at −20 °C for up to 1 week.
13. Detergent extraction buffer: 20 mM HEPES pH 7.6, 7.5 mM $MgCl_2$, 0.2 mM EDTA, 30 mM NaCl, 1 M urea, 1 % NP-40, with protease/phosphatase inhibitor mix. Store at −20 °C for up to 1 week.
14. 0.4 N H_2SO_4.
15. Trichloroacetic acid.
16. Acetone.
17. 1 M Tris (pH unadjusted).

2.2 Protein Extraction and Electrophoresis Components

1. BCA protein assay (catalogue number 23227) (Thermo Scientific).
2. 5× Laemmli buffer: 60 mM Tris pH 6.8, 25 % glycerol, 5 % β-mercaptoethanol, a few flakes bromophenol blue. Store extra at −20 °C for months. Store working volume at room temperature for months.
3. Nitrocellulose membrane, 0.45 μM (catalogue number 162-0115) (BioRad).
4. Histone H2A antibody (catalogue number sc-8648) (Santa Cruz).
5. Adenine nucleotide transporter antibody (catalogue number sc-9299) (Santa Cruz).
6. Nitrile gloves (catalogue number 19-130-1597B) (Fisher).
7. Oriole (catalogue number 161-0496) (BioRad).
8. Stainless steel surgical blade (catalogue number 371211) (Bard-Parker).

2.3 Digestion and LC/MS/MS Components

Preparation of any solution that will be used in the digestion or LC/MS/MS steps should be prepared using a hairnet in addition to lab coat and gloves. Special care should be employed to prevent contamination using tips and tubes which have been sealed and when possible, using solvents dedicated for MS use that have been kept contaminant-free.

1. 1.5 ml Low Adhesion Microcentrifuge Tubes (catalogue number 1415-2600) (USA Scientific).
2. 0.65 ml Prelubricated Microcentrifuge Tubes (catalogue number 3206) (Costar).
3. HPLC water.
4. Acetonitrile for LC-MS.
5. 50 mM NH_4HCO_3, 50 % acetonitrile. Store at room temperature for a month.
6. 10 mM DTT/10 mM TCEP: 10 mM dithiothreitol, 10 mM Tris (2-carboxyethyl)phosphine hydrochloride. Make fresh.
7. 100 mM iodoacetamide. Make fresh.
8. 0.02 μg/μl trypsin (Promega, catalogue number V5111) in 50 mM NH_4HCO_3. Make fresh on ice.
9. 0.02 μg/μl chymotrypsin (Roche, catalogue number 11 418 467 001) in 50 mM NH_4HCO_3. Make fresh on ice.
10. 5 % formic acid. Store at room temperature for weeks.
11. 0.1 % formic acid, 50 % acetonitrile. Store at room temperature for weeks.
12. HPLC vials and inserts (refer to your system for proper models).
13. Mobile phase A: 0.1 % formic acid, 2 % acetonitrile. Store at room temperature for weeks.
14. Mobile phase B: 0.1 % formic acid, 80 % acetonitrile. Store at room temperature for weeks.
15. Reversed phase column (catalogue number PFC7515-B14-10) (New Objective).
16. Xcalibur software (Thermo Scientific).

3 Methods

3.1 Heart Isolation for Whole Heart Lysate

Carry out all steps on ice unless otherwise indicated.

1. Fill a glass dounce with 2 ml whole heart lysate buffer and place on ice along with a beaker of PBS. Anesthetize adult mouse (8–12 weeks) with isoflurane and sacrifice by cervical dislocation. Work quickly to remove the heart, rinse it in cold PBS, and place in dounce. Homogenize on ice, being careful

not to create bubbles, until there is no more visible tissue. Collect the lysate in a 2 ml centrifuge tube.

2. Sonicate for 10–15 s, 3–6 times. Put sample on ice between sonications. Centrifuge at 4 °C for 5 min at 16,200 × *g*. Remove the supernatant and store at −80 °C. The supernatant is the whole heart lysate fraction.

3.2 Nuclear Isolation from Whole Heart

Carry out all steps on ice unless otherwise indicated.

1. Fill a glass dounce with 2 ml lysis buffer and place on ice along with a beaker of PBS. Anesthetize adult mouse (8–12 weeks) with isoflurane and sacrifice by cervical dislocation. Work quickly to remove the heart, rinse it in cold PBS, and place in dounce. Homogenize on ice, being careful not to create bubbles, until there is no more visible tissue. Pass lysate through a 100 μm strainer and collect the lysate in a 2 ml centrifuge tube.
2. Centrifuge at 4 °C for 5 min at 1,500 × *g*. Meanwhile, prepare a 2 ml tube with 1 ml of sucrose buffer on ice. After centrifugation, remove the supernatant and store at −80 °C. This is the cytosolic fraction. Resuspend the pellet in 500 μl cold lysis buffer and gently place on top of the sucrose buffer using a transfer pipette. Be careful not to disturb the sucrose pad or mix the two layers.
3. Centrifuge at 4 °C for 10 min at 2,400 × *g*. Remove the sucrose pad along with the lipid layer on top and rinse the pellet in 200 μl cold PBS/EDTA. This is the nuclei pellet.

3.3 Nuclear Isolation from Adult Isolated Cardiomyocytes

Isolation of high-quality cardiac myocytes is one of the most important factors for successful experimentation. The protocols for adult cardiac myocyte isolation vary between different laboratories and depend on the downstream experiments. Here, we describe a succinct protocol using Langendorff perfusion apparatus. For more details on isolation of cardiac myocytes, please see these protocols (22–24).

1. Isolate whole heart from animal. The perfusion system should be prepared and warmed up before starting to remove heart from animal.
2. Connect aorta to the system and perfuse with perfusion buffer to remove blood from the heart, followed by collagenase to digest tissue. Maintain the temperature of perfusate at 36–37 °C.
3. Cut the ventricles in a dish and mince into small pieces using a scalpel. Triturate with a plastic transfer pipette to dissect tissue.
4. Filter the cell suspension through a 100 μm Nylon cell strainer into 15 ml tubes. Sediment cells by gravity for 8–10 min followed by gentle centrifugation for 1 min at 180 × *g*. Wash pellet in PBS twice. Carry out the next steps on ice unless otherwise indicated.

5. Resuspend pellet in 1 ml ice-cold lysis buffer and incubate on ice for 10 min.
6. Centrifuge at 4 °C for 5 min at 1,500 × *g*. Meanwhile, prepare a 2 ml tube with 1 ml of sucrose buffer on ice. After centrifugation, remove the supernatant and store at −80 °C. This is the cytosolic fraction.
7. Resuspend the pellet in 500 μl cold lysis buffer and gently place on top of the sucrose buffer using a transfer pipette. Be careful not to disturb the sucrose pad or mix the two layers.
8. Centrifuge at 4 °C for 10 min at 2,400 × *g*. Remove the sucrose pad along with the lipid layer on top and rinse the pellet in 200 μl cold PBS/EDTA. This is the nuclei pellet.

3.4 Whole Nuclei Preparation

Carry out all steps on ice unless otherwise indicated.

1. Resuspend the nuclei pellet from Subheading 3.2 (whole heart) or 3.3 (isolated myocytes) in 300 μl TRIS, SDS, EDTA buffer.
2. Sonicate for 10–15 s, 3–6 times. Put sample on ice between sonications. Centrifuge at 4 °C for 5 min at 16,200 × *g*. The pellet should be small. If it is not, additional sonication may be necessary. Remove the supernatant and store at −80 °C. The supernatant is the whole nuclei fraction.

3.5 Nucleoplasm and Chromatin Fractionation

Carry out all steps on ice unless otherwise indicated.

1. Resuspend the nuclei pellet from section 3.2 (whole heart) or 3.3 (isolated myocytes) in 200 μl detergent extraction buffer and vortex for 10 s twice, then place on ice for 10 min.
2. Centrifuge at 4 °C for 5 min at 16,200 × *g*. Remove the supernatant and store at −80 °C. This is the nucleoplasm. Rinse the pellet with 200 μl cold PBS/EDTA and resuspend it in 300 μl TRIS, SDS, EDTA buffer.
3. Sonicate for 10–15 s, 3–6 times. Put sample on ice between sonications. Centrifuge at 4 °C for 5 min at 16,200 × *g*. The pellet should be small. If it is not, additional sonication may be necessary. Remove the supernatant and store at −80 °C. The supernatant is the chromatin fraction.

3.6 Acid Extraction (See Note 1)

Carry out all steps on ice unless otherwise indicated.

1. Resuspend the nuclei pellet from 3.2 (whole heart) or 3.3 (isolated myocytes) in 200 μl detergent extraction buffer and vortex for 10 s twice, then place on ice for 10 min.
2. Centrifuge at 4 °C for 5 min at 16,200 × *g*. Remove the supernatant and store at −80 °C. This is the nucleoplasm. Rinse the pellet with 200 μl cold PBS/EDTA and resuspend it in 400 μl 0.4 N H_2SO_4. Vortex sample until clumps are dissolved and rotate at 4 °C for 30 min or overnight.

3. Centrifuge at 4 °C for 10 min at 16,000 × *g*. Remove the supernatant to a new tube and add to it 132 μl of trichloroacetic acid, drop by drop. Invert between drops and place on ice for 30 min.
4. Centrifuge at 4 °C for 10 min at 16,000 × *g*. Discard the supernatant. The supernatant is the acid insoluble fraction. Rinse the pellet with 200 μl of ice-cold acetone. Centrifuge at 4 °C for 10 min at 16,000 × *g*. Repeat the wash by rinsing the pellet with 200 μl of ice-cold acetone and again centrifuging at 4 °C for 10 min at 16,000 × *g*. Discard the supernatant and air-dry the pellet.
5. Resuspend the pellet in 100 μl of TRIS, SDS, EDTA buffer. Use 1 M Tris stock (pH unadjusted) to set the pH to 8. Sonicate for 15 min in a water bath (partially filled with ice) to resuspend the protein. Store at −80 °C. This is the acid-extracted fraction.

3.7 Protein Extraction, Electrophoresis, and Digestion

The following steps can be performed at room temperature unless otherwise indicated. When working with samples that will be run on the mass spectrometer, it is important to prevent keratin contamination. The experimenter should wear a lab coat, gloves, and hairnet and work in an area free of dust. Similarly, use tips and tubes that have been kept covered and wash with soap and water the gel running apparatus and box the gel will be stained in. Spray down any surface the gel will touch with 75 % ethanol and wipe with a Kimwipe. Finally, use HPLC grade solvents that have been kept clean of contamination.

1. Remove samples from −80 °C and thaw on ice. Using the bicinchoninic acid (BCA) protein assay, determine the protein concentration of each sample. Dilute samples using 5× Laemmli buffer. (Buffer must be at least 1/5 of the final volume, but can be more.) Depending on the fraction, we usually dilute our samples to 0.5–5 μg/μl. Boil the samples at 100 °C for 10 min and then return to ice. Once diluted in Laemmli buffer, samples can be stored at −20 °C.
2. Run equal amounts of each sample on a one-dimensional gel. Depending on your intentions, 20 μg of sample is a good starting amount. After running, the gel can be transferred to a nitrocellulose membrane and blotted to verify success of the fractionation (see Note 2) or used for mass spectrometric analysis (subsequent steps).
3. Transfer gel to a clean, light tight box. Avoid touching the gel and instead use a squirt bottle with deionized water to loosen the gel off of the glass plate and into the box. Add enough Oriole stain to cover (use nitrile gloves when handling Oriole) and leave on shaker for at least 90 min. The gel can also be

stained over night at 4 °C; however, a small amount may precipitate out leaving tiny overexposed flecks on the gel. After staining, the Oriole should be disposed of properly and the gel can be kept at 4 °C in deionized water.

4. Image the gel using a UV light box. Spray off the surface with 75 % ethanol and wipe down with a Kimwipe several times before placing your gel on it. It is best to take images at multiple exposures. Choose an exposure that best reflects what you will see when you cut the gel, to mark and label the bands that will be cut. For studies of the total proteome, we cut each lane into 25 2-mm bands. Cut out each band using a small razor and further section it into three equal pieces, width-wise. Collect the three pieces into a labeled 1.5 ml tube (it is best to use "low bind" tubes for digestion). Store pieces at −20 °C.
5. Digest the gel pieces (see Note 4). Wash gel plugs twice in 50 mM NH_4HCO_3/50 % acetonitrile (ACN), dehydrate with ACN, and dry in a SpeedVac. Reduce with 10 mM DTT/10 mM TCEP at 56 °C for 30 min. Wash and dehydrate the gel plugs as before and alkylate in 100 mM iodoacetamide for 22 min in the dark. Wash, dehydrate, and SpeedVac as before. Digest in 0.02 μg/μl trypsin solution in 50 mM NH_4HCO_3 overnight at 37 °C (see Note 3 for enzymatic considerations). Halt the digestion with 5 % formic acid solution and collect the supernatant in a 0.65 ml low bind tube. Further extract peptides in 0.1 % formic acid/50 % ACN twice, collecting the supernatant after each, and SpeedVac the combined supernatant down to approximately 30 μl. For quantitative analysis, be careful to concentrate each sample equally. Transfer the samples to HPLC vials. Store the samples and the left over gel plugs at −20 °C. For an extended protocol see our Jove publication (PMID: 19455095).

3.8 LC/MS/MS

1. Inject 10 μl of sample for each run using a nano-flow method optimized for a range of peptides. We use a linear gradient from 5 %B to 50 %B over 60 min to 95 %B over the next 15 min, and then we hold at 95 %B for 10 min. We use a 200 nl/min flow rate through a New Objective, reversed phase, C18 column (75 μm i.d.). For MS/MS, use a high mass accuracy mass spectrometer and acquire date in a data-dependent mode. We use a Thermo Orbitrap to fragment the six most abundant parent ions.
2. We recommend repeating this process for at least three biological replicates (samples from different animals) and two technical replicates (the same sample analyzed an additional time by LC/MS/MS). Gross changes in peptide abundance between experimental conditions can be observed in the total ion chromatogram (TIC) which displays each peptide detected as a three-dimensional peak Fig. 3.

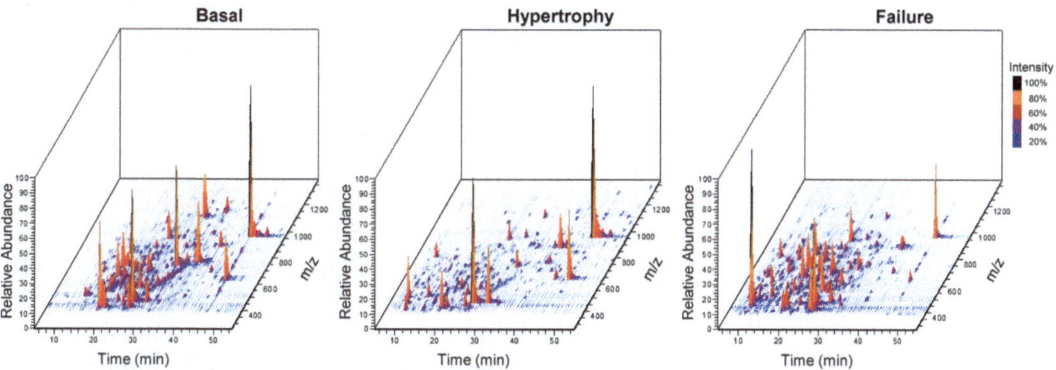

Fig. 3 3D rendering of total ion chromatogram. The total ion chromatogram (TIC) from an LC/MS run can be viewed in three dimensions to display all peptides detected as individual peaks. The 3D-TIC of acid-extracted chromatin proteins from basal (*left panel*), hypertrophic (*middle panel*), and failing (*right panel*) mouse myocardium illustrates the changes in protein abundance that occur during disease progression. Figures generated using Qual Browser embedded in Xcalibur software. *X*-axis (time), *Y*-axis (intensity), and *Z*-axis (*m/z*)

3.9 Database Searching

1. Identify peptides by searching the spectra against the UniProt protein database of choice via a search algorithm (such as SEQUEST or MASCOT) integrated into commercially available software such as BioWorks and Xcalibur or publically available software such as PROWL, X! Tandem, and SpectraST. More details on database searching algorithms can be found in another protocol article (25).
2. Consider modifying search parameters to allow for cysteine carbamidomethylation and methionine oxidation, two common modifications created during the sample processing.
3. Calculate a false-positive rate using reverse database searching. A reverse database is created by reversing protein sequences in the original database.
4. Filter protein identifications to only accept matches of a threshold confidence. We recommend the following parameters to start with: Xcorr > 3 (+2), >4 (+3), >5 (+4); deltaCN > 0.1, consensus score ≥20, mass tolerance 2 Da for parent ion, mass tolerance of 0.5 Da for product ion, at least two unique peptides per protein and no more than two missed cleavages. Manually inspect all spectra used for protein identification and reject identifications when less than two peptides are identified (see Note 5).

3.10 Label-Free Quantitation

Label-free quantitation requires consistent sample prep and LC/MS/MS conditions as well as the use of a high mass accuracy mass spectrometer. While quantitative comparisons using metabolic labeling combine samples into a single run and compare spectra from that same run, label-free quantitation requires samples be run individually and spectra compared across runs. Thus, the following quantitation relies on having taken effort to ensure high reproducibility in the previous steps (see Note 6).

1. Use individual peptide signals to determine the relative abundance of specific peptides between samples (i.e., healthy and diseased or treated and untreated) using label-free quantitation. Available programs include Census (Prof. Yates' group) (26), Elucidator (Microsoft) (27), SIEVE (Thermo Scientific) (28), and Scaffold (Proteome Software) (29). Briefly, these programs will align data across runs, normalize signal intensity, and quantify peptide peaks through spectral counting or by measuring peak area to generate peptide abundance ratios between samples.
2. Connect peptide abundance to protein identification by employing search algorithms (MASCOT, SEQUEST, X! Tandem) in the above programs.
3. Assess accuracy and reproducibility through ANOVA analysis of the data. We recommend generating a PCA plot to compare different biological replicates (different experimental samples) and technical replicates (running the same sample on the mass spectrometer multiple times). As already mentioned, these replicates are crucial to label-free quantitation, and a PCA plot will allow you to confirm reproducible grouping of data, with technical replicates most closely clustered. Additionally, close clustering of biological replicates not only confirms accuracy but also determines the presence of proteomic differences between treatment groups. One representative result is shown in Fig. 4.

3.11 Bioinformatic Analysis

Depending on the different purposes of the project, many bioinformatics tools are available for various analyses. One interesting question to ask is where these genes coding for identified proteins are located in the genome. In this section we will map the proteins identified by MS to chromosomes and visualize the location of these genes within the genome. We will then conduct gene ontology analysis for these proteins to build a gene ontology tree with enriched terms. Most of the procedures are straightforward and do not require program coding. Refer to Table 1 for the description of these tools.

1. Convert UniProt identifiers to Ensembl Genomes or RefSeq nucleotide identifiers using ID Mapping (see Note 7).
2. Extract gene annotation information from UCSC Genome Browser. Go to UCSC Genome Browser and select mouse as species. Upload gene identifiers from previous step to the Table Browser and download the gene structure information in BED format.

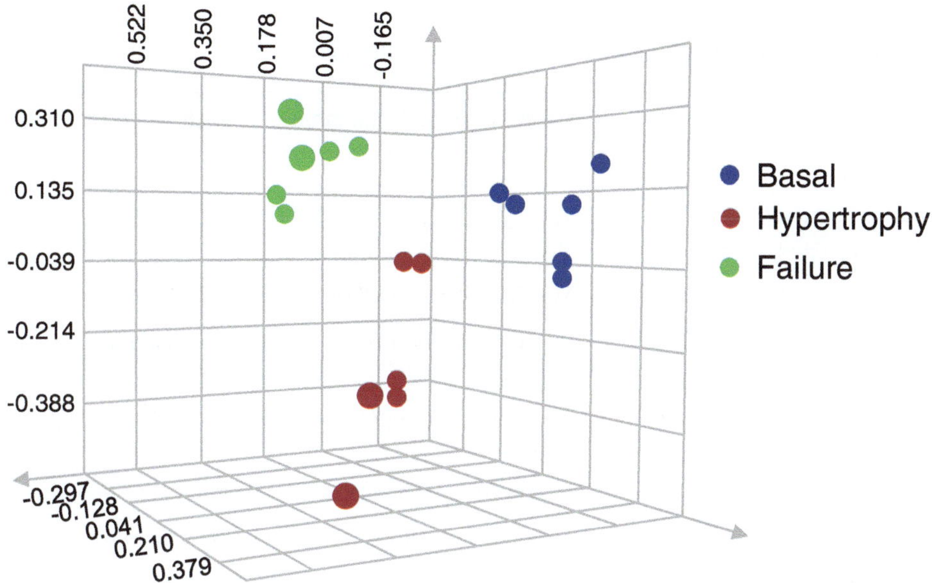

Fig. 4 Analysis of variance and principal component analysis. Cardiac chromatin fractions from three basal mice (*blue*), three mice in cardiac hypertrophy (*red*), and three mice in cardiac failure (*green*) were run on a one-dimensional protein gel. Proteins of 75–150 kDa were cut and digested. Each sample was analyzed twice by LC/MS/MS for a technical replicate. The data was searched and protein abundance determined by label-free quantitation. ANOVA analysis was used to generate a PCA plot where axes represent peptide intensity and each dot represents a single replicate. Clustering by disease state reveals distinct proteomic changes occurring during cardiac hypertrophy and failure and confirms the ability of this methodology to reproducibly detect these changes

3. Visualize the location of these genes in chromosomes. Submit the BED file generated from the previous step to UCSC Genome Browser as a custom track.
4. Build gene ontology tree with enriched gene ontology terms. Convert UniProt identifiers into Mouse Genome Informatics (MGI) identifiers using ID Mapping. Submit MGI IDs to GO Term Finder and choose "MGI (Generic GO slim)" for GO Slim, if mouse sample is used. Genes can be clustered by GO terms using GO Term Mapper.

4 Notes

1. Acid extraction allows a greater enrichment for histone proteins as compared to chromatin fractionation, as it selects for proteins that are tightly bound to the DNA. Performing analysis using both fractions allows for a more complete characterization.
2. Before carrying out mass spectrometric analysis, assess the purity of your fractionation by running a gel and transferring

Table 1
Publicly available tools and databases for bioinformatic analysis

Name (URL)	Function (refer to the manual of each tool for more details)
Algorithms	
ID Mapping (www.uniprot.org/?tab=mapping)	Map identifiers to or from UniProtKB
PICR (www.ebi.ac.uk/Tools/picr/)	Map identifiers based on sequence identity
UCSC Genome Browser (genome.ucsc.edu)	Visualize the location of genes within genome
ProtParam (web.expasy.org/protparam/)	Compute various physiochemical parameters for a given protein
PROSITE (prosite.expasy.org)	Determine protein domains and families
DAVID Bioinformatics Resources (david.abcc.ncifcrf.gov)	Annotate large list of genes based on gene ontology
GENE ONTOLOGY (GO) Tools (go.princeton.edu)	GO Term Finder and GO Term Mapper. Find significant GO terms shared by a list of genes
Clover (zlab.bu.edu/clover/)	Identify overrepresented functional sites within a set of DNA sequences
Cytoscape (www.cytoscape.org)	Analyze and visualize complex protein-protein interaction networks
Databases	
Database of Interacting Proteins (dip.doe-mbi.ucla.edu/)	Database for experimentally determined interactions between proteins
Ensembl (www.ensembl.org)	Genome database for vertebrates and other eukaryotic species
InterPro (www.ebi.ac.uk/interpro/)	Database for protein families, domains, regions, and sites
KEGG PATHWAY (www.genome.jp/kegg/pathway.html)	Database of molecular interaction and reaction networks
MotifMap (motifmap.ics.uci.edu)	Comprehensive maps of candidate regulatory elements encoded in the genomes of model species

to a nitrocellulose membrane. Probe for a nuclear protein (we use H2A) and a mitochondrial protein (we use adenine nucleotide transporter). Electron microscopy of the nuclei pellet (last steps of Subheading 3.2 and 3.3) can also verify enrichment of intact nuclei.

3. We typically digest in trypsin 18–24 h at 37 °C. A digestion markedly shorter than this could be incomplete, likewise going longer than 24 h runs the risk of nonspecific cleavage at other residues. Because trypsin cleaves at the C-terminal end of lysines and arginines, the acidic tails of histones are cut into peptides that are often too small to be detected by the mass

spectrometer. Substituting chymotrypsin for trypsin for the low molecular weight bands is an option to get around this without sacrificing the reproducibility of trypsin. Digest in 0.02 μg/μl chymotrypsin solution in 50 mM NH_4HCO_3 overnight at room temperature for approximately 20 h. The specificity of cleavage, and thus the peptides that will be generated, will vary with timing and can be adjusted depending on the specific goals of the experiment.

4. The protocol included is for in gel digestion as the fractions we are analyzing are very complex. One-dimensional separation by mass via electrophoresis and subsequent cutting of bands greatly reduces the complexity of the sample by distributing the proteins of one fractionation across multiple samples (bands), thereby increasing the ability of the mass spectrometer to detect lower abundance peptides. However, for less complex fractionations (such as acid extractions from a homogenous cell population) or for samples which have undergone subsequent purification (e.g., via HPLC), the one-dimensional separation may not be necessary and the sample could be directly digested in solution without running a gel or cutting bands. When working with a new protein sample, assess the complexity by running a gel and staining with Oriole. Determine if in-solution digestion is possible based off of the number and intensity of bands. For samples with a large number of bands of varying intensity, in gel digestion would be preferable. However, in-solution digestion could be an option for samples with only a few bands of equal intensity, including an equally or more intense band at the molecular weight of your protein of interest. Note that among other adjustments for in-solution digestion, the concentration of enzyme and the length of digestion will vary.

5. With trypsin, and even chymotrypsin, digestion (see Note 4), histone proteins will often be cleaved into only one or a few detectable peptides (due to their small size and amino acid composition). Thus, we recommend searching low molecular weight bands twice to increase the ability to identify histones in these samples. First, search using the above recommended protocol. Then, re-search the data removing the requirements for two unique peptides and for the Consensus Score, since this score also takes into account the number of detected peptides and may be low as a result. This will generate a list of possible histone proteins which must be confirmed by manual inspection of their spectra. Briefly, the identified peptide(s) for each should be examined for three key things. First, you should confirm the presence of the parent ion. Second, you should confirm that the majority of the high abundant and medium abundant daughter ions are assigned to your peptide. Third,

you should look individually at the B and Y ions by charge. Working across a peptide from N terminal to C terminal, there are a series of peptide bonds that could have broken, each one generating a B and Y ion of a different mass. As a rule of thumb, a good fragmentation will give at least five ions of the same charge (e.g., B2+) in a row. A final consideration for histones is the potential for the single identified peptide to match to more than one variant. In this case, knowledge of the sample and what other histone peptides were identified, as well as other methods may be necessary.

6. Even when samples are prepared together and run using the same buffers on the same day, small variations in injection, ionization, and fragmentation must be accounted for. Several means of normalizing data between runs exist. Peptide elution profiles can be normalized through alignment algorithms. Additionally, signal intensity can be normalized to either background noise, a known analyte in your sample, or a known standard that was added during sample preparation. In all cases, biological and technical replicates are still necessary for determining statistical significance.
7. Identifier conversion between protein database and gene database is used extensively in this section. However, it should be noted that one gene identifier may be mapped to multiple protein identifiers, and vice versa. PICR tool provided by EBI maps identifiers according to the similarity between protein and gene sequences and is useful when only sequence is available. However, it is slower than ID Mapping and may take several hours or days to map if a long list of identifiers are provided.

Acknowledgments

The Vondriska lab is supported by grants from the National Heart, Lung, and Blood Institute of the NIH and the Laubisch Endowment at UCLA. EM is recipient of the Jennifer S. Buchwald Graduate Fellowship in Physiology at UCLA, HC is the recipient of an American Heart Association Pre-doctoral Fellowship, and SF is the recipient of an NIH K99 Award.

References

1. Franklin S et al (2011) Specialized compartments of cardiac nuclei exhibit distinct proteomic anatomy. Mol Cell Proteomics 10:703
2. Franklin S et al (2012) Quantitative analysis of chromatin proteome reveals remodeling principles and identifies HMGB2 as a regulator of hypertrophic growth. Mol Cell Proteomics, 11: M111.014258
3. Lieberman-Aiden E et al (2009) Comprehensive mapping of long-range interactions reveals folding principles of the human genome. Science 326:289–293

4. van Steensel B, Dekker J (2010) Genomics tools for unraveling chromosome architecture. Nat Biotechnol 28:1089–1095
5. Kornberg RD (1974) Chromatin structure: a repeating unit of histones and DNA. Science 184:868–871
6. Zhang Z, Pugh BF (2011) High-resolution genome-wide mapping of the primary structure of chromatin. Cell 144:175–186
7. Rada-Iglesias A et al (2010) A unique chromatin signature uncovers early developmental enhancers in humans. Nature 470:279–283
8. Barski A et al (2007) High-resolution profiling of histone methylations in the human genome. Cell 129:823–837
9. Cuddapah S et al (2011) Genomic profiling of HMGN1 reveals an association with chromatin at regulatory regions. Mol Cell Biol 31:700–709
10. Schones DE, Zhao K (2008) Genome-wide approaches to studying chromatin modifications. Nat Rev Genet 9:179–191
11. Kouzarides T (2007) Chromatin modifications and their function. Cell 128:693–705
12. Schreiber SL, Bernstein BE (2002) Signaling network model of chromatin. Cell 111: 771–778
13. Matouk CC, Marsden PA (2008) Epigenetic regulation of vascular endothelial gene expression. Circ Res 102:873–887
14. Ho L, Crabtree GR (2010) Chromatin remodeling during development. Nature 463:474–484
15. Haberland M, Montgomery RL, Olson EN (2009) The many roles of histone deacetylases in development and physiology: implications for disease and therapy. Nat Rev Genet 10: 32–42
16. Narlikar GJ, Fan HY, Kingston RE (2002) Cooperation between complexes that regulate chromatin structure and transcription. Cell 108:475–487
17. Bai L, Santangelo TJ, Wang MD (2006) Single-molecule analysis of RNA polymerase transcription. Annu Rev Biophys Biomol Struct 35: 343–360
18. Cairns BR (2009) The logic of chromatin architecture and remodelling at promoters. Nature 461:193–198
19. Heineke J, Molkentin JD (2006) Regulation of cardiac hypertrophy by intracellular signalling pathways. Nat Rev Mol Cell Biol 7:589–600
20. Rajabi M, Kassiotis C, Razeghi P, Taegtmeyer H (2007) Return to the fetal gene program protects the stressed heart: a strong hypothesis. Heart Fail Rev 12:331–343
21. Razeghi P et al (2001) Metabolic gene expression in fetal and failing human heart. Circulation 104:2923–2931
22. Louch WE, Sheehan KA, Wolska BM (2011) Methods in cardiomyocytes isolation, culture, and gene transfer. J Mol Cell Cardiol 51:288–298
23. Schluter KD, Schreiber D (2005) Adult ventricular cardiomyocytes: isolation and culture. Methods Mol Biol 290:305–314
24. O'Connell TD, Rodrigo MC, Simpson PC (2007) Isolation and culture of adult mouse cardiac myocytes. Methods Mol Biol 357:271–296
25. Nesvizhskii AI, Vitek O, Aebersold R (2007) Analysis and validation of proteomic data generated by tandem mass spectrometry. Nat Methods 4:787–797
26. Park SK, Venable JD, Xu T, Yates JR 3rd (2008) A quantitative analysis software tool for mass spectrometry-based proteomics. Nat Methods 5:319–322
27. Lomenick B et al (2009) Target identification using drug affinity responsive target stability (DARTS). Proc Natl Acad Sci USA 106:21984–21989
28. Kamleh A et al (2008) Metabolomic profiling using Orbitrap Fourier transform mass spectrometry with hydrophilic interaction chromatography: a method with wide applicability to analysis of biomolecules. Rapid Commun Mass Spectrom 22:1912–1918
29. Searle BC (2010) Scaffold: a bioinformatic tool for validating MS/MS-based proteomic studies. Proteomics 10:1265–1269

Chapter 8

Proteomic Analysis of Interstitial Aortic Valve Cells Acquiring a Pro-calcific Profile

Millioni Renato, Elisa Bertacco, Cinzia Franchin, Giorgio Arrigoni, and Marcello Rattazzi

Abstract

Cell-driven processes are now considered of relevance for the pathogenesis of aortic stenosis. In particular, during calcific valve degeneration, interstitial valve cells (VIC) resident in the leaflet can acquire an osteogenic/pro-calcific profile and actively contribute to matrix mineralization. The proteomic study described in this chapter is undertaken to investigate modifications in the proteome of bovine aortic VIC acquiring a calcifying phenotype. This approach can be useful to clarify cellular pathways involved in VIC pro-calcific differentiation and identify innovative therapeutic targets.

Key words Aortic valve calcification, Interstitial valve cells, 2-DE analysis, MS analysis

1 Introduction

Calcific aortic valve stenosis is the most common acquired valve disease and represents the third cause of cardiovascular death in western countries (1). Although aortic stenosis prevalence is progressively increasing due to aging of the population, most of the mechanisms driving calcific degeneration of aortic valve leaflets are still unknown, and no medical therapies are available to either slow or prevent calcific valve degeneration. The traditional view of valve calcification as a progressive, ineluctable event has been recently questioned by evidence showing the importance of a balance between promoting and inhibiting factors and the relevance of cell-driven processes (2). Although dystrophic calcification represents the main pathological finding, aspects of mature lamellar bone and endochondral bone formation have been described in the calcified aortic valves. Moreover, analyses of calcified leaflets have shown the expression of bone-related proteins (such as osteopontin, osteocalcin (OC), alkaline phosphatase activity (ALP), OPG, RANKL, BMPs), chondro-/osteoblast-specific transcription factors (such as Runx2/Cbfa-1 and Sox9), and mediators involved in osteogenic

Fernando Vivanco (ed.), *Heart Proteomics: Methods and Protocols*, Methods in Molecular Biology, vol. 1005,
DOI 10.1007/978-1-62703-386-2_8,

programs (such as Wnt/Lrp5/beta-catenin pathway) (3). On the whole, these data suggest that processes resembling those involved in bone remodeling can take place during aortic valve calcification. Recent epidemiological and basic science studies suggest that inflammatory molecules, modified lipids, infectious agents, and circulating levels of calcium and phosphate could act as potential promoters of valve calcification. Some of these factors have been shown to promote the acquisition of an osteogenic profile by interstitial valve cells (VIC). These are the most prevalent type of cell in the valve and play a crucial role in maintaining the structural integrity of the leaflets. However, VIC biology, pathobiology, and phenotype-related functions are poorly understood. We recently identified a peculiar subpopulation of bovine VIC characterized by a fibroblast-like phenotype that expresses osteoblast-like markers (ALP and OC) and promotes collagen–matrix calcification in response to endotoxin (LPS) and elevated phosphate (Pi) levels (4). Proteomic analysis of VIC acquiring a calcifying profile can be useful to recognize molecular pathways involved in cell pathological differentiation and might help to identify novel therapeutic targets. In this chapter we describe the materials and methods used to conduct a proteomic study in VIC acquiring a pro-calcific profile.

2 Materials

All solutions and equipments used for cell culture must be sterile. During cell culture media preparation, use proper sterile technique, work in a laminar flow hood, and pre-warm solutions before use. Prepare all solutions and solvents for two-dimensional electrophoresis (2-DE) and mass spectrometry (MS) analysis using ultrapure water. Always wear protective gloves when working with acrylamide, IPG strips, or surfaces that come into contact with acrylamide solutions. All reagents must be of MS grade. Trypsin must be at least of sequencing grade. Store reagents according to manufacturer's instructions. Solutions are kept at room temperature unless otherwise specified. MS analysis can be performed with any mass spectrometer with MS/MS capabilities. Concentrations and amounts of reagents and solutions are optimized for MALDI-TOF/TOF 4800 Plus from AB Sciex (Foster City, CA, USA). Stainless steel MALDI plates must be compatible with the instrument used.

2.1 Cell Culture Media Formulations

1. Growth medium: Dulbecco's Modified Eagle Medium (DMEM) containing 4.5 g/l glucose plus FBS 20 %, 100 U/ml penicillin, and 100 μg/ml streptomycin.
2. Treatment medium: DMEM containing 4.5 g/l glucose plus FBS 5 %, 100 U/ml penicillin, and 100 μg/ml streptomycin supplemented with LPS (100 ng/ml; *E. coli*).
3. Cellular lysis buffer Tris–HCl 12 mM, pH 7.4, DTT 1 mM, and a cocktail of protease inhibitors (Complete, Roche).

2.2 Components and Reagent Solutions for 2-DE

1. Components: (a) First dimension: IPGphor, IPGphor strip holders, immobilized pH gradient (IPG) strips (GE Healthcare), (b) Second dimension: Ettan DALTsix gel caster and vertical electrophoresis apparatus, including the power supply and the thermostatic circulator (GE Healthcare), (c) transmittance scanner and 2-DE image analysis software.
2. IPG strip rehydration stock solutions: urea stock solution: 8 M urea, 2 % CHAPS, 0.5 % ampholyte buffer, 0.002 % bromophenol blue, 25 ml; urea–thiourea stock solution: 7 M urea, 2 M thiourea, 2 % CHAPS, 0.5 % ampholyte buffer, 0.002 % bromophenol blue; store aliquots at −20 °C.
3. SDS equilibration buffer: 50 mM Tris–HCl pH 8.8, 6 M urea, 30 % glycerol, 2 % SDS, 0.002 % bromophenol blue; store aliquots at −20 °C. DTT (100 mg for 10 ml of SDS equilibration buffer) or iodoacetamide (250 mg for 10 ml of SDS equilibration buffer) should be added before use.
4. Laemmli SDS electrophoresis buffer: 25 mM Tris base, 192 mM glycine, 0.1 % SDS. The pH of this solution should not be adjusted.
5. SDS-PAGE polyacrylamide solution: acrylamide mix 29:1, 12 % T, Tris–HCl 0.375 M, APS 0.1 %, SDS 0.1 %, TEMED 0.04 %, pH 8.8. This solution must be prepared just before use (see Note 1).
6. Agarose sealing solution: 25 mM Tris base, 192 mM glycine, 0.1 % SDS, 0.5 % agarose. Heat on a heating stirrer until agarose is completely dissolved. Do not allow the solution to boil over.
7. Colloidal Coomassie brilliant blue (CBB) staining solutions: 0.1 % w/v CBB G250, 25 % v/v methanol, 5 % v/v acetic acid; CBB destaining solution: 5 % v/v methanol and 7,5 % v/v acetic acid.

2.3 Components and Solutions for In-Gel Digestion and MALDI-TOF/TOF Mass Spectrometry

Concentrations of solutions and amount of reagents reported have been calculated to allow the digestion and analysis of 100 spots from 2-DE.

1. Components: (a) 0.6 ml Maxymum Recovery microcentrifuge tubes for in-gel protein digestion and peptide extraction (Axygen, Union City, CA, USA) (see Note 2); (b) stainless steel MALDI plate; (c) stainless steel scalpels to excise gel spots; (d) SpeedVac vacuum concentrator; (5) microcentrifuge; (6) thermoblock or thermal bath set at 37 °C.
2. TFA 0.1 % solution: 0.1 % trifluoroacetic acid (TFA) in water. 10 ml of solution is required. Measure 10 ml of ultrapure water (possibly with a 5 ml pipette or in a graduated cylinder). Add 10 μl of TFA (see Notes 3 and 4) and transfer into a 25 ml glass bottle.

3. Ammonium bicarbonate solution: ammonium bicarbonate (NH_4HCO_3) 50 mM. pH adjustment is not needed. 10 ml of solution is required. Solution is stable for a few days at room temperature (better to prepare just before use).
4. Destaining solution: 50 % acetonitrile (ACN) in ammonium bicarbonate (NH_4HCO_3) 50 mM. 18 ml of the solution is required.
5. Matrix dissolving solution: 30 % ACN, 0.1 % TFA in water. 1.0 ml of solution is required.
6. Matrix solution: α-cyano-4-hydroxycinnamic acid (CHCA) 5 mg/ml in 0.1 % TFA, 30 % ACN. Dissolve the matrix with an appropriate amount of matrix dissolving solution to obtain a final concentration of 5 mg/ml. About 2 μl of matrix solution is needed for each 2-DE spot (see Note 5).
7. Trypsin solution: porcine modified trypsin (sequencing grade, Promega) 12.5 ng/μl in ammonium bicarbonate solution (NH_4HCO_3 50 mM). About 1.0 ml of this solution is required. Trypsin dissolved in ammonium bicarbonate can be stored at –20 °C. Do not refreeze after thawing.

3 Methods

3.1 VIC Isolation and Treatment

1. Obtain bovine hearts (15 months of age) from a slaughterhouse within 30 min from slaughter (see Note 6).
2. Place immediately the whole hearts in growth medium formulation added with Amphotericin B (0.5 μg/ml). Keep samples in ice during transportation to the laboratory.
3. Obtain interstitial valve cells (BVIC) from the noncoronary leaflet of aortic valve using an explant method. After scraping out fibrosa and ventricularis endothelial layers, mince aortic leaflet in fragments of about 2–3 mm^3, digest with type I collagenase (125 U/ml; Sigma), elastase (8 U/ml; Fluka, Germany), and soybean trypsin inhibitor (0.375 mg/ml; Sigma) for 30 min at 37 °C. Collect fragments in a Tissue Culture Petri dish (Falcon BD Bioscience, San Diego, CA) and incubate at 37 °C, 5 % CO_2 in growth medium formulation. Usually, after 7–10 days primary BVIC spread from predigested fragments. Cultures are visualized by phase contrast microscopy.
4. After reaching 80–90 % confluence, split BVIC into Tissue Culture 100 mm Petri dishes at the density of 10^4 cells/cm^2 (see Note 7). The day after plating, BVIC become adherent. Change media every third day.
5. Once confluence is reached, treat the cells with the treatment medium containing LPS or control for 12 days. Change media every third day (see Note 8).

3.2 Protein Extraction

1. Wash cells three times with PBS to remove any traces of serum, calcium, and magnesium that would inhibit the activity of the reagents.
2. For each 100 mm Tissue Culture Petri dish, add 0.250–0.300 ml of lysis buffer. Gently shake the plate to get complete coverage of the cell layer. Lyse cells by scraping the dish, collect lysate, and keep it on ice. Submit the lysate to three cycles of freeze–thawing in liquid nitrogen and sonication in ice.
3. Fractionate the sample by ultracentrifugation (100,000 × g for 1 h at 10 °C). Solubilize the pellet, membrane-enriched protein fraction (MF), in the urea–thiourea stock solution. Concentrate and desalt the supernatant, cytosol-enriched protein fraction (CF), through ultrafiltration (Microcon-Amicon YM-3, Millipore Corporation).
4. Quantify the proteins with a colorimetric assay (Bradford, BCA or similar).
5. Add a constant amount of solubilized proteins (300 μg) to 450 μl of the urea solution for CF and urea–thiourea stock solution for MF. Vortex (see Note 9).

3.3 First Dimension: Isoelectric Focusing

1. Pipette 450 μl of sample-containing rehydration solution into the ceramic strip holders (see Note 10).
2. Lay down the IPG strips gel-side down into the rehydration solution. Gently lift and lower the strip and slide it back and forth along the holder to assure a complete and even wetting. Remove any large air bubbles (see Note 11). The electrodes must be in contact with the acrylamide of the strip. Overlay with mineral oil (see Note 12) and place the strip holders over the cooling plate of the IPGphor, on the electrode contact areas.
3. IPG gel strips are rehydrated overnight, and then IEF is performed until the desired total product time × voltage applied for each strip is reached (see Note 13). IPG strip containing the sample can be stored at −20 °C for several days.
4. To prepare focused proteins for migration into the SDS-PAGE gel, incubate the IPG strips for 15 min in the SDS equilibration buffer containing 1 % DTT and then for 15 min in the same solution containing 4 % w/v iodoacetamide (IAA) instead of DTT (see Note 14).

3.4 SDS-PAGE Casting and IPG Strip Loading

1. Following the manufacturer instructions, assemble the gel caster on a level bench. Add APS and TEMED to the polyacrylamide solution just before pouring the solution in the caster. Pour the gels to within 1 cm of the top of the glass plates. A stacking gel is not necessary.
2. Hold one end of the equilibrated IPG strip with forceps and place it across the exposed top part of the higher glass plates.

Slide the IPG strip between the glass plates with the help of a thin spatula.

3. Seal the IPG strip in place: for each IPG strip, melt agarose solution in a heating magnetic stirrer and slowly pipette it across the length of the IPG strip (see Note 15). Allow a minimum of 5 min for the agarose to solidify.
4. Prepare the Ettan DALTsix gel vertical electrophoresis apparatus following manufacturer's instructions and begin electrophoresis using the Laemmli SDS electrophoresis buffer (see Note 16).
5. Once the dye front reaches the end of the gel, turn off the power supply, disassemble the electrophoretic apparatus, and open the glass plates by using a spatula. The gel will remain attached to one of the glass plates. After removing agarose and the strip, rinse the gel with water and carefully transfer it to a clean vessel for CBB staining and destaining.

3.5 Gel Staining and Image Analysis

1. Stain the gel with CBB at room temperature for at least 3 h with gentle agitation.
2. After removing CBB by aspiration, cover the gel with the destaining solution. Use gentle agitation and change destaining solution several times until the spots are visible and the background staining is low.
3. Gel images can be acquired by using transmittance scanners (with an optical density of 4), charge-coupled device cameras, or laser imaging devices, depending on the protein labeling or staining techniques used for spot visualization (see Note 17).
4. The spot volumes, after background subtraction, are expressed as a numeric value of optical density (spot intensities). Spot intensities of gel replicates are automatically normalized by the 2-DE image analysis software allowing comparison between gels.
5. After completion of spot matching, use the normalized intensity values of individual spots to compare protein levels between groups (treated vs. untreated) (see Note 18).

3.6 Protein Spot Excision and In-Gel Digestion

We report a digestion protocol not requiring reduction and alkylation of cysteine residues that were performed during sample separation for SDS-PAGE. Keratin contamination is a common inconvenient when analyzing proteins from 1D or 2D gels by mass spectrometry. Store the gels covered and handle them with great care always using powder-free gloves (powder contamination interferes with MS analysis).

1. Store intact gels in 5 % acetic acid at 4 °C using clean vessels until spots are excised. Protein spots of interest for MS analysis can be excised manually using a scalpel or, alternatively, a clean

pipette tip and placed into 0.6 ml Maxymum Recovery microcentrifuge tubes (Axygen). Carefully mark the tubes; spots can be stored indefinitely at −20 °C.

2. Wash gel spots with 50 μl destaining solution (50 % ACN/25 mM NH_4HCO_3) for 10–15 min under constant agitation. The gel sample should turn opaque. Centrifuge shortly and dispose supernatant. Repeat the procedure twice. At this point, normally gel spots should be colorless.
3. Dry spots under vacuum in a SpeedVac for 10–15 min. Drying the spots maximizes the access of the enzyme into the gel upon rehydration (see Note 19).
4. Add trypsin solution to each gel spot. Normally 5 μl of solution per millimeter square of gel should be used. For average spot sizes, 10 μl of solution is generally enough. Leave gel spots to rehydrate in ice (or at 4 °C) for 30 min.
5. Check if there is still liquid inside the tubes or if the gel adsorbed all the solution. In this case, add a few microliters of ammonium bicarbonate solution to cover the gel pieces. If after rehydration a large excess of solution is still present, remove it from the tube, so as to have enough liquid to cover the gel piece.
6. Digest samples at 37 °C overnight or at least for 3–4 h.

3.7 Extraction of Peptide Mixtures and Spotting onto the MALDI Target Plate

If conducted manually (i.e., without the support of robotics devices), we recommend to perform this procedure on a limited number of samples (20–30). The extraction of peptide mixtures must be conducted with great care to avoid mix-up of the samples. For optimum results, use only freshly prepared matrix solution.

1. Mark a new set of 0.6 Maxymum Recovery tubes, matching the tubes containing the gel spots.
2. Add 20–30 μl of 75 % ACN, 0.1 % TFA to each gel spot and incubate for 20 min under constant agitation. Then transfer the solution into the corresponding tube (a new pipette tip must be used for each sample). Repeat the procedure twice. At this point gel spots should appear completely dehydrated and hard.
3. Reduce the volume of each sample (about 70–100 μl at this point) in SpeedVac down to about 5–10 μl (see Note 20). Samples can be stored at −20 °C until the MS analysis is performed.
4. Pipette 1 μl of sample into a clean 0.6 Maxymum Recovery tube and add 1 μl of matrix solution. Mix well by pipetting up and down. Spot 0.8 μl of mixture onto a MALDI plate (see Notes 21 and 22). After spotting all samples, let them dry completely, leaving the plate in a dark place for 10–15 min (see Notes 23 and 24).

3.8 MS Analysis and Database Search

Mass spectrometry analysis is a complex procedure that can be conducted with very different methods, instruments, and informatics tools and that is impossible to summarize here.

The following method refers to the MS analysis conducted with a MALDI-TOF/TOF 4800 Plus (AB Sciex) and will not be very specific because procedures and instrumental parameters depend on the characteristic of the instrument used. Procedures will be very different if samples are analyzed with nanoESI-MS/MS instruments coupled online with a LC system. Tuning and calibrating the instrument over the required mass range, both in MS and MS/MS mode, is important to obtain high-confidence identifications (see Note 25).

1. We usually perform a data-dependent analysis by programming the instrument to acquire a MS spectrum per each sample, select the ten most intense ions, and, on these, perform MS/MS analysis (see Note 26).
2. After the acquisition, raw data files are converted in a format that is readable by search engines. At this point spectra are usually transformed by a series of operations such as peak picking, deisotoping, and noise filtering (see Note 27).
3. Files are analyzed with a search engine (like Mascot, SEQUEST, X!tandem, or similar) against a specific database. It is necessary to define some information like enzyme specificity, mass tolerance at the MS and MS/MS level, maximum number of allowed missed cleavages, and static and dynamic protein modifications. It is important to perform the analysis also against a randomized or inverted database or use other statistical tools to evaluate the false discovery rate for the identifications (see Note 28).
4. For samples derived from 2-DE, the results obtained by the search engines are usually easy to interpret since ideally one or only very few proteins should be identified per sample. Search engines generally return results associated to a probabilistic score that can be used to evaluate the confidence of identifications. Sometimes it is important to manually evaluate the obtained results, overall when a protein has been identified with a low score or with a very limited number of peptides.

4 Notes

1. A 12 % T acrylamide gel is suitable for the separation of proteins with molecular weight (MW) ranging from 180 to 10 kDa. Porosity gradient gels with a downward gradient of increasing acrylamide percentage allow better resolution compared to homogeneous single percentage gels. Since the reproducible preparation of gradient gel casting is complex and time consuming, we suggest using commercially available precast gradient gels.

2. Maxymum Recovery tubes are treated to prevent binding of peptides to the tube surface.

3. Trifluoroacetic acid (TFA) is a very strong and fuming acid. Handle with great care under a fume hood, wearing gloves and goggles. TFA is stored at 4 °C; before opening the bottle, wait till the reagent has reached room temperature (preserve TFA from adsorbing humidity). Do not insert plastic pipette tips directly into the bottle of TFA (plastic can dissolve and contaminate concentrated TFA). Transfer few drops of pure TFA in a small glass container and then pipette the quantity needed. We suggest to prepare a stock solution of 50 % TFA (store at −20 °C in a glass container) which is less dangerous to handle and can be used without reaching room temperature. If 50 % TFA is prepared, quantities to be used must be doubled.

4. In case the MS analysis is conducted using an instrument interfaced with an electrospray (ESI) source, TFA (that suppresses ionization in ESI sources) should be replaced with equal amount of formic acid (FA) in all solutions.

5. CHCA must be of high purity (MS grade) to obtain reproducible and good quality results. CHCA is light sensitive and should be stored in the dark. Matrix solution must be prepared fresh just before MS analysis. Let the bottle reach room temperature before opening the lid.

6. VIC can be obtained from different animal species. Bovine VIC (BVIC), used in this case, are characterized by high proliferative rate in vitro. Similar to BVIC, a good rate of cell proliferation can be obtained with porcine VIC, a cell type commonly used to investigate valve biology as well. In vitro studies conducted by using smaller animals (such as rats and mice) are severely hampered by the low number of VIC obtained. Human VIC are mainly collected from aged/pathological valves and usually show a low proliferation rate in culture. Although human cells can furnish valuable information about the profile of VIC present in the diseased valves, they might present some limits for studying normal tissue physiology and/or the initial phase of pathological processes.

7. The same experiment can be conducted after cloning the VIC by using a dilutional approach. Investigation on clonal cells allows to perform the proteomic analysis on specific VIC subpopulations. In particular, this approach enables to study changes happening in the proteomic profile during clonal VIC pro-calcific differentiation and/or comparing the proteomes of cells harboring different calcifying potential (4).

8. In the present protocol, we used LPS as inducer of calcifying profile in VIC. This pathological stimulus can be substituted by other calcification promoters such as inorganic phosphate, inflammatory cytokines, or oxidized lipids. However, before

conducting the proteomic study, the researchers need to confirm the calcifying potential of the cells treated with the calcification inducer. In particular, it should demonstrate the acquisition of an osteoblast-like profile (i.e., expression of ALP and OC) and in vitro deposition of calcium by the VIC.

9. Although several standard 2-DE protocols have been published, these methods need to be adapted and further optimized for the type of sample to be analyzed. In general, we recommend the following: (a) use urea and nonionic or zwitterionic detergents to solubilize proteins; (b) Triton X100 is poorly efficient when used with urea alone, while it is much more efficient in urea–thiourea and particularly useful for the solubilization of hydrophobic proteins (5); (c) desalt the sample or prepare sample with final salt concentration less than 10 mM; (d) samples containing urea should not be heated (<37 °C) in order to avoid the protein carbamylation by isocyanate which covalently modifies lysine residues, inducing changes in isoelectric point. The amount of proteins loaded in the first dimension depends on strip length, pH range, and detection method used. The method described here is optimized for 24 cm long IPG strips, 4–7 pH range, and CBB staining. Sample concentration must be high enough to avoid excessive dilution of rehydration stock solution (no more than 1/8).

10. Samples can be loaded either through addition to the rehydration solution or by loading into the rehydrated IPG strip via sample cups. We prefer the first approach because it is simpler and eliminates the risk of precipitate formation at the application point of cup loading.

11. Reproducibility of pH gradient profiles is limited by batch-to-batch preparation variability of the IPG slab gel from which the individual strips are obtained. Not consecutively numbered IPG strips may derive from different master gels and should preferably not be used in the same quantification experiment, because this leads to a high increase of the inter-gel spot variances (6, 7).

12. Addition of mineral oil is mandatory to minimize sample evaporation and urea crystallization during IEF.

13. We recommend applying low voltages (30–50 V) during IPG strip rehydration to facilitate the entry of high MW proteins into gel matrix. The optimum focusing time to achieve the best IEF quality and reproducibility is the time needed for all the proteins to reach the steady state. A low volt–hours product will result in poor focusing, whereas overfocusing can induce excessive water exudation at the IPG gel surface (reverse electroendosmotic flow) with formation of distorted protein patterns and horizontal streaks at the basic side of 2D gel. Optimum focusing time should be empirically established for each combination of protein sample, protein loading, pH

range, and length of strip used. In general, the IEF of longer strip and narrow pH range requires higher value of total volt–hours product.

14. IAA is used to alkylate any free DTT, thus preventing its migration through the second-dimension gel. DTT migration can in fact result in artifact known as point streaking (vertical lines in silver-stained gels). IAA also alkylates the sulfhydryl groups of the proteins preventing their reoxidation during the SDS-PAGE run and the formation of vertical spot streaking.
15. The strip should remain on the surface of the slab gel. Avoid trapping air bubbles between strip and the second-dimension gel. The gel face of the strip should not touch the opposite glass plate.
16. To avoid streaking and minimize loss of proteins, protein transfer from the first to the second dimension should be performed slowly (<5 V/cm). The choice of voltage depends on the time of application and generally is between 60 and 120 V.
17. A proper setting of image digitalization parameters is fundamental. Generally, it is recommended to acquire gel images into TIFF files with 16 bit of grayscale depth at 300 d.p.i. resolution.
18. Post-experimental 2-DE variability, mainly due to improper use of 2-DE image analysis software, can be sufficient to mask changes in the levels of the protein under investigation (8, 9). In particular, we recommend to check manually all the detected spots to minimize the number of false positives (i.e., artifacts such as bubbles, imperfections in the gel, or dye residues) and to perform a separate analysis of the spots that have been manually segmented (10).
19. After being dried in the SpeedVac, spots will be smaller and hard, sometimes even difficult to see inside the tube if the original gel spots were very small. The drying process and the use of gloves might build up static energy to which dried gel spots are very sensitive; therefore, extract tubes from the SpeedVac with great care to avoid losing the gel spots.
20. Completely dried samples can favor binding of peptides to the tube surface. If samples are completely dry, resuspend them with 10 μl of ACN 20 % in 0.1 % TFA solution. If samples will be analyzed by LC-MS, do not use ACN and resuspend them with 10 μl of 0.1 % TFA solution.
21. Samples can be spotted onto the matrix also using different methods. For an overview of the different spotting procedures and sample preparations, refer to Kussmann and Roepstorff (11). The method we normally use gives generally very good and reproducible results, with a thin homogeneous layer of sample/matrix optimal for an automated data acquisition.
22. It is a good practice to spot each sample in duplicate or triplicate. The crystallization process is influenced by many factors, and from time to time, sample–matrix co-crystallization turns

out to be unideal. Moreover, during the MALDI analysis, matrix and sample are destroyed by the laser; therefore, having the same sample in more than one spot onto the plate could be useful if the first analysis is not successful due to a not ideal crystallization or if extra data need to be collected.

23. The presence of a high percentage of ACN in the final sample solution allows for a fast co-crystallization of the matrix–sample mixture. This leads to a highly homogeneous layer of small crystals that is particularly useful when MS and MS/MS data are acquired in a completely automated way. In this case it is important to have the least possible variation of signals when laser is fired at different positions across the spot.
24. The matrix is light sensitive; therefore, it is a good practice to keep the MALDI plates (once the samples have been spotted on them) in the darkness to avoid matrix degradation. Best results are obtained when samples are spotted just before the MS analysis is performed.
25. If the instrument supports the application, it can be a good practice to mix an internal standard with the matrix solution to obtain an internal calibration for each sample. Any synthetic peptide with a molecular weight possibly within a 1,200–1,800 Da mass range can be used as internal standard provided that it is spiked into the solution in the appropriate amount. The operator must establish the correct amount of standard since this depends on the specific sequence of the peptide chosen and on its ionization efficiency. Internal calibration is particularly useful when a large number of samples are analyzed in a single MALDI plate. In this case, calibrating the instrument only with an external calibration might lead to a significant variation of mass accuracy across the positions of the MALDI plate. Moreover, large number of samples means also long acquisition times; therefore, overall with TOF analyzers, a substantial variation of mass accuracy and precision could be observed during the day.
26. We found that acquiring ten MS/MS spectra per sample usually results in high-confidence protein identifications. This number can be increased, but matrix–sample spots tend to degrade quite quickly, and the acquisition of more spectra might lead to poor results while significantly increasing the acquisition time.
27. The conversion tools usually require the setting of a certain number of parameters. These depend on the specific tool used and will not be discussed here. However, it is very important to set these parameters with correct values. Converting the raw MS and MS/MS spectra with not optimized parameters can result in less confident protein identifications, since the way in which the files are converted into a readable format affects the final scores of the search engines.

28. To set the correct parameters for the search engines is of absolute importance. Using wrong parameters, like a wrong mass tolerance for peptides or fragment ions, can lead to a complete failure of the analysis or significantly reduce the score associated to an identified protein. Also the database used for the analysis must be correctly chosen, since the number of sequences in the database influences the final score of identified proteins: very small databases will result in unreliable identifications, while unnecessarily large databases can result in low-score identifications. It is always a good choice to use a database relative to the particular specie from which the samples derive.

Acknowledgments

The authors wish to thank Veneto Banca Holding for funding the acquisition of the MALDI-TOF/TOF. This work was supported by the University of Padova (CPDA082784/08 to GA, CPDA083710/08 to M.R.), and FORIBICA.

References

1. Goldbarg SH, Elmariah S, Miller MA, Fuster V (2007) Insights into degenerative aortic valve disease. J Am Coll Cardiol 50:1205–1213
2. Rajamannan NM, Evans FJ, Aikawa E, Grande-Allen KJ, Demer LL, Heistad DD, Simmons CA, Masters KS, Mathieu P, O'Brien KD, Schoen FJ, Towler DA, Yoganathan AP, Otto CM (2011) Calcific aortic valve disease: not simply a degenerative process: a review and agenda for research from the National Heart and Lung and Blood Institute Aortic Stenosis Working Group. Executive summary: calcific aortic valve disease-2011 update. Circulation 124:1783–1791
3. Miller JD, Weiss RM, Heistad DD (2011) Calcific aortic valve stenosis: methods, models, and mechanisms. Circ Res 108:1392–1412
4. Rattazzi M, Iop L, Faggin E, Bertacco E, Zoppellaro G, Baesso I, Puato M, Torregrossa G, Fadini GP, Agostini C, Gerosa G, Sartore S, Pauletto P (2008) Clones of interstitial cells from bovine aortic valve exhibit different calcifying potential when exposed to endotoxin and phosphate. Arterioscler Thromb Vasc Biol 28:2165–2172
5. Luche S, Santoni V, Rabilloud T (2003) Evaluation of nonionic and zwitterionic detergents as membrane protein solubilizers in two-dimensional electrophoresis. Proteomics 3:249–253
6. Taylor RC, Coorssen JR (2006) Proteome resolution by two-dimensional gel electrophoresis varies with the commercial source of IPG strips. J Proteome Res 5:2919–2927
7. Fuxius S, Eravci M, Broedel O, Weist S, Mansmann U, Eravci S, Baumgartner A (2008) Technical strategies to reduce the amount of "false significant" results in quantitative proteomics. Proteomics 8:1780–1784
8. Millioni R, Puricelli L, Sbrignadello S, Iori E, Murphy E, Tessari P (2012) Operator- and software-related post-experimental variability and source of error in 2-DE analysis. Amino Acids 42:1583–1590
9. Millioni R, Miuzzo M, Sbrignadello S, Murphy E, Puricelli L, Tura A, Bertacco E, Rattazzi M, Iori E, Tessari P (2010) Delta2D and Proteomweaver: performance evaluation of two different approaches for 2-DE analysis. Electrophoresis 31:1311–1317
10. Millioni R, Sbrignadello S, Tura A, Iori E, Murphy E, Tessari P (2010) The inter- and intra-operator variability in manual spot segmentation and its effect on spot quantitation in two-dimensional electrophoresis analysis. Electrophoresis 31:1739–1742
11. Kussmann M, Roepstorff P (2000) Sample preparation techniques for peptides and proteins analyzed by MALDI-MS. Methods Mol Biol 146:405–424

Chapter 9

Differential Protein Expression Analysis of Degenerative Aortic Stenosis by iTRAQ Labeling

Sergio Alonso-Orgaz, Tatiana Martin-Rojas, Enrique Calvo, Juan Antonio López, Fernando Vivanco, and María G. Barderas

Abstract

Degenerative aortic stenosis is the most common worldwide cause of valve replacement. While it shares certain risk factors with coronary artery disease, it is not delayed or reversed by reducing exposure to risk factors (e.g., therapies that lower lipids). Therefore, it is necessary to better understand its pathophysiology for preventive measures to be taken. In order to identify proteins that are involved in AS, we designed a simple method for protein extraction, digestion, labeling, identification, and differential protein expression analysis of both normal and stenotic aortic valve to identify biomarkers of this pathology, which may facilitate early diagnosis and treatment.

Key words Human aortic valve, Stenosis, iTRAQ labeling

1 Introduction

Degenerative aortic valve stenosis (AS) is the leading cause for valve replacement in Western countries and the third leading cause of cardiovascular disease (1). Since degenerative AS increases with age, the aging process of Western population forecasts a significant increase in AS prevalence in future decades; thus, the medical and economic burden is most likely to increase (2). The principal morphologic characteristics consist in valve thickening and bone formation. These morphological hallmarks are the result of active regulated cell processes, consisting in inflammation, cellular infiltration, lipid metabolism deregulation, extracellular matrix remodeling with increasing fibrosis, valve thickening, as well as angiogenesis and calcification (3–5). Although it shares many risk factors with atherosclerosis, recent experimental evidence, and clinical trials, it appears that atherogenesis is not pivotal to the pathogenesis of AS, underlining the need to better understand this disease (6, 7).

Additional research about the molecular basis of aortic valve stenosis may contribute to a better understanding of the mechanisms

Fernando Vivanco (ed.), *Heart Proteomics: Methods and Protocols*, Methods in Molecular Biology, vol. 1005, DOI 10.1007/978-1-62703-386-2_9, © Springer Science+Business Media New York 2013

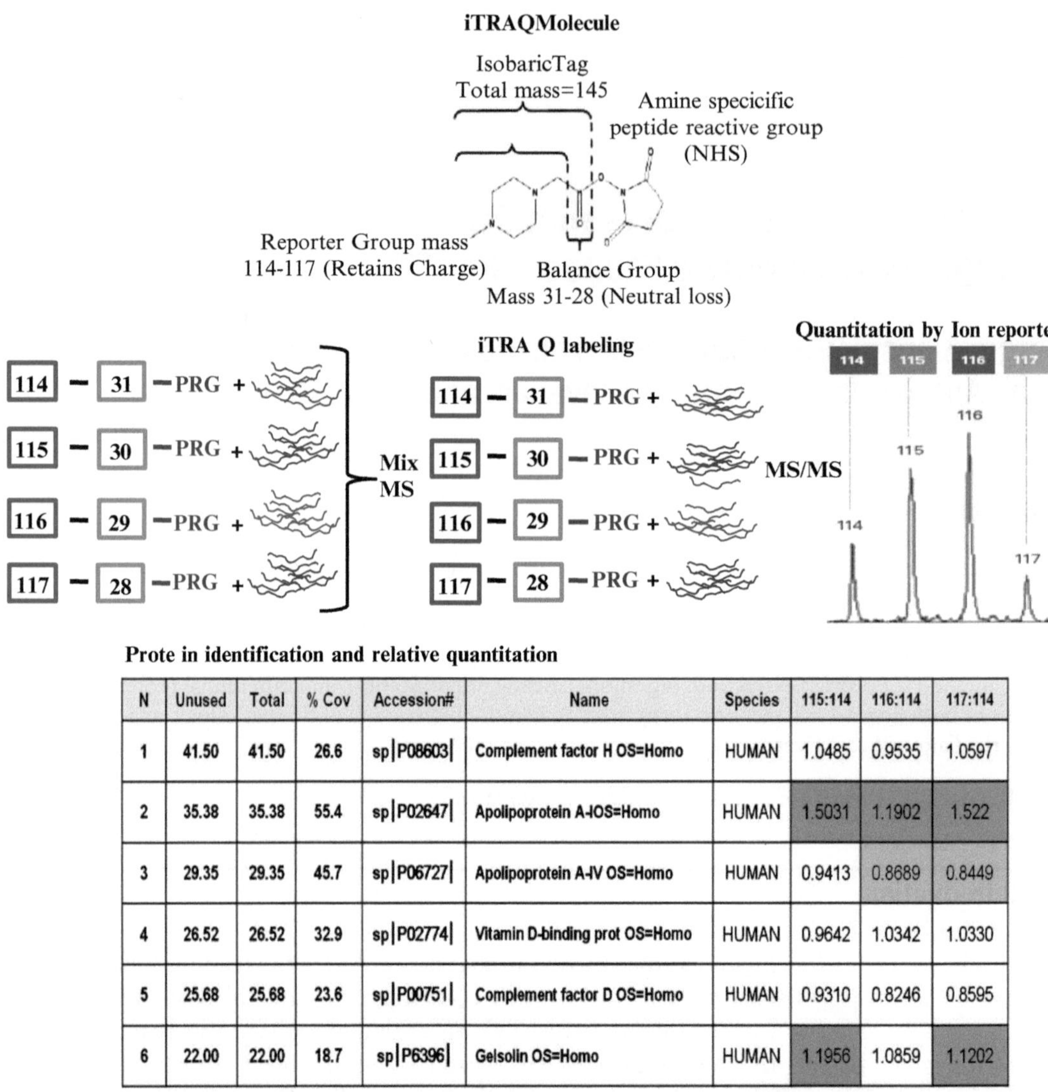

Prote in identification and relative quantitation

N	Unused	Total	% Cov	Accession#	Name	Species	115:114	116:114	117:114
1	41.50	41.50	26.6	sp|P08603|	Complement factor H OS=Homo	HUMAN	1.0485	0.9535	1.0597
2	35.38	35.38	55.4	sp|P02647|	Apolipoprotein A-IOS=Homo	HUMAN	1.5031	1.1902	1.522
3	29.35	29.35	45.7	sp|P06727|	Apolipoprotein A-IV OS=Homo	HUMAN	0.9413	0.8689	0.8449
4	26.52	26.52	32.9	sp|P02774|	Vitamin D-binding prot OS=Homo	HUMAN	0.9642	1.0342	1.0330
5	25.68	25.68	23.6	sp|P00751|	Complement factor D OS=Homo	HUMAN	0.9310	0.8246	0.8595
6	22.00	22.00	18.7	sp|P6396|	Gelsolin OS=Homo	HUMAN	1.1956	1.0859	1.1202

Fig. 1 Schematic diagram illustrating the iTRAQ molecule, iTRAQ-labeling process, and protein identification and quantitation

involved in cellular and structural remodeling and thus help build the basis for the identification of new treatments that can help address this disease. An understanding of the pathogenesis of this disease would be greatly improved with the identification of the differential protein expression that occurs between aortic valves with and without stenosis.

The amine-specific isobaric peptide tags for relative and absolute quantification (iTRAQ) of proteins by mass spectrometry in different samples is one major advance in quantitative proteomics (8, 9). The iTRAQ method is based on the differential covalent labeling of peptides from proteolytic digests with one of four iTRAQ reagents (114, 115, 116 and 117), resulting in the incorporation of 144.1 Da to peptide N-termini and lysine residues. Peptides with different tags are indistinguishable by mass but can be differentiated by collision-induced dissociation (CID) (normally applied during MS peptide sequencing) through release of a reporter ion, each of which has a different mass (114.1, 115.1, 116.1 or 117.1 Da). The analysis of the intensity of reporter ions allows the simultaneous sequencing and quantification of labeled peptides (Fig. 1).

This chapter provide protocols that facilitate from the initial tissue processing to differential protein expression analysis in the study of AS.

2 Materials

2.1 Equipment

1. Milli-Q System (Millipore, Bedford, MA).
2. Thermomixer comfort (Eppendorf).
3. Ultrasonic bath.
4. Vortex.
5. Centrifuge (5840R Eppendorf).
6. SpeedVac.
7. Nano-HPLC (Ultimate 3000, Dionex).
8. LTQ-Orbitrap mass spectrometer (Thermo Fisher, San José, CA, USA).
9. Mortar.
10. Proteome Discoverer 1.0 software (Thermo Fisher).

2.2 Reagents

1. PBS, pH 7.4.
2. RPMI medium.
3. Liquid nitrogen (Liquid N_2).
4. Protein-extraction buffer-1: Tris–HCl 10 mM pH 7.5, 500 mM NaCl, 0.1 % Triton X-100, 1 % β-mercaptoethanol, 1 mM PMSF.
5. Protein-extraction buffer-2: 7 M urea, 2 M Thiourea, 4 % CHAPS.
6. Mini dialysis kit, 1 kDa cutoff (cat: 80-6483-75, GE Healthcare).
7. Tris buffer.
8. Bradford-Lowry reagent (cat: 500-0006, Bio-Rad).
9. 2-D Clean-Up kit (cat: 80-6484-51, GE Healthcare).
10. Triethylammonium bicarbonate (TEAB), 1 M (cat: T7408, Sigma).
11. Urea.

12. Reducing reagent: Tris(2-carboxyethyl)phosphine (TCEP), 50 mM, included in iTRAQ kit.
13. Cysteine-blocking reagent: Methyl methanethiosulfonate (MMTS), 200 mM, included in iTRAQ kit.
14. Sequencing grade modified trypsin (20 μg). (Cat: V511A, Promega).
15. iTRAQ Multiplex (4-plex) kit (cat: 4352135, AB Sciex).
16. Ethanol absolute.
17. pH paper (cat: 921 60, Macherey-Nagel).
18. Milli-Q water.
19. C-18 reversed phase (RP) nano-column (100 μm I.D. and 12 cm, Teknokroma).
20. Acetonitrile, HPLC grade.

3 Methods

3.1 Protein Tissue Sample Obtention

The valve tissues were processed according to our protocol described previously (10). The procedure is schematized in Fig. 2.

Heart valves were obtained from AS patients who had undergone aortic valve replacement due to severe degenerative AS. Control macroscopically normal valves were obtained within 4 h of death from autopsies of subjects who had died due to other causes than cardiovascular disease and who had no history of coronary artery disease or diabetes mellitus (see Note 1). Once in the laboratory, all manipulations were conducted at 4 °C:

1. Wash valve samples three times in phosphate buffered saline (PBS) to eliminate blood contaminants.
2. Frozen aortic valve tissue in liquid nitrogen and homogenized using a mortar.
3. Resuspend 0.2 g of the powder tissue in 400 μL of protein-extraction buffer-1 (11).
4. Vortex and sonicate three times, for 2 min, to facilitate protein dissolution.
5. Centrifuge the homogenate at 21,000 × g for 15 min at 4 °C, to precipitate the membranes and tissue debris. Collect the supernatant (supernatant E1) containing most of the soluble proteins. Store at 4 °C.
6. Add 400 μL of protein-extraction buffer-2 (12), to solubilize the pellet. Vortex and sonicate three times, for 2 min, to facilitate protein dissolution.
7. Centrifuge the new homogenate at 21,000 × g for 15 min at 4 °C. Collect the supernatant and store at 4 °C. This second supernatant (supernatant E2) is rich in hydrophobic proteins, mainly membrane proteins.

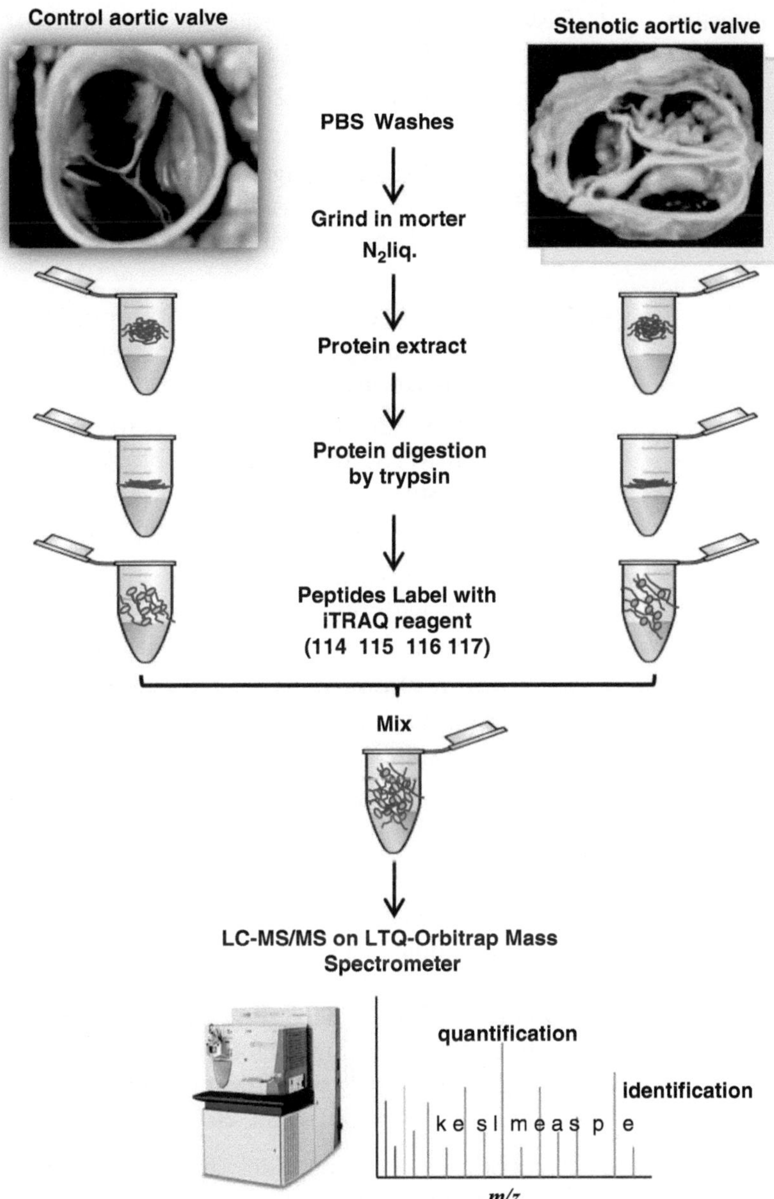

Fig. 2 Flowchart representing the analysis of human AS valve iTRAQ analysis (iTRAQ labeling and MS/MS analysis)

8. Mix the two supernatants (E1 + E2).
9. Dialyze the supernatants mixed against 2 mM Tris buffer overnight (12–16 h) at 4 °C, to remove any substances that could modify the protein quantification. Determine protein concentration by Bradford-Lowry method (13).
10. Precipitate 100 μg of each aortic valve extracts. Clean using 2-D Clean-Up kit, according to manufacturer's protocol. Now, the aortic valve extracts are ready to protein digestion.

3.2 Protein Digestion

1. Dissolve 100 μg of each aortic valve lysate in 20 μL of dissolution buffer (TEAB 0.5 M, pH 8.5, urea 8 M).
2. Vortex and sonicate, for 2 min, to facilitate protein dissolution, and spin.
3. Add 2 μL of reducing agent (TCEP), to reduce the disulfide bonds of the proteins, vortex, and spin. (Do not use DTT or 2-mercaptoethanol because it generates thiol groups.) (See Note 2).
4. Incubate at 37 °C for 30 min at 600 rpm (it is important not to exceed this temperature in order to reduce the risk of protein carbamylation). Spin the samples.
5. Alkylate the sample by adding 1 μL of cysteine-blocking reagent (MMTS) that reversibly blocks the cysteine group. Vortex and spin.
6. Incubate at room temperature for 10 min.
7. Dilute the sample with 100 μL, to dilute urea concentration below 2 M, with 500 mM TEAB. It is important for the protein digestion; trypsin is not active at urea 2 M.
8. Reconstitute Promega sequencing grade trypsin (20 μg) with 20 μL Milli-Q water to prepare a 1 μg/μL solution of trypsin. Vortex and spin.
9. Digest protein sample with 2.5 μL of sequencing grade trypsin (ratio 1:40, w/w) (see Note 3). Vortex and spin. Incubate samples at 37 °C overnight (12–16 h).
10. Spin samples.
11. Dry the sample in a centrifugal vacuum concentrator.

3.3 iTRAQ-Labeling Protocol

Follow iTRAQ labeling according to manufacturer's protocol (AB Sciex). In order to maximize labeling efficiency, the volume of the sample digest must be less than 50 μL and the sample concentration must be at 1–5 μg/μL (see Note 4).

1. Reconstitute dried peptide samples with 30 μL of TEAB, 0.5 M, and pH 8.5.
2. Vortex and sonicate, for 2 min, to facilitate peptide sample dissolution, and spin.
3. Bring each vial of iTRAQ reagent (114, 115, 116, and 117) to room temperature.
4. Add 70 μL of ethanol to each iTRAQ reagent vial being used; the percentage of ethanol in the reaction tube must be at least 70 % v/v. Vortex for 1 min and spin.
5. Transfer the contents of one iTRAQ reagent vial to one sample tube. Label control samples with 2 iTRAQ reagent, i.e., 114 and 116. Label stenosis samples with another iTRAQ reagents, i.e., 115 and 117. Vortex and spin.

6. Check the pH of each sample. Make sure it is between 7.8 and 8.5 (see Note 5). The final concentration of ethanol must be 70% v/v.
7. Incubate the tubes in the dark at room temperature for 1 h.
8. Add 100 μL of Milli-Q water to each tube to quench the iTRAQ reaction. Incubate at room temperature for 30 min.
9. Combine the contents of all iTRAQ reagent-labeled sample tubes into one tube. Vortex and spin.
10. Dry completely the tube containing all the combined iTRAQ mixes in a centrifugal vacuum concentrator.

Now, the iTRAQ-labeling peptides from both samples control and stenotic valves are ready to identification and relative quantitation (Fig. 3).

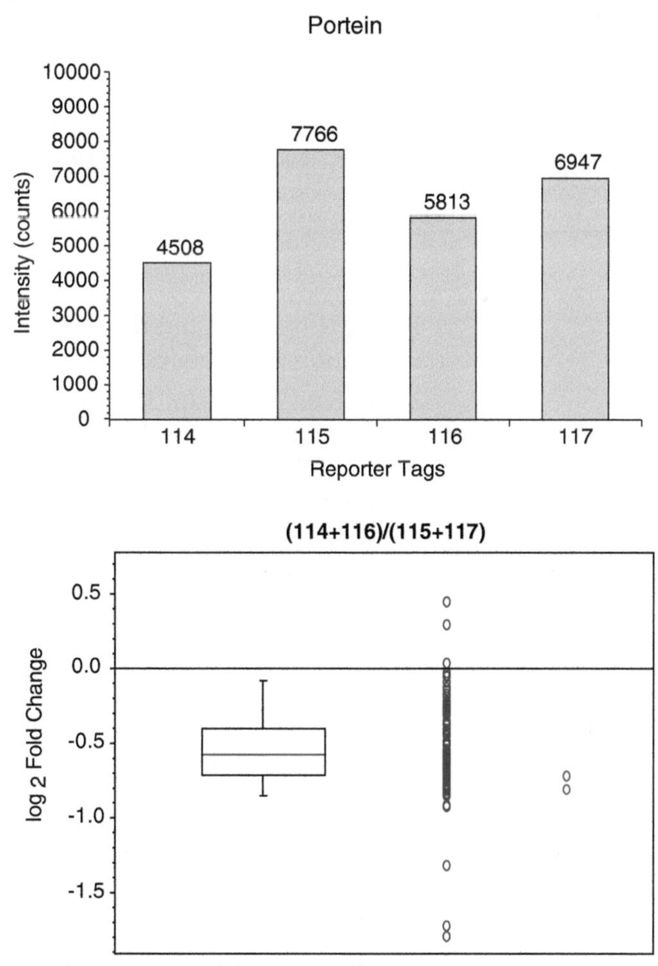

Fig. 3 Representative report of a protein differentially expressed. Control samples are labeled with 114 and 116 iTRAQ reagents and AS samples with 115 and 117 iTRAQ reagents

3.4 Mass Spectrometry and Protein Quantitation

Subject the four-plex peptide mixture to nano-liquid chromatography coupled to mass spectrometry for peptide separation in order to improve the protein identification and quantitation.

Inject the peptides mixed onto a C-18 reversed phase (RP) nano-column (100 μm I.D. and 12 cm, Teknokroma). Analyzed in a continuous acetonitrile gradient consisting of 0–43 % B in 140 min and 50–90 % B in 1 min (B = 95 % acetonitrile, 0.5 % acetic acid). Use a flow rate of 300 nL/min to elute peptides from the RP nano-column to an emitter nanospray needle for real-time ionization and peptide fragmentation on an LTQ-Orbitrap mass spectrometer. An enhanced FT-resolution spectrum (resolution = 60,000) followed by the MS/MS spectra from most intense five parent ions are analyzed along the chromatographic run (180 min). Set the dynamic exclusion at 0.5 min.

For protein identification, search fragmentation spectra against MSDB database using the Mascot™ program. Allow two mixed cleavages, and set an error of 15 ppm or 0.8 Da for full MS or MS/MS spectra searches, respectively. Perform the identifications by Proteome Discoverer 1.0 software (Thermo Fisher). Set the decoy database search for FDR analysis at 0.05 by applying corresponding filters. Specify phosphorylation in Serine, Threonine, or Tyrosine as variable modification.

4 Notes

1. RPMI medium and the temperature at 4 °C are used for preservation of biological material obtained by surgery.
2. DTT or 2-mercaptoethanol generates thiol groups that interfere with cysteine blocking and then may interfere with the iTRAQ™ reagents. If your sample contains these substances, perform acetone or 2-D Clean-Up precipitation to clean up the sample.
3. The appropriate relationship with the digestion in solution is at 1:40 w/w. On the other hand, acetonitrile can be added to a final volume of 25 %, since the presence of an organic component improve the digestion process.
4. In order to maximize labeling efficiency, the volume of the sample digest must be less than 50 μL and the protein concentration must be at 1–5 μg/μL. This peptide concentration range is the optimal for labeling the iTRAQ reagent. If the volume of the sample digest is greater than 50 μL, dry the sample in a centrifugal vacuum concentrator, then reconstitute with 30 μL dissolution buffer.
5. The percentage of ethanol in the reaction tube must be at least 70 % v/v. If ethanol percentage is less than the 70 %, the iTRAQ reagent is hydrolyzed and do not label correctly.

6. The pH must be between 7.8 and 8.5, because the iTRAQ reaction occurs at basic pH. If your sample is at pH below that indicated, then add up to 5 μL of TEAB 0.5 M to get the pH at or above 7.8. You must simultaneously add ethanol as needed to keep the final concentration of ethanol at 70 % v/v.

Acknowledgments

This work was supported by grants from the Instituto de Salud Carlos III (FIS PI070537, PI11/02239, PI11/01401), Fondos Feder, Redes temáticas de Investigación Cooperativa en Salud (RD06/0014/1015), and Fundación para la Investigación Sanitaria de Castilla-La Mancha (FISCAM PI2008-08, PI2008-28, PI2008-52).

References

1. Cowell SJ, Newby DE, Boon NA, Elder AT (2004) Calcific aortic stenosis: same old story? Age Ageing 33:538–544
2. Nkomo VT, Gardin JM, Skelton TN, Gottdiener JS, Scott CG, Enriquez-Sarano M (2006) Burden of valvular heart diseases: a population based study. Lancet 368:1005–11
3. Wallby L, Janerot-Sjoberg B, Steffensen T, Broqvist M (2002) T lymphocyte infiltration in non-rheumatic aortic stenosis: a comparative descriptive study between tricuspid and bicuspid aortic valves. Heart 88:348–351
4. Rabuş MB, Kayalar N, Sareyyüpoğlu B, Erkin A, Kirali K, Yakut C (2009) Hypercholesterolemia association with aortic stenosis of various etiologies. J Card Surg 24:146–150
5. Soini Y, Salo T, Satta J (2003) Angiogenesis is involved in the pathogenesis of nonrheumatic aortic valve stenosis. Hum Pathol 34:756–763
6. Chan KL, Teo K, Dumesnil JG, Ni A, Tam J, ASTRONOMER Investigators (2010) Effect of Lipid lowering with rosuvastatin on progression of aortic stenosis: results of the aortic stenosis progression observation: measuring effects of rosuvastatin (ASTRONOMER) trial. Circulation 121:306–314
7. van der Linde D, Yap SC, van Dijk AP, Budts W, Pieper PG, van der Burgh PH et al (2011) Effects of rosuvastatin on progression of stenosis in adult patients with congenital aortic stenosis (PROCAS Trial). Am J Cardiol 108: 265–271
8. Chaerkady R, Harsha HC, Nalli A, Gucek M, Vivekanandan P, Akhtar J et al (2008) Quantitative proteomic approach for identification of potential biomarkers in hepatocellular carcinoma. J Proteome Res 7:4289–4298
9. Polisetty RV, Gautam P, Sharma R, Harsha HC, Nair SC, Gupta MK et al (2012) LC-MS/MS analysis of differentially expressed glioblastoma membrane proteome reveals altered calcium signaling and other protein groups of regulatory functions. Mol Cell Proteomics 11:M111.013565
10. Gil-Dones F, Martin-Rojas T, Lopez-Almodovar LF, de la Cuesta F, Darde VM, Alvarez-Llamas G et al (2010) Valvular aortic stenosis: a proteomic insight. Clin Med Insights Cardiol 4:1–7
11. Gil-Dones F, Martín-Rojas T, López-Almodovar LF, Juárez-Tosina R, de la Cuesta F, Alvarez-Llamas G et al (2010) Development of an optimal protocol for the proteomic analysis of stenotic and healthy aortic valves. Rev Esp Cardiol 63:46–53
12. Barderas MG, Tuñón J, Dardé VM, De la Cuesta F, Durán MC, Jiménez-Nacher JJ et al (2007) Circulating human monocytes in the acute coronary syndrome express a characteristic proteomic profile. J Proteome Res 6:876–886
13. Bradford MM (1976) A rapid and sensitive method for the quantitation of microgram quantities of protein utilizing the principle of protein-dye binding. Anal Biochem 72:248–254

Chapter 10

Characterization of Zebrafish Cardiac Proteome Using Online pH Gradient SCX–RP HPLC–MS/MS Platform

Jiang Zhang, Kevin A. Lanham, Warren Heideman, Richard E. Peterson, and Lingjun Li

Abstract

Two-dimensional HPLC coupled with tandem MS (MS/MS) has become a mainstream technique in the shotgun proteomics for large-scale identification of proteins from biological samples. This powerful technology provides speed, sensitivity, and dynamic range which are essential to probe complex peptide mixtures from proteomic samples. Herein we present a pH gradient SCX–RP 2D HPLC–MS/MS method designed to improve the peptide resolution and protein identification from complex proteomic samples. The comparison between the pH gradient SCX–RP 2D HPLC method and traditional salt gradient SCX–RP method was presented. A two-step sample prefractionation method utilizing microwave-assisted tryptic digestion to improve the identification of insoluble proteins was also introduced. This novel 2D HPLC–MS/MS method was applied to the heart proteomic sample of the zebrafish, *Danio rerio*, to provide comprehensive cardiac proteomic profiling of this important model organism for cardiovascular and environmental toxicology studies.

Key words pH gradient, SCX–RP, 2D HPLC, Mass spectrometry, Proteome, Cardiac, Zebrafish

1 Introduction

Cardiovascular disease is the leading cause of human death and morbidity, affecting a large population in the United States, Europe, and other regions in the world. The prevalence of this disease and its dramatic socioeconomic cost underscore an urgent need to understand the underlying cellular and molecular mechanisms of the disease. The zebrafish, *Danio rerio*, is a well-established vertebrate animal model and has recently gained increasing research interest as a cardiovascular model due to its similarity to other vertebrate and mammalian species in the cardiovascular development and pathology (1). Compared with other animal models, the zebrafish provides a unique advantage for large-scale genetics combined with exceptional visualization of early stage development and physiological processes. The zebrafish has also

Fernando Vivanco (ed.), *Heart Proteomics: Methods and Protocols*, Methods in Molecular Biology, vol. 1005, DOI 10.1007/978-1-62703-386-2_10,

been used as a valuable animal model in environmental toxicology to evaluate the cardiac toxicity of environmental chemicals on the vertebrate cardiovascular development (2, 3).

Recent development in proteomics has provided a powerful tool in the investigation of the molecular mechanism of cardiovascular disease. A mainstream technology in the shotgun proteomics is the two-dimensional liquid chromatography (2D HPLC) coupled with tandem mass spectrometry (MS/MS), which has been widely used to separate complex tryptic peptide mixtures and perform large-scale protein identifications from complex biological samples (4–6). Compared with traditional gel electrophoresis-based methods, 2D HPLC–MS/MS provides high-speed, enhanced resolution and sensitivity, which have catalyzed the rapid growth of proteomic research in the recent decades. 2D HPLC has been most commonly implemented in a strong cation exchange (SCX)/reversed-phase (RP) format, where a salt buffer is used to elute peptide or protein mixtures based on charge and the fractions from the SCX separation are further separated in the RP phase based on analyte hydrophobicity. Despite its good orthogonality with the RP HPLC, SCX separation especially the salt gradient method suffers from problems with low peptide and protein resolution, which limits the overall resolution and peak capacity of 2D HPLC and therefore the downstream peptide and protein identification by mass spectrometry. Additionally, mobile phase containing high salt concentrations affects the electrospray ionization of peptides and proteins, promotes the formation of adduct peaks, and reduces the sensitivity for peptide and protein analysis (7). Therefore, improving the resolution and MS compatibility of 2D HPLC is of critical importance to enable more sensitive and comprehensive characterization of proteomic samples.

Recently, pH gradient SCX–RP has emerged as an attractive alternative approach to the traditional salt gradient SCX–RP HPLC method. In this method, a buffer solution of increasing pH is used to elute proteins in the SCX dimension, and the resulted fractions are further separated by reversed-phase HPLC. This method separates proteins and peptides based on their isoelectric points rather than ionic strength. Therefore, it provides higher resolution than the method based on the salt concentration gradient elution. This pH gradient-based ion exchange separation was first introduced in a chromatofocusing format by Sluyterman and coworkers (8–10), where the pH gradient was delivered by polymeric Ampholyte buffers. Several other groups further studied this method by using low molecular weight buffer to replace polyampholytes and deliver more reproducible linear pH gradient, therefore providing improved protein separation and reduced the protein–polyampholyte complex association (11–15). In addition, reduced salt concentration also results in improved MS compatibility. Recently, this method has been applied to tryptic peptide samples to provide improved peptide

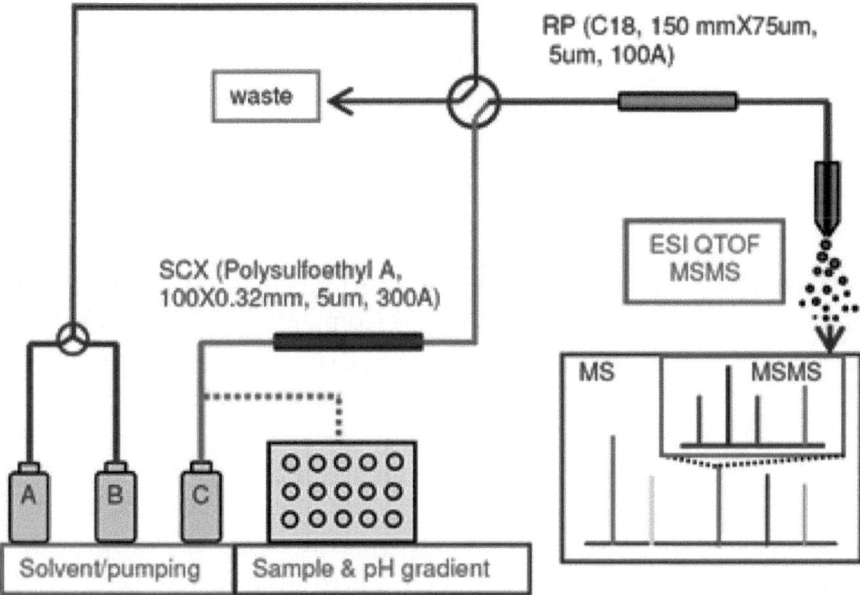

Fig. 1 A schematic outline of pH gradient SCX–RP HPLC–MS/MS setup. The first dimension used a SCX column, and the digested zebrafish heart proteomic samples are first loaded onto the SCX dimension from pump C and autosampler. The pH gradient is delivered from the autosampler, and the fractionated peptides are eluted to the RP column. Each peptide fraction is subsequently separated by a 120 min RP gradient. The eluted peptides from RP separation are then analyzed by ESI MS/MS analysis

separation and protein identification in proteomic studies. Herein we present an efficient pH gradient SCX–RP HPLC platform implemented on a capillary LC system coupled with QTOF MS/MS for comprehensive cardiac proteome profiling in the zebrafish. Figure 1 depicts the setup of this 2D HPLC–MS/MS platform. Details of this method are discussed below.

2 Materials

All solutions are prepared with deionized water, and reagents used are analytical grade, unless indicated otherwise.

2.1 Zebrafish Heart Protein Extraction and Digestion

1. Tricane (MS-222, Sigma-Aldrich, St. Louis, MO, USA).
2. Tris-buffered saline: 100 mM NaCl, 50 mM Tris. 0.5 mM EDTA, 1 mM DTT, adjust pH to 7.6 with HCl, store at 4 °C (reagents from Sigma-Aldrich, ACS grade).
3. Protease inhibitor cocktail tablet (Roche Applied Science, Mannheim, Germany).
4. Dithiothreitol (DTT), iodoacetic acid (IAA).
5. Sequencing grade trypsin (Promega, Madison, WI, USA).

6. Ammonium bicarbonate, formic acid.
7. Fine forceps (Fine Science Tools, Foster City, CA).
8. Dounce tissue homogenizer (Fisher Scientific, Inc., Pittsburgh, PA, USA).
9. Ultrasonic sonicator (Qsonica LLC., Newtown, CT, USA).
10. Microwave oven (General Electric, Fairfield, CT, USA).

2.2 2D HPLC–MS/MS Instrumentation

1. SCX–RP HPLC was performed on a Waters CapLC system with autosampler (Waters Corp., Milford, MA, USA) (see Note 1).
2. SCX column (320 μm × 10 cm, PolySULFOETHYL A, 5 μm, 100 Å, Microtech Scientific, Vista, CA, USA).
3. RP column (75 μm × 15 cm, C_{18}, 5 μm, 300 Å, Column Technology, Fremont, CA, USA). RP trap (C_{18}, 300 μm × 1 cm, 5 μm, 300 Å, LC Packings, Sunnyvale, CA, USA).
4. ESI MS/MS experiments were performed on a Waters Micromass Q-TOF Micro mass spectrometer (Waters Corp., Milford, MA, USA), equipped with a nanoelectrospray ionization source (nESI) (see Note 2).
5. Acetonitrile (HPLC grade).
6. SCX pH gradient buffers: ten discrete pH buffers were prepared with 10 mM citric acid, ammonium hydroxide, and trimethylamine. pH were adjusted to 2.0, 2.5, 3.0, 3.5, 4.0, 4.5, 5.0, 5.5, 6.0, 7.0, 8.0.
7. RP mobile phase A: 0.1 % formic acid in double deionized water (v/v).
8. RP mobile phase B: 0.1 % formic acid in acetonitrile (v/v).

3 Methods

3.1 Heart Dissection and Protein Extraction

All animal procedures are conducted according to the established guidelines for the care and use of laboratory animals approved by the National Institutes of Health.

1. The zebrafish (AB wild type) are bred and embryos raised in the laboratory according to the procedures described by Westerfield (16).
2. Fifteen adult zebrafish are sacrificed by anesthetization with 0.168 mg/ml Tricane solution.
3. Extract the hearts by microdissection with fine forceps and carefully remove the connective tissue. The procedure was performed in a Petri dish on dry ice to keep low temperature during microdissection (see Note 3).
4. Rinse the dissected hearts with cold Tris buffer to remove the blood.

5. Immediately snap-freeze the dissected hearts in liquid nitrogen and place the hearts into a microcentrifuge tube in a dry ice bucket.
6. The dissected hearts are homogenized using a 1 ml Dounce tissue homogenizer with 0.5 ml Tris buffer with protease inhibitor cocktail (see Note 4). (For the quantity of the protease inhibitor, refer to the manufacturer's guideline.)
7. Further sonicate the protein homogenate in an ice bath with a microsonicator tip for five cycles, with 1 min for each cycle and 30 sec. interval.
8. The homogenized protein extract is then centrifuged at 16,000 × *g* for 10 min. This results in the separation of the supernatant that contains soluble proteins and the pellet portion that contains insoluble proteins. Collect both fractions into a 1 ml microcentrifuge tube. Figure 2 illustrates the two-step protein fractionation and digestion method.

3.2 Protein Digestion

1. Denature the supernatant protein fraction with 8 M urea (see Note 5).
2. The denatured protein solution is reduced with 20 μl 100 mM DTT solution in the dark for 20 min.
3. After the reduction, the protein solution is alkylated with 20 μl 150 mM iodoacetic acid in a water bath for 20 min.
4. Dilute the protein solution to reduce the urea concentration to 1 M.
5. Add trypsin to the protein solution at a 1:25 ratio and digest overnight at 37 °C in a water bath (see Note 6).
6. Concentrate the protein digest solution by SpeedVac to a reduced volume of 200 μl. Add 0.1 % formic acid to adjust pH to 2 (see Note 7).
7. Centrifuge the concentrated digest solution at 16,000 × *g* for 10 min. Save the supernatant portion at −80 °C before LC–MS/MS analysis.
8. Treat the pellet with 1 ml 50 % formic acid with sonication for 15 min.
9. Irradiate the homogenate in a conventional microwave oven at half power (~0.6 kW) with 30 sec. intervals for 2 min (see Note 8).
10. Centrifuge the homogenate at 16,000 × *g* for 5 min and collect the supernatant.
11. Concentrate the supernatant with SpeedVac and reduce the supernatant volume to 200 μl (see Note 9).
12. Add 100 mM ammonium bicarbonate to adjust the pH to 7.4.

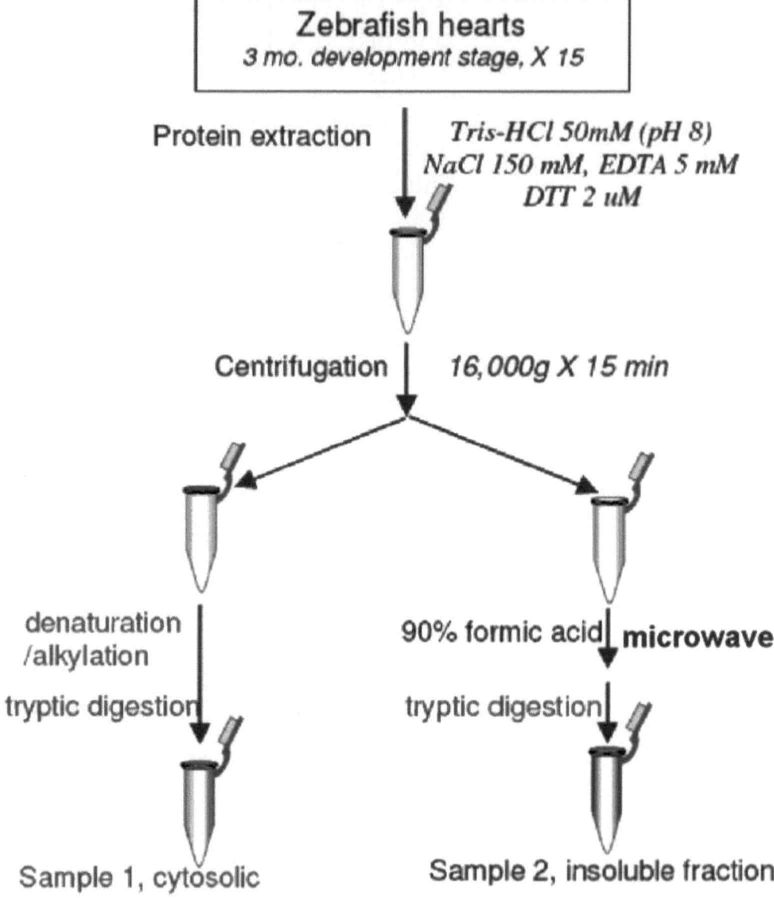

Fig. 2 A schematic outline of the two-step sample fractionation and digestion method. A total of 15 adult zebrafish hearts are extracted with Tris–HCl buffer and centrifuged, resulting in soluble and pellet portions. The soluble protein fraction undergoes conventional tryptic digestion procedures. The insoluble proteins in the pellet are acid hydrolyzed in 50 % formic acid assisted by microwave irradiation. The resulted solution is further digested by trypsin. Both fractions are subsequently analyzed by SCX–RP HPLC–ESI MS/MS analysis for peptide and protein identification from the zebrafish heart tissue

13. Add trypsin at a 1:50 ratio and digest overnight at 37 °C in water bath (see Note 10).
14. The digested sample is frozen at −80 °C until LC–MS/MS analysis.

3.3 Online pH Gradient SCX–RP HPLC–MS/MS Analysis

1. Load 60 μl cytosolic protein digests to the SCX column using the autosampler on the capillary HPLC system (see Note 11).
2. For SCX separation, apply ten sequential pH gradient buffers (from pH 2 to pH 8) to the SCX column delivered through the autosampler and auxiliary pump C with a flow rate of 5 μl/min.

Each pH gradient elution is delivered for 5 min, and the eluted peptide fraction is transferred onto the RP trap column.

3. For each SCX fraction, apply a 120 min RP gradient for the RP separation. A typical RP gradient profile is as follows: 0–5 min, 5 % mobile phase B; 5–85 min, ramp to 35 % B; 85–105 min, ramp to 95 % B; 105–110 min, 95 % B; 110–115 min, revert to 5 % B; and 115–120 min, 5 % B. The mobile phase gradient is delivered at a flow rate of approximately 200 nl/min after split flow.
4. The eluted peptides are analyzed by ESI MS/MS on a Waters Micromass Q-TOF Micro mass spectrometer using the following conditions: capillary voltage 3.2 kV, sample cone voltage 35 V, and extraction cone voltage.
5. 1 V, source temperature 100 °C. The data-dependent analysis mode parameters: MS survey scans range *m/z* 400–2,000, 1.0 s scan time. MS/MS scans are acquired on the three most abundant precursor ions, each MS/MS scan is acquired from *m/z* 50–2,000 for 1.9 s, with interscan time 0.1 s; fragmentation is performed on 2+, 3+, and 4+ ions, with an intensity threshold of 15 counts/s. A dynamic exclusion of 120 s is used to minimize redundant identifications; singly charged ions are also excluded for MS/MS analysis.
6. The collision energy is set according to the ion charge state and mass-to-charge ratio and varied from 16 to 55 eV.

3.4 Data Analysis

1. Convert the raw MS/MS data to peak list (.pkl) files using MassLynx software (version 2.0, Waters) with the following parameters: spectral smoothing, Savitzky–Golay, 3 channels × 2, and centroid (80 % top).
2. Download a local protein database for the zebrafish from the NCBI protein server for the taxonomy entry 7955 (*Danio rerio*). A reversed database is created and appended to the forward database for the decoy database search and evaluation of the false-positive rate.
3. Submit the peak list files to Mascot search engine (Matrix Science, UK) for protein identification (see Note 12). The database search parameters are set as follows: precursor ion tolerance 300 ppm, product ion tolerance 0.6 Da for monoisotopic mass, respectively; peptide charge states include 2+, 3+, and 4+; enzymatic specificity, trypsin with two maximum missed cleavages allowed; static modification, carbamidomethyl cysteine; and variable modifications, N-acetylation and Met-oxidation. The level of confidence for peptide identifications is based on the Mascot assignment of identity for peptides (see Note 13).
4. The isoelectric points of identified peptides and proteins are calculated via the algorithm on the Expert Protein Analysis System (EXPASY) proteomic server (http://us.expasy.org).

4 Notes

1. Other HPLC systems capable of 2D HPLC may also be used for the 2D HPLC separation.
2. Other mass spectrometers capable of automated acquisition of tandem mass spectra (e.g., ion trap mass spectrometers) may be used in place of Q-TOF mass spectrometer.
3. The connective tissue should be carefully removed to ensure the endogenous protein identification from the heart tissue. Similarly the heart tissue should be rinsed to remove blood proteins.
4. The protease inhibitor is helpful to reduce the nonspecific enzymatic activity during the cell lysing and improve the subsequent tryptic digestion and protein identification.
5. The protein denaturation may also be performed with other methods such as addition of organic solvents or quick heating by water bath.
6. The protein concentration in the extract can be determined by a protein assay (e.g., Bradford assay, BCA assay) to estimate the quantity of trypsin.
7. It is essential to adjust the sample pH to the equilibrium condition to ensure proper sample binding and separation on the SCX column. Adding acid to the digest may cause some peptide precipitation from the solution, and therefore the peptide solution needs to be centrifuged to remove the precipitate before the sample loading to the column.
8. Irradiation time needs to be carefully adjusted based on the power of the microwave oven to avoid sample overheating and loss. A water bath is helpful to maintain a proper control in the heating.
9. It is helpful to maximize the protein concentration to ensure efficient digestion by relatively small amount of trypsin used in the digestion.
10. The trypsin quantity can be estimated based on the protein concentration measured by a protein assay.
11. Multiple loading may be used if the sample loop does not have sufficient loading volume.
12. A license is needed to access the full function of Mascot search engine for protein database search. Other open source search engines may also be used, such as OMSSA (http://pubchem.ncbi.nlm.nih.gov/omssa).
13. Other peptide and protein validation tools, such as PeptideProphet (http://peptideprophet.sourceforge.net) and ProteinProphet (http://proteinprophet.sourceforge.net), may also be used for proper validation of identified peptides and proteins.

Acknowledgements

The authors wish to thank Dr. Chia-Hui Shieh for providing the SCX column and technical assistance. We also thank the Mass Spectrometry Facility at the University of Wisconsin-Biotechnology Center and Dr. Amy Harms for access to the Mascot server for protein database search. This work was supported by the University of Wisconsin Sea Grant Institute.

References

1. Bakkers J (2011) Zebrafish as a model to study cardiac development and human cardiac disease. Cardiovasc Res 91:279–288
2. Carney SA et al (2006) Understanding dioxin developmental toxicity using the zebrafish model. Wiley. Hoboken, New Jersey, USA p 7–18
3. Heideman W et al (2005) Zebrafish and cardiac toxicology. Cardiovasc Toxicol 5:203–214
4. Washburn MP, Wolters D, Yates JR (2001) Large-scale analysis of the yeast proteome by multidimensional protein identification technology. Nat Biotechnol 19:242–247
5. Zhang J et al (2010) Characterization of the adult zebrafish cardiac proteome using online pH gradient strong cation exchange-RP 2D LC coupled with ESI MS/MS. J Sep Sci 33:1462–1471
6. Dowell JA et al (2008) Comparison of two-dimensional fractionation techniques for shotgun proteomics. Anal Chem 80:6715–6723
7. Garcia MC (2005) The effect of the mobile phase additives on sensitivity in the analysis of peptides and proteins by high-performance liquid chromatography-electrospray mass spectrometry. J Chromatogr B Analyt Technol Biomed Life Sci 825:111–123
8. Sluyterman LAA, Wijdenes J (1978) Chromatofocusing-isoelectric-focusing on ion-exchange columns. 2. Experimental verification. J Chromatogr 150:31–44
9. Sluyterman LAA, Wijdenes J (1981) Chromatofocusing. 4. Properties of an agarose polyethyleneimine ion-exchanger and its suitability for protein separations. J Chromatogr 206:441–447
10. Sluyterman LAA, Wijdenes J (1981) Chromatofocusing. 3. The properties of a DEAE-agarose anion-exchanger and its suitability for protein separations. J Chromatogr 206:429–440
11. Kang XZ, Frey DD (2003) High-performance cation-exchange chromatofocusing of proteins. J Chromatogr A 991:117–128
12. Kang XZ, Frey DD (2004) Chromatofocusing of peptides and proteins using linear pH gradients formed on strong ion-exchange adsorbents. Biotechnol Bioeng 87:376–387
13. Pepaj M et al (2006) Fractionation and separation of human salivary proteins by pH-gradient ion exchange and reversed phase chromatography coupled to mass spectrometry. J Sep Sci 29:519–528
14. Pepaj M et al (2006) Two-dimensional capillary liquid chromatography: pH Gradient ion exchange and reversed phase chromatography for rapid separation of proteins. J Chromatogr A 1120:132–141
15. Shan L, Anderson DJ (2002) Gradient chromatofocusing. Versatile pH gradient separation of proteins in ion-exchange HPLC: characterization studies. Anal Chem 74:5641–5649
16. Westerfield M (2000) The zebrafish book, A guide for the laboratory use of zebrafish (Danio rerio), 4th edn. University of Oregon Press, Eugene, OR

Chapter 11

Proteomic Analysis of Brain Mitochondrial Proteome and Mitochondrial Complexes

Ana Lopez-Campistrous and Carlos Fernandez-Patron

Abstract

We describe various complementary techniques to achieve multidimensional mitochondrial proteome fractionation and analysis. Previously described methods for 2D-DIGE/mass spectrometry and 1D-SDS-PAGE/Western techniques and protein complex analysis by BN-PAGE/Western and sucrose gradient ultracentrifugation/SDS-PAGE/mass spectrometry are optimized to characterize the brain mitochondrial proteome. This approach allows for a comprehensive identification of mitochondrial proteomic differences between health and disease conditions.

Key words Proteomics, Brain, Mitochondria, Sodium-dodecyl sulfate polyacrylamide gel electrophoresis (SDS-PAGE), Isoelectric focusing (IEF), Tandem mass spectrometry (MS/MS), Blue native (BN) electrophoresis, Sucrose gradient ultracentrifugation

1 Introduction

An emerging common hallmark of many human diseases, including neurodegenerative, cardiovascular conditions, and hypertension, is the dysfunction of mitochondria (1, 2). Mitochondrial dysfunction in the central nervous system has been extensively investigated for several neurodegenerative diseases, including vascular dementia, Alzheimer's, Huntington's, and Parkinson's disease (3, 4). Mitochondrial dysfunction is characterized by decreased expression of mitochondrial components and transcription factors involved in mitochondrial biogenesis as well as defects in the assembly of respiratory complexes (5, 6).

By combining various complementary techniques, we achieved multidimensional proteome fractionation and analysis of the brain mitochondria. We optimized previously described methods for 2D-DIGE/mass spectrometry and 1D-SDS-PAGE/Western techniques and protein complex analysis by BN-PAGE/Western and sucrose gradient ultracentrifugation/SDS-PAGE/mass spectrometry (7–9). Furthermore, we combined these methods with a novel

Fernando Vivanco (ed.), *Heart Proteomics: Methods and Protocols*, Methods in Molecular Biology, vol. 1005,
DOI 10.1007/978-1-62703-386-2_11,

complex assembly analysis based on mass spectrometric abundance indexes to characterize the brain mitochondrial complexes. The resultant strategy is applicable to identifying protein complex assembly differences in health vs. disease.

2 Materials

To avoid contamination of the samples prior to mass spectrometry analysis, prepare all solutions, including the SDS running buffer, using HPLC-grade water and analytical-grade reagents. Filter all solutions through 0.45 μm filter. Also, carefully wash all electrophoretic equipment, glass plates, separators, and tanks, with ethanol followed with ultrapure water rinses. Strip holders should be cleaned with nonionic detergents only, using dedicated vessels and brushes and with ultrapure water.

2.1 1D and 2D SDS Polyacrylamide Gel Components

1. Monomer stock solution: 40 % Acrylamide/bisacrylamide solution 29:1 (BioRad). Store at 4 °C.
2. 4× Resolving gel buffer: 1.5 M Tris–HCl pH 8.8; in 1,000 mL: 181.5 g Tris base, 750 mL HPLC-grade water, adjust to pH 8.8 with HCl and add HPLC-grade water to 1,000 mL. Store at 4 °C.
3. Ammonium persulfate: Prepare a fresh 10 % solution in HPLC-grade water.
4. TEMED: ultrapure (Fisher Bioreagents). Store at 4 °C.
5. SDS Running Buffer: 25 mM Tris base, 192 mM Glycine, 0.1 % SDS.

2.2 Buffers

1. Tris/sucrose buffer: 10 mmol/L Tris–HCl pH 7.80, 0.25 mol/L sucrose, 2 mmol/ L EDTA.
2. Tris/EDTA buffer: 100 mmol/L Tris–HCl pH 7.5, 1 mmol/L EDTA, 1 mmol/L PMSF.
3. CHAPS lysis buffer: 8 mol/L urea, 4 %w/v CHAPS, 2 %v/v IPG buffer pH 3–10 NL (GE Healthcare). Store 1 mL aliquots at −20 °C. Add 2 mg/mL DTT before use.
4. CHAPS labeling buffer: 10 mmol/L Tris–HCl pH 8.5, 5 mmol/L magnesium acetate, 8 mol/L urea, and 4 % CHAPS. Store 1 mL aliquots at −20 °C.
5. DIGE-sample buffer: 8 mol/L urea, 4 % CHAPS, 2 % (v/v) IPG buffer pH 3–10 NL. Store 1 mL aliquots at −20 °C. Add 2 mg/mL DTT before use.
6. Strip equilibration buffer: 50 mmol/L Tris–HCl pH 8.8, 6 mol/L urea, 30 % (v/v) glycerol, 2 % (w/v) SDS. Store 15 mL aliquots at −20 °C. Add 10 mg/mL DTT before use.

7. Agarose sealing solution: 0.5 % Agarose, 1× SDS running buffer, traces of bromophenol blue. Prepare fresh.
8. Sucrose gradient buffer: 10–80 % (w: v) sucrose dissolved in 100 mmol/L Tris–HCl, 0.05 % NP-40, 0.1 % *n*-dodecyl b-D-maltoside, 1 mmol/L EDTA.
9. SDS loading buffer: 10 mmol/L Tris–HCl, pH 6.8, 20 % glycerol, 5 % SDS, 0.05 % bromophenol blue, and 10 % .β-mercaptoethanol.

2.3 BN-PAGE

1. Cathode buffer: 50 mM tricine, 7.5 mM imidazole, 0.02 % Coomassie Blue G-250, pH 7.0.
2. Anode buffer: 25 mM imidazole, pH 7.0.
3. BN-PAGE sample buffer: 100 mmol/L Tris–HCl pH 7.5, 1 mmol/L EDTA, 1 mmol/L PMSF, and 2 mmol/L protease inhibitor cocktail, 1 % NP-40 and 0.1 % *n*-dodecyl b-D-maltoside.
4. Colloidal Coomassie blue G-250 stock: 5 % solution in 500 mmol/L aminocaproic acid.
5. Transfer buffer: 50 mmol/L Tris–HCl pH 8.3.
6. Antibodies: Antibodies that recognize mitochondrial subunits III and IV were purchased from Mitosciences (Oregon). Antibodies against 24 kDa subunit of complex I were a kind gift from Prof. John Walker (Cambridge).

3 Methods

3.1 Brain Mitochondrial Proteome Analysis

3.1.1 Preparation of Mitochondria from Rat Brains

1. Harvest the brain tissue and snap-freeze in liquid nitrogen; pulverize while still frozen, or store frozen tissue (liquid nitrogen) until use (see Note 1).
2. Resuspend at 1 g/mL Tris/sucrose buffer (10 mmol/L Tris–HCl pH 7.80, 0.25 mol/L sucrose, 2 mmol/L EDTA) prior to differential centrifugation.

3.1.2 Differential Centrifugation

1. First, centrifuge at 200 × *g* for 15 min. Collect the supernatant.
2. Centrifuge again at 750 × *g* for 10 min.
3. Centrifuge resultant supernatant again at 20,600 × *g* for 15 min.
4. Discard supernatant; keep resultant pellet.
5. Wash the pellet twice in the Tris/sucrose buffer supplemented with 0.5 mmol/L PMSF.

3.1.3 Proteome Analysis with 2D-DIGE

1. Resuspend the whole mitochondrial pellet (step 4, Subheading 3.1.2) in 140 μL of CHAPS lysis buffer (8 mol/L urea, 4 %w/v CHAPS, 2 %v/v IPG buffer pH 3–10 NL, and 2 mg/mL DTT).

2. After a freeze/thaw cycle, incubate samples on ice for 30 min followed by 5 min incubation in the ultrasound bath (Branson) at 4 °C. Centrifuge samples at 14,000 × *g* for 10 min at 4 °C.
3. Sample clean up: precipitate the supernatants using the Protein CleanUp Kit (GE Healthcare, or a TCA precipitation protocol). This step primarily eliminates contaminants that interfere with labeling and isoelectric focusing steps (i.e., nucleic acids and membrane-bound phospholipids). Do not overdry pellets since they become hard to resuspend. Resuspend the protein pellets in 100 μL of CHAPS labeling buffer (10 mmol/L Tris–HCl pH 8.5, 5 mmol/L magnesium acetate, 8 mol/L urea, and 4 % CHAPS).
4. Determine protein concentration before proceeding with sample labeling (ProteinQuant Kit (GE Healthcare) or other compatible method).
5. Label a total of 30 μg of protein from each sample (e.g., control vs. treatment) with 120 pmol of Cy3 or Cy5 Fluors in the dark for 30 min. Terminate the reaction by adding lysine at a final concentration of 10 nmol.
6. An internal standard for the labeling is produced by mixing equal amounts of proteins from all the samples contained in the experiment and labeled with Cy2 (see Note 2).
7. Add an equal volume of DIGE-sample buffer (8 mol/L urea, 4 % CHAPS, 2 %v/v IPG buffer pH 3–10 NL, and 2 mg/mL DTT) to each reaction; incubate on ice for 15 min.
8. Proceed to mix the samples, for every gel in the experiment mix: one sample labeled with Cy3, other with Cy5 and the internal standard (Cy2-labeled).
9. Load samples on to 3–10 NL, 24 cm IEF strips (see Note 3). Focus at 8 kV until a reading of 76 kV/h is reached.
10. Equilibrate strips prior to SDS-PAGE in 15 mL of equilibration buffer (50 mmol/L Tris–HCl pH 8.8, 6 mol/L urea, 30 %v/v glycerol, 2 %w/v SDS, 10 mg/mL DTT) for 15 min on a rocking platform.
11. Position strips on 13 % SDS-PAGE gels and perform the electrophoretic run at 0.2 W per gel for 1 h, then at 1 W per gel overnight at 10 °C until the bromophenol blue dye front reaches the gel bottom (see Note 4).

3.1.4 Image Acquisition and Analysis of 2D-DIGE Data

1. Visualize CyDye-labeled proteins by scanning gels using a compatible fluorescence imager (e.g., Typhoon™ 9400 Imager, GE Healthcare).
2. Scan Cy2 images using a 488 nm laser and an emission filter of 520 BP 40. Cy3 images are scanned using a 532 nm laser and an emission filter of 580 nm BP 30. Cy5 images are scanned

using a 633 nm laser and an emission filter of 670 nm BP 30, at 100 μm resolution. If needed crop images prior to analysis using Image Quant V 5.2 (GE Healthcare) or compatible software.

3. For the image analysis of the CyDye-stained gels, you will need a dedicated software (e.g., DeCyder 2D, GE Healthcare) or a software capable of cross-stain analysis (e.g., Progenesis Workstation, Nonlinear Dynamics).
4. After performing automatic spot detection and background subtraction, manually screen the gels to delete artifacts and to correct matching if needed.
5. Perform spot volume normalization. As default procedure, some software normalize to the volumes in the internal standard (Cy2-stained) samples. Spot volumes can also be normalized to the average volumes of the untreated samples in one specific group (e.g., control group). Choose the normalization method that best suits your experiment.
6. Proteins that are identified as differentially expressed are best excised from a preparative gel prior to identification by MALDI-TOF-TOF and/ or LC/MS/MS (see Note 5).

3.2 Mitochondrial Complex Fractionation

3.2.1 Sucrose Gradient Ultracentrifugation

1. Resuspend the washed pellet from step 4 in Subheading 3.1.2 in 100 μL of Tris/EDTA buffer (100 mmol/L Tris–HCl pH 7.5, 1 mmol/L EDTA, 1 mmol/L PMSF) supplemented with 2 mmol/L protease inhibitor cocktail, 1 % NP-40, and 0.1 % *n*-dodecyl b-D-maltoside.
2. Layer the resulting mixture of protein complexes on top of a 3 mL linear gradient from 10 to 40 % (w: v) sucrose dissolved in 100 mmol/L Tris–HCl, 1 mmol/L EDTA 0.05 % NP-40, 0.1 % *n*-dodecyl b-D-maltoside (see Note 6).
3. Separate the solubilized mitochondrial complex mixture by ultracentrifugation (128,000 × *g*) for 16.5 h at 4 °C, in a Beckman Coulter Optima L-90K Ultracentrifuge with an SW 41Ti rotor.
4. Collect fractions by carefully pipetting from the top of the tube (i.e., light end of the sucrose gradient) into vials. Snap-freeze and stored at −80 °C until use.

3.2.2 1D-SDS-PAGE of Sucrose Gradient Separated Proteins

To separate complex components, load and subject sucrose gradient fractions to SDS-PAGE separation under reducing conditions:

1. Add SDS loading buffer (10 mmol/L Tris–HCl, pH 6.8, 20 % glycerol, 5 % SDS, 0.05 % bromophenol blue, and 10 % mercaptoethanol) in a 1:4 ratio to each fraction aliquoted.
2. Heat the mixture at 100 °C for 5 min; load 25 μL of mixture per well on a 10 % polyacrylamide SDS-PAGE separating gel with 4.5 % stacking gel.

3. Carry electrophoresis in a Vertical Electrophoresis System for 75 min at 200 V, iv. Use exactly the same procedure for each experimental group in the study (i.e., control and treatment).
4. Stain the gels with Colloidal Coomassie G-250 (see Note 5). We use the in-house developed software (Gelviewer 2006), to determine gel regions where protein amount (as determined by Coomassie staining intensity) is different in two different experimental groups:
5. Briefly, the gel surface is divided using a grid, where each column corresponds to a fraction and the rows divide the gels into 20 distinct molecular weight intervals (previous analysis had shown that finer gridding of the gels studied up to 60 molecular weight intervals did not reveal new regions of different protein distribution).
6. Next, Coomassie staining intensity of resultant rectangles is measured, and the MW regions of the gels with large log10 differences (i.e., control: treatment >0.5) are subjected to mass spectrometric analysis for identification of the proteins in these regions (Fig. 1).

3.3 Mass Spectrometry-Based Identification of Proteins

Tryptic digestion of proteins of interest can be performed manually or using an automated station:

3.3.1 In-Gel Digestion of Proteins

1. First, wash excised protein bands with 50 mmol/L ammonium bicarbonate, then 50 %v/v acetonitrile in water followed by 100 %v/v acetonitrile for dehydration.
2. Reduce and alkylate in 10 mmol/ L DTT and 50 mmol/L iodoacetamide, respectively.
3. Digest overnight with trypsin in 50 mmol/L ammonium bicarbonate pH 8.0 at 37 °C.
4. Extract tryptic peptides using sequential steps of 1 % (v/v) formic acid, 2 % (v/v) acetonitrile, and 50 % (v/v) acetonitrile.

3.3.2 Analysis of Peptide Extracts Using a MALDI-TOF/TOF System

1. Spot samples for analysis on the Ultraflex TOF/TOF (1.2 μL of the peptide extract) directly on a 600 μm spot size anchorchip plate.
2. Allow the sample to dry and spot 0.6 μL of a 2 mg/mL solution (2:1 acetonitrile: 2.5 % trifluoroacetic acid in water) alpha-cyano-4-hydroxycinnamic acid matrix on top.
3. Use an accelerating voltage of 25 kV for peptide mass fingerprinting (PMF).
4. Perform external calibration using $(M+H)^+$ ions of angiotensin I, angiotensin II, substance P, bombesin, and adrenocorticotropic hormones (clip 1–17 and clip 18–39).
5. Use the data accumulated from 200 consecutive laser shots to produce each spectrum.

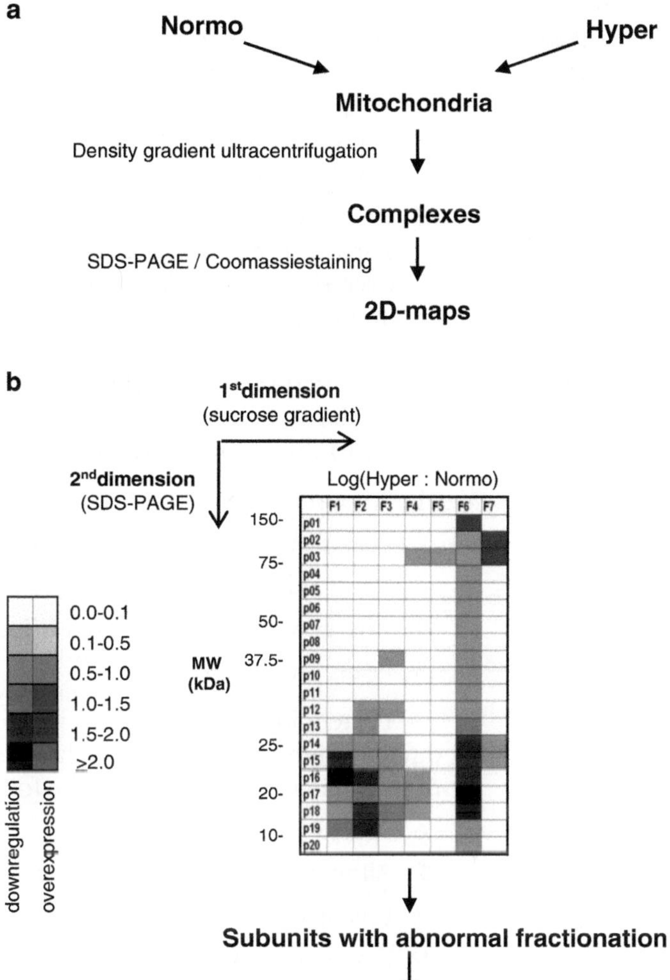

Fig. 1 Schematic representation of the analysis of mitochondrial complexes by sucrose gradient/1D SDS/PAGE approach. (**a**) Diagram of the strategy used for analysis of mitochondrial complexes by sucrose gradient ultracentrifugation/ SDS-PAGE /MS/MS. (**b**) Representation of 2D complex fractionation maps showing the logarithm10 of the ratio of Coomassie staining intensity in indicated states. Log (ratio) ranges are indicated using a color scale. Positive and negative values have the meaning of an overexpression or downregulation, respectively. Regions of the resultant 2D maps with different protein staining intensity are subsequently processed for identification by MS/MS

6. To additionally analyze samples using LIFT-TOF/TOF MS/MS from the same target, choose a maximum of five precursor ions per sample for MS/MS analysis. In the TOF1 stage, all ions are accelerated to 8 kV to promote metastable fragmentation then lifted by 19 kV to high potential energy in the LIFT cell.

3.3.3 Analysis of Peptide Extracts by Electrospray Mass Spectrometry Coupled with a Capillary HPLC

1. Separate tryptic peptides using a linear (water/acetonitrile) gradient (0.2 % formic acid) on a PicoFrit reversed-phase capillary column (5 μm BioBasic C18, 300 Å pore size, 75 μm ID × 10 cm, 15 μm tip) (New Objectives, MA, USA), with an in-line PepMap column (C18, 300 μm ID × 5 mm) (LC Packings, CA, USA) that serves as a loading/desalting column.
2. Run the PicoFrit column at 300 nL/min using a gradient starting at 8 % acetonitrile and increasing to 60 % acetonitrile over 40 min. Use data-dependent acquisition for the four most intense precursor ions.

3.3.4 Interpretation of the Peptide Mass Spectra

For the interpretation of the mass spectra, we use the Mascot software (Matrix Science Ltd, London, UK) to create peak lists and adhere to the following set of steps:

1. Smoothing is not applied.
2. Peak to noise criteria of 2 for peak picking.
3. Centroids are calculated at peak height of 50 %.
4. Charge states are calculated using the Q-TOF survey scan.
5. Peaks are de-isotoped.

Protein identification from the generated MS/MS data is conducted using the Mascot search engine (Mascot Daemon 2.1.03, Matrix Science, UK). We use the following parameters:

6. Specify the enzyme as trypsin.
7. Allow for one missed cleavage.
8. Precursor mass accuracy of ±0.6 Da.
9. Fragment ion mass accuracy of ±0.6 Da.
10. Fixed and variable modifications: carbamidomethyl (C) and oxidation (M), respectively, (we also include deamidation (NQ) as a variable modification for proteins separated by 2D-PAGE).
11. Perform data-dependent acquisitions on peptides with a charge state of 1 (MALDI-TOF results) or 2 or 3 (for Q-TOF analysis).
12. Perform protein identification from the MS/MS by searching the NCBI nonredundant database (NCBInr_22.fasta, 3,651,628 sequences, 1,255,333,329 residues, 35,247 sequences after taxonomy filter).
13. Select the taxonomy parameter according to your samples and experiments. In our experiments, we kept the taxonomy parameter open to entries from rodents as the main goal of our study was to detect potential differences in the abundance of known subunits of mitochondrial complexes from rat brain.
14. Only peptides with a score threshold high enough to warrant protein identity are used in the identification of the proteins.

gi|27717677
SHR control P20 F6
MS/MS Fragmentation of TAESSAVAATK
Found in gi|27717677, PREDICTED: similar to NADH oxidoreductase [Rattus norvegicus]

Match to Query 5: 1034.528608 from(518.271580,2+) intensity(4938.0000)
10: Sum of 10 scans in range 1098 (rt=23.3645, f=2, i=66) to 1107 (rt=23.5588, f=2, i=75) [\\qtof\qtof19\JAN2004.PRO\Data\6p032-64.raw]

From data file \\qtof\qtof19\JAN2004.PRO\Data\6p032-64.raw

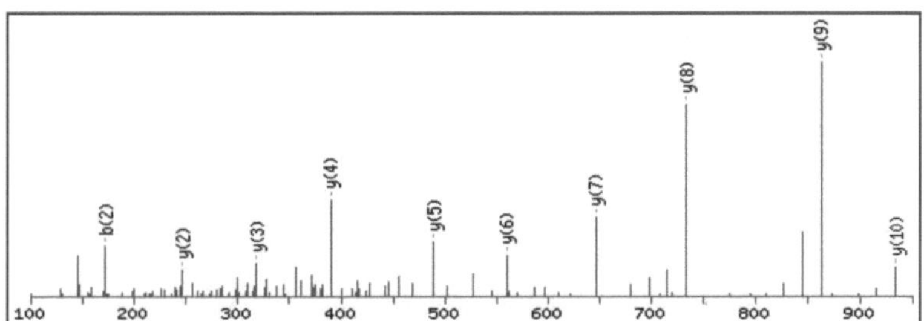

Monoisotopic mass of neutral peptide Mr(calc) : 1034.52
Fixed modifications: Carbamidomethyl (C)
Ions Score: 83 **Expect:** 3.6e-007
Matches (Bold Red): 10/98 fragment ions using 11 most intense peaks

#	b	b^{++}	b^0	b^{0++}	Seq.	y	y^{++}	y*	y^{*++}	y^0	y^{0++}	#
1	102.05	51.53	84.04	42.53	T							11
2	**173.09**	87.05	155.08	78.04	A	**934.48**	467.75	917.46	459.23	916.47	458.74	10
3	302.13	151.57	284.12	142.57	E	**863.45**	432.23	846.42	423.71	845.44	423.22	9
4	389.17	195.09	371.16	186.08	S	**734.40**	367.71	717.38	359.19	716.39	358.70	8
5	476.20	238.60	458.19	229.60	S	**647.37**	324.19	630.35	315.68	629.36	315.18	7
6	547.24	274.12	529.23	265.12	A	**560.34**	280.67	543.31	272.16	542.33	271.67	6
7	646.30	323.66	628.29	314.65	V	**489.30**	245.16	472.28	236.64	471.29	236.15	5
8	717.34	359.17	699.33	350.17	A	**390.23**	195.62	373.21	187.11	372.22	186.62	4
9	788.38	394.69	770.37	385.69	A	**319.20**	160.10	302.17	151.59	301.19	151.10	3
10	889.43	445.22	871.42	436.21	T	**248.16**	124.58	231.13	116.07	230.15	115.58	2
11					K	147.11	74.06	130.09	65.55			1

Fig. 2 Protein identification based on the mass spectra of a single peptide. Despite having apparent low scores, these scores are enough to give a statistically significant identification of the protein. The peptide fragmentation produced nine consecutive y ions, which together with the parent-ion mass made protein identification unequivocal

15. For proteins identified using a single peptide assignment, only peptides with statistically significant scores that also produced at least five consecutive y or b ions, which together with the parent-ion mass make protein identification unequivocal, should be considered (Fig. 2; also see Note 7).

3.4 Protein Complex Analysis Using BN-PAGE/Western Analysis

This method, with minor modifications, has been described earlier (11–13).

3.4.1 BN/PAGE Complex Separation

1. Resuspend the washed mitochondrial pellet from step 4 in Subheading 3.1.1 in BN-PAGE sample buffer (100 mmol/L Tris–HCl pH 7.5, 1 mmol/L EDTA, 1 mmol/L PMSF and 2 mmol/L protease inhibitor cocktail, 1 % NP-40, and 0.1 % *n*-dodecyl b-D-maltoside). Determine the protein concentration using the Bradford assay.
2. Shortly before starting the BN-PAGE, add the colloidal Coomassie blue G-250 stock solution (5 % in 500 mmol/L aminocaproic acid) to adjust to a detergent: Coomassie ratio of 4:1 (g/g).
3. Conduct the BN-PAGE separation on 4–15 % gradient polyacrylamide gel, run for 10 h (see Note 8). Run the gels at a fixed voltage (100 V) and temperature (4 °C).

3.4.2 Western Analysis After BN/PAGE

1. Following separation by BN-PAGE, incubate the gel for 15 min in standard SDS running buffer supplemented with 10 % beta-mercaptoethanol and prewarmed at 60 °C.
2. Rinse the gel for 30 s in transfer buffer (50 mmol/L Tris–HCl, pH 8.3).
3. Electrotransfer proteins onto a nitrocellulose membrane.
4. Block the membrane and probe using commercially available antibodies against the subunits of mitochondrial respiratory complexes (Fig. 3).

4 Notes

1. Snap-freezing the tissues allows for effective pulverization of the whole tissue sample. The pulverized tissue can then be weighted and resuspended at the desired concentration.
2. All the experimental samples are mixed to create the internal standard, for example, if your experiment has four control and four treatment samples, all eight samples are mixed in equal quantities, 30 μg of protein of the total protein mix is then labeled with the internal standard for every gel that will be run in the experiment.
3. Isoelectrofocusing strips are produced commercially and are available in different sizes and different ranges of pH gradients. It is recommended that you test run your samples to select the size and pH gradient that allows for the best protein separation.
4. It is important to remove all air bubbles between the strip and the top of the SDS/PAGE gel. Boil the agarose solution and stir

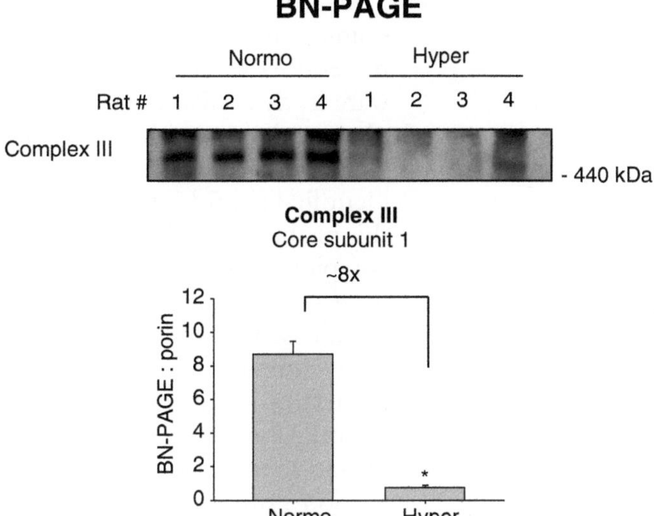

Fig. 3 Mitochondrial respiratory complexes display assembly defects in hypertension. To determine whether mitochondrial respiration rates were affected in hypertension (Hyper) vs. normotension (Normo), equal loads of whole-brain mitochondria protein preparations (100 μg of total protein) were subjected to Blue native (BN)-PAGE followed by Western analysis with antibodies against marker subunits. Band intensities of the mitochondrial complex III were normalized to those of prohibitin, an inner mitochondrial membrane marker of which the abundance was unchanged in hypertension vs. normotension (as measured by 2D-PAGE)

occasionally to dissolve. Seal the strip in place by immediately pipetting the agarose solution on top of it. To further minimize variations, run all gels simultaneously.

5. To maximize the probabilities of protein identification, a preparative 2D gel is loaded with up to 1 mg of protein, stained with Colloidal Coomassie G-250, and used to excise the protein spots that will be identified by MS/MS. Colloidal Coomassie G-250 is our protein stain of choice prior to MS/MS, owing to its simplicity, reproducibility, and high sensitivity (~10–50 ng protein/band). Stain the gels for exactly 16 h in colloidal Coomassie blue in 20 % methanol and destain in 25 % methanol for 24 h.
6. To achieve linearity, we prepare the gradient using an ABI-173A HPLC gradient system. A discrete gradient can also be obtained by carefully layering the sucrose solutions manually.
7. After mass spectrometric analysis, we find that many different proteins tend to comigrate in 1D-SDS-PAGE and not all proteins are individually detected by the Coomassie stain. Further, some proteins are found at more than one molecular weight, consistent with precursor forms, covalently (e.g., disulfide bridge) stabilized complexes, or degradation fragments.

Thus, the notion of protein abundance based on Coomassie staining intensity alone can be ambiguous. When we examine the emPAI values (10) for the detected proteins (using software freely available at http://xome.hydra.mki.co.jp), we also observe that individual protein identifications have their individual emPAI values, which depend on the number of identified peptides for each identified protein. To circumvent this problem, we first focus on known components of protein complexes of interest (i.e., mitochondrial respiratory complexes), which allow us to assess protein complex fractionation based on the co-fractionation of diverse subunits of the same complex. Second, for each of these subunits, we use the highest of its emPAI values in each fraction. If fractionation of emPAI values of one or more subunits of a complex differs in one state vs. another state (e.g., hypertension vs. normotension), this would indicate that integrity of the parent complex may differ as well. The more subunits of a complex that exhibit different fractionation in one state vs. another state, the more likely the complex is to be perturbed that state.

8. Several options exist to obtain BN-PAGE gradient gels. They can be produced in-house either by casting them manually or by the use of a gradient maker. They are also commercially available in precast form from a number of companies (i.e., Life Technologies).

References

1. Arrell DK, Elliott ST, Kane LA, Guo Y, Ko YH, Pedersen PL, Robinson J, Murata M, Murphy AM, Marban E, Van Eyk JE (2006) Proteomic analysis of pharmacological preconditioning: novel protein targets converge to mitochondrial metabolism pathways. Circ Res 99:706–714
2. Bernal-Mizrachi C, Gates AC, Weng S, Imamura T, Knutsen RH, DeSantis P, Coleman T, Townsend RR, Muglia LJ, Semenkovich CF (2005) Vascular respiratory uncoupling increases blood pressure and atherosclerosis. Nature 435:502–506
3. Keeney PM, Xie J, Capaldi RA, Bennett JP Jr (2006) Parkinson's disease brain mitochondrial complex I. J Neurosci 26:5256–5264
4. Swerdlow RH, Parks JK, Cassarino DS, Shilling AT, Bennett J, Jr P, Harrison MB, Parker WD Jr (1999) Characterization of hybrid cell lines containing mtDNA from Huntington's disease patients. Biochem Biophys Res Commun 261:701–704
5. Ogilvie I, Kennaway NG, Shoubridge EA (2005) A molecular chaperone for mitochondrial complex I assembly is mutated in a progressive encephalopathy. J Clin Invest 115:2784–2792
6. Nijtmans LG, Henderson NS, Attardi G, Holt IJ (2001) Impaired ATP synthase assembly associated with a mutation in the human ATP synthase subunit 6 gene. J Biol Chem 276:6755–6762
7. Kislinger T, Ox B, Kannan A, Chung C, Hu P, Ignatchenko A, Scott MS, Gramolini AO, Morris Q, Hallett MT, Rossant J, Hughes TR, Frey B, Emili A (2006) Global survey of organ and organelle protein expression in mouse: combined proteomic and transcriptomic profiling. Cell 125:173–186
8. Taylor SW, Warnock DE, Glenn GM, Zhang B, Fahy E, Gaucher SP, Capaldi RA, Gibson BW, Ghosh SS (2002) An alternative strategy to determine the mitochondrial proteome using sucrose gradient fractionation and 1D PAGE on highly purified human heart mitochondria. J Proteome Res 1:451–458
9. Taylor SW, Fahy E, Ghosh SS (2003) Global organellar proteomics. Trends Biotechnol 21:82–88

10. Ishihama Y, Oda Y, Tabata T, Sato T, Nagasu T, Rappsilber J, Mann M (2005) Exponentially modified protein abundance index (emPAI) for estimation of absolute protein amount in proteomics by the number of sequenced peptides per protein. Mol Cell Proteomics 4:1265–1272
11. Schagger H, Cramer WA, von Jagow G (1994) Analysis of molecular masses and oligomeric states of protein complexes by blue native electrophoresis and isolation of membrane protein complexes by two-dimensional native electrophoresis. Anal Biochem 217:220–230
12. Wittig I, Schagger H (2005) Advantages and limitations of clear-native PAGE. Proteomics 5:4338–4346
13. Swamy M, Siegers GM, Minguet S, Wollscheid B, Schamel WW (2006) Blue native polyacrylamide gel electrophoresis (BN-PAGE) for the identification and analysis of multiprotein complexes. Sci STKE 2006(345):pl4

Chapter 12

Oxidative Modifications of Mitochondria Complex II

Liwen Zhang*, Patrick T. Kang*, Chwen-Lih Chen, Kari B. Green, and Yeong-Renn Chen

Abstract

Increased superoxide ($O_2^{\bullet-}$) and nitric oxide (NO) production is a key mechanism of mitochondrial dysfunction in myocardial ischemia/reperfusion injury. In the complex II, oxidative impairment, decreased protein S-glutathionylation, and increased protein tyrosine nitration at the 70 kDa subunit occur in the post-ischemic myocardium (Zhang et al., Biochemistry 49:2529–2539, 2010; Chen et al., J Biol Chem 283:27991–28003, 2008; Chen et al., J Biol Chem 282: 32640–32654, 2007). To gain the deeper insights into ROS-mediated oxidative modifications relevant in myocardial infarction, isolated complex II is subjected to in vitro oxidative modifications with GSSG (to induce cysteine S-glutathionylation) or $OONO^-$ (to induce tyrosine nitration). Here, we describe the protocol to characterize the specific oxidative modifications at the 70 kDa subunit by nano-LC/MS/MS analysis. We further demonstrate the cellular oxidative modification with protein nitration/S glutathionylation with immunofluorescence microscopy using the antibodies against 3-nitrotyrosine/glutathione and complex II 70 kDa polypeptide (AbGSC90) in myocytes under conditions of oxidative stress.

Key words Mitochondria, Complex II, Nitration, S-glutathionylation, Protein disulfide linkage, S-sulfonation

1 Introduction

Mitochondrial dysfunction in ischemia-reperfusion (I/R) injury is caused by oxidative stress. Alterations of physiological redox status in the mitochondria of the post-ischemic heart include hyperoxygenation (1), overproducing $O_2^{\bullet-}$ and $O_2^{\bullet-}$-derived oxidants, and the consequence electron transfer chain functional impairment. Myocardial ischemia also alters NO metabolism, which includes increasing NO production and subsequent peroxynitrite formation in the post-ischemic heart (2, 3). The mitochondrial redox pool is enriched in reduced glutathione (GSH) with a high physiological

*Both authors contributed equally to this work.

Fernando Vivanco (ed.), *Heart Proteomics: Methods and Protocols*, Methods in Molecular Biology, vol. 1005, DOI 10.1007/978-1-62703-386-2_12,

concentration (in the mM range). Overproduction of reactive oxygen species (ROS) increases the ratio of oxidized glutathione (GSSG) to GSH. In the post-ischemic heart, accumulated GSSG associating with GSH depletion has been observed in the myocardium and mitochondrial preparation (4). Redox alteration involved in ROS production, NO metabolism, and GSH pool depletion in mitochondria can induce oxidative posttranslational modification of electron transport chain. Specifically, increasing protein tyrosine nitration (complexes I–III) and protein S-glutathionylation (complex I) has been marked in the mitochondria of post-ischemic heart, providing deep insights into disease pathogenesis (2–5).

In vitro protein tyrosine nitration and S-glutathionylation followed with LC/MS/MS has been emerged as a powerful approach to map the specific sites of oxidative modifications in the complex II (2, 3, 6). This approach is also capable of extending the research scope of investigating other important modifications such as protein radical formation (2, 6), disulfide bond formation (2, 7), and protein S-sulfonation (2).

The development and availability of specific antibodies against the 70 kDa subunit of the complex II (3, 8), 3-nitrotyrosine, and GSH has facilitated the study of oxidative modifications of the complex II in myocytes and myocardium using immunofluorescence microscopy and immunoprecipitation techniques. Through the approach, we are able to detect enhanced protein nitration of complex II under the conditions of hypoxia/reoxygenation.

2 Materials

2.1 Equipment

1. Olympus fluorescence microscopy (Olympus American Co., model IX-71, Center Valley, PA) with 40× objective (Olympus UApo/340).
2. Nano-liquid chromatography–tandem mass spectrometric analysis is performed on a Thermo Fisher LTQ mass spectrometer equipped with a nanospray source operated in positive ion mode. The LC system is a Dionex UltiMate 3000 Thermo Fisher. A 5 cm 75 μm ID BioBasic C18 column packed directly in the nanospray tip is used for chromatographic separation.

2.2 General Reagents and Supplies

1. Peroxynitrite ($OONO^-$) is purchased from Cayman Chemical (Ann Arbor, MI). BCNU (carmustine, a potent glutathione reductase inhibitor), GSH, NADH, sodium dithionite, and all chemicals are purchased from Sigma-Aldrich unless indicated otherwise. The monoclonal antibodies against 3-nitrotyrosine or GSH are purchased from Upstate Biotechnology, Inc. (Lake Placid, NY) and ViroGen Corp. (Watertown, MA), respectively. The polyclonal antibodies against the 70 kDa subunit

(AbGSC90) of complex II are generated in house (3, 8). Secondary antibodies conjugated with fluorochrome are from Invitrogen, Life Technologies Corporation (Grand Island, NY).

2. Gelatin/fibronectin pre-coating agent: autoclave 0.02 % gelatin (0.1 g gelatin into 500 mL water), dilute 500 μL of fibronectin (received as 0.1 % solution) in 99.5 mL of 0.02 % gelatin to make the gelatin/fibronectin pre-coating agent, and stored at 4°C up to 2 weeks.
3. Supplemented Claycomb medium (complete growth medium of HL-1 myocytes): Claycomb medium, 10 % fetal bovine serum, 100 U/mL–100 μg/mL penicillin–streptomycin, 2 mM L-glutamine, and 100 μM norepinephrine (Sigma-Aldrich, St. Louise, MO), and store at 4°C up to 2 weeks (9).
4. Hank's balanced salt solution (HBSS): sodium chloride (8 g), potassium chloride (400 mg), potassium phosphate monobasic (KH_2PO_4, 60 mg), glucose (1 g), sodium phosphate dibasic (Na_2HPO_4, 47.9 mg), sodium bicarbonate (350 mg), to a final volume of 1 L with water, and store at 4°C.

2.3 SDS-Polyacrylamide Gel Electrophoresis

1. NuPAGE® Novex® 4–12 % Bis-Tris precast mini gels 1.0 mm is purchased from Invitrogen (Carlsbad, CA).
2. Sample buffer (4×): Tris base (0.682 g), Tris–HCl (0.666 g), glycerol (4 g), SDS (0.8 g), EDTA (6 mg), Serva Blue G250 (0.75 mL of 1 % solution), Phenol Red (0.25 mL of 1 % solution) dissolved in 10 mL ultrapure water. The pH of 1× sample buffer should be ~8.5. Do not use acid or base to adjust pH.
3. MES SDS Running buffer (20×): MES [2-(*N*-morpholino) ethane sulfonic acid, 97.6 g, 1 M], Tris Base (60.6 g, 1 M), SDS (10.0 g), EDTA (3.0 g), dissolved in 500 mL of distilled water. Do not adjust pH with base. The pH of 1× running buffer should be ~7.3.
4. Pre-stained full-range Rainbow™ molecular weight markers are purchased from GE Healthcare (Fairfield, CT).

2.4 Western Blotting for Protein S-Glutathionylation and Protein Nitration

1. Transfer buffer (20×): Bicine (500 mM), Bis-Tris (500 mM), EDTA (20.5 mM), chlorobutanol (1 mM) prepared with ultrapure water. 1× buffer should be pH 7.2.
2. GE Nitrocellulose Pure Transfer Membranes (GE Healthcare), and TRANS-BLOT® paper, 15×20 cm (Bio-Rad).
3. Tris-buffered saline with Tween-20 (TTBS): 25 mM Tris–HCl, pH 7.4, 137 mM NaCl, 0.1 % Tween-20.
4. Blocking buffer: 5 % (w/v) nonfat dry milk in TTBS.
5. Antibody dilution buffer: TTBS supplemented with 5 % nonfat dry milk.

6. Primary antibody: anti-3-nitroptyrosine polyclonal antibody, anti-GSH monoclonal antibody, and AbGSC90.
7. Secondary antibody: Amersham ECL™-horseradish peroxidase linked anti-rabbit IgG antibody (GE Healthcare).
8. Amersham Enhanced Chemiluminescent ECL™ Western Blotting Detection Reagents (GE Healthcare).
9. Amersham Hyperfilm ECL™ (8 × 10 in.) (GE Healthcare).

2.5 Mass Spectrometry to Identify the Specific Site of Oxidative Modification

1. Staining solution: 0.25 % Coomassie Brilliant Blue R-250 in solution containing 45.4 % (v/v) methanol and 9.2 % (v/v) acetic acid.
2. Fixing solution: solution containing 40 % (v/v) methanol and 10 % (v/v) acetic acid.
3. Destaining solution: solution containing 0.5 % (v/v) ethanol and 0.75 % (v/v) acetic acid.
4. Washing solution: solution containing 50 % (v/v) methanol and 10 % (v/v) acetic acid.
5. Other buffer/reagents used for the digestion: 100 mM ammonium bicarbonate buffer, acetonitrile, dithiothreitol solution (5 mg/mL, in 50 mM ammonium bicarbonate buffer) and iodoacetamide (15 mg/mL, in 50 mM ammonium bicarbonate buffer).
6. Enzymes: sequencing grade trypsin (Promega, Madison, WI): 20 ng/μL in 50 mM ammonium bicarbonate solution; chymotrypsin (Roche Diagnostics, Indianapolis, IN): 25 ng/μL in 50 mM ammonium bicarbonate solution.
7. Extraction solution: solution containing 50 % (v/v) acetonitrile and 5 % (v/v) formic acid.
8. Montage In-Gel Digestion Kit (Millipore, Bedford, MA).

3 Methods

3.1 Detection of Complex II Protein S-Glutathionylation and Protein Nitration from the Post-ischemic Heart

3.1.1 Immunoprecipitation (Ip) of Complex II 70 kDa FAD-Binding Subunit from the Tissue Homogenates of Post-ischemic Myocardium

1. Myocardium is excised from the risk and non-ischemic regions of the post-ischemic heart (6) (see Fig. 1a).
2. The excised myocardium is homogenated in the HEPES buffer (pH 7.2 containing 250 mM sucrose, 1 mM EDTA, protease/phosphatase inhibitors), and then centrifuged at 600 × *g* for 20 min. The supernatant is collected as tissue homogenate, and the protein concentration is adjusted to 2 mg/mL with RIPA buffer (containing 1× protease inhibitor, 100 μM PMSF, and 10 μg leupeptin/aprotinin).
3. 2 μL of AbGSC90 (11.1 mg/mL) is mixed with Rabbit IP Matrix (80 μL, Santa Cruz Biotechnology, San Diego, CA), and agitated at 4°C for 1 h using rotary shaker.

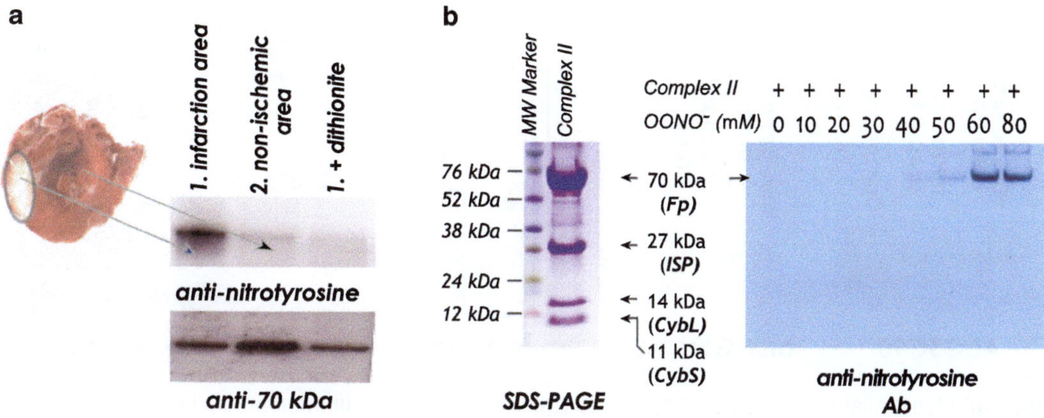

Fig. 1 (**a**) The rat heart model of in vivo myocardial ischemia/reperfusion and TCC staining of infarct region in the post-ischemic myocardium. Myocardial tissue homogenates from non-ischemic and infarct (risk) regions are subjected to immunoprecipitation with AbGSC90 and subsequently subjected to SDS-PAGE and immunoblotted with anti-3-nitrotyrosine (*upper panel*) and anti-70 kDa (*lower panel*) antibodies. (**b**) *Left panel*, SDS-PAGE of $OONO^-$-treated complex II and stained by Coomassie blue. *Right panel*, isolated complex II is subjected to in vitro protein tyrosine nitration. Protein (1 μM, based on heme *b*) is incubated with various concentrations of $OONO^-$ (0–80 μM) at 37°C for 1 h. Excess $OONO^-$ is removed by uric acid (1 mM). The $OONO^-$-treated complex II is subjected to SDS-PAGE and then immunoblotting with anti-3-nitrotyrosine antibody

4. Tissue homogenate (2 mg/mL, 0.5 mL) is then added to the AbGSC90-linked beads, and agitated at 4°C overnight using rotary shaker.
5. Spin the mixture at 300 × *g* for 5 min by microfuge, and discard the supernatant.
6. Wash the beads with PBS (1 mL) four times, and remove the PBS as much as possible after the fourth washing.
7. For the detection of complex II S-glutathionylation: The beads are resuspended in 30 μL of PBS, and digested with 1× sample buffer without dithiothreitol (DTT) addition at 70°C for 10 min. In a parallel negative control experiment, PBS-suspended beads should be digested with 1× sample buffer with DTT (25 mM) addition to eliminate the signal of S-glutathionylation (Fig. 2a).
8. For the detection of complex II tyrosine nitration: The beads are resuspended in 30 μL of PBS, and digested with 1× sample buffer with DTT addition at 70°C for 10 min. In a separate negative control experiment, PBS-suspended beads should be digested with 1× sample buffer with DTT and sodium dithionite (25 mM) addition to eliminate the signal of protein nitration (see Note 1).
9. The beads are removed by brief spin, and the sample of supernatant is loaded and subjected to SDS-polyacrylamide gel electrophoresis (SDS-PAGE).

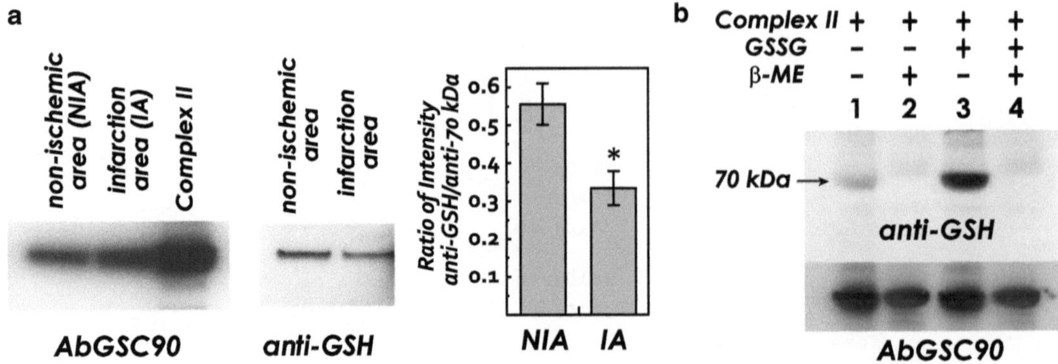

Fig. 2 (**a**) Myocardial tissue homogenates (300 μg) are subjected to immunoprecipitation with AbGSC90, and subsequently subjected to SDS-PAGE and immunoblotted with anti-70 kDa and anti-GSH antibodies. (**b**) Complex II is isolated from the bovine heart. *Lane 1*, native complex II (50 pmol, based on heme b) is subjected to SDS-PAGE and followed by immunoblotting with anti-GSH (*upper panel*) and anti-70 kDa antibodies (*lower panel*). *Lane 2*, native complex II pretreated with β-ME (1 %) prior to SDS-PAGE. *Lane 3*, glutathionylated complex II. *Lane 4*, glutathionylated complex II pretreated with β-ME (1 %) prior to SDS-PAGE

3.1.2 SDS-PAGE

1. These instructions assume the use of XCell SureLock™ Mini-Cell gel electrophoresis system and Novex® Pre-Cast Gel of 4–12 % Bis-Tris ZOOM Gel 1.0 mm, IPG well. They are easy to operate and provide highly reproducible experiment.
2. Prepare 1× running buffer by diluting 80 mL of 10× running buffer to ~800 mL with distilled water.
3. Complete the assembly of the precast gel and XCell SureLock Mini-Cell unit, and connect to a power supply.
4. Add running buffer to the upper buffer chamber and lower buffer chamber.
5. Load the sample of tissue homogenates (100 μg) in a well, including one well for 5 μL of pre-stained rainbow marker.
6. Gel is run at room temperature (R.T.) at 190 V for 50 min [current 100–125 mA/gel (start); 60–80 mA/gel (end)].
7. Proteins of the gel are transferred to nitrocellulose membrane (50 V and 1 h).
8. Blocking the membrane with 5 % dry milk (Bio-Rad, CA) in TTBS at R.T. for 1 h.
9. Add monoclonal anti-GSH Ab (primary Ab, 500× dilution) or polyclonal anti-3-nitrotyrosine Ab (primary Ab, 2,000× dilution) to the membrane in blocking solution, and agitate at 4°C for overnight.
10. Wash the membrane with TTBS 10 min for three times.
11. Add the secondary Ab (Sheep HRP conjugated Ab against mouse IgG from GE Healthcare, 3,000× dilution) in blocking buffer, and agitate at R.T. for 1 h.

12. Discard the secondary Ab in blocking buffer, and wash the membrane twice with TTBS and twice in TBS.
13. Membrane is ready for the visualization by using ECL Western blotting detection agent following manufacturer's protocol.

3.2 Detection of Cellular Protein S-Glutathionylation and Protein Nitration by Immunofluorescence Microscopy

3.2.1 Detecting Protein Glutathionylation of BCNU-Treated HL-1 Cells by Immunofluorescence Microscopy

1. Mouse myocytes (HL-1 cell line) are cultured in supplemented Claycomb medium in gelatin/fibronectin pre-coated polystyrene tissue culture flask at 37°C in the presence of 5 % CO_2 with standard cell culturing technique (9).
2. HL-1 myocytes are seeded at a density of 2×10^5 cells per well on fibronectin pretreated coverslip (BD BioCoat) in 6-well tissue culture plate overnight prior to conduct fluorescence imaging experiments (see Note 2).
3. HL-1 myocytes are subjected to BCNU (80 μM, 4 h) treatment in growth medium at 37°C (see Note 3).
4. Aspirate growth medium, briefly rinse cells with PBS (see Note 4).
5. Fixation: flood cells with 3.7 % paraformaldehyde in PBS at R.T. for 15 min.
6. Remove paraformaldehyde solution; wash with PBS, 10 min for three times. After fixation, samples should be kept in shaker.
7. Incubate cells in 0.3 % Triton X-100 and 5 % goat serum in PBS for 1 h for permeabilization and blocking.
8. Add primary antibodies (anti-GSH monoclonal Ab and AbGSC90) into the blocking solution, and incubate at R.T. for 2 h or at 4°C overnight.
9. Remove primary antibody solution; wash with PBS, 10 min for three times (see Note 5).
10. Block with 5 % goat serum in PBS for 20 min.
11. Add secondary antibodies into the blocking solution; incubate at R.T. for 1 h (see Note 6).
12. Discard secondary antibody solution; wash with PBS, 10 min for three times.
13. Mount the coverslips with cells on a glass slide with antifade mounting medium, VECTASHIELD with DAPI (see Note 7).
14. Apply nail polisher along the edge of the coverslips. After the drying of the nail polisher, specimens are ready for the fluorescence microscopy detection.
15. Images are processed and overlaid with QCapture Pro software (QImaging, Surrey, BC, Canada) (see Fig. 3).

3.2.2 Detecting Protein Nitration of Post-hypoxic Myocytes by Immunofluorescence Microscopy

1. Myocytes (using HL-1 myocytes or H9c2 cardiac myoblasts, confluent cells with >90 % viability) in glucose-free medium are placed on coverslips in 35-mm sterile dishes at a density of 10^5 cells/dish (see Note 8).

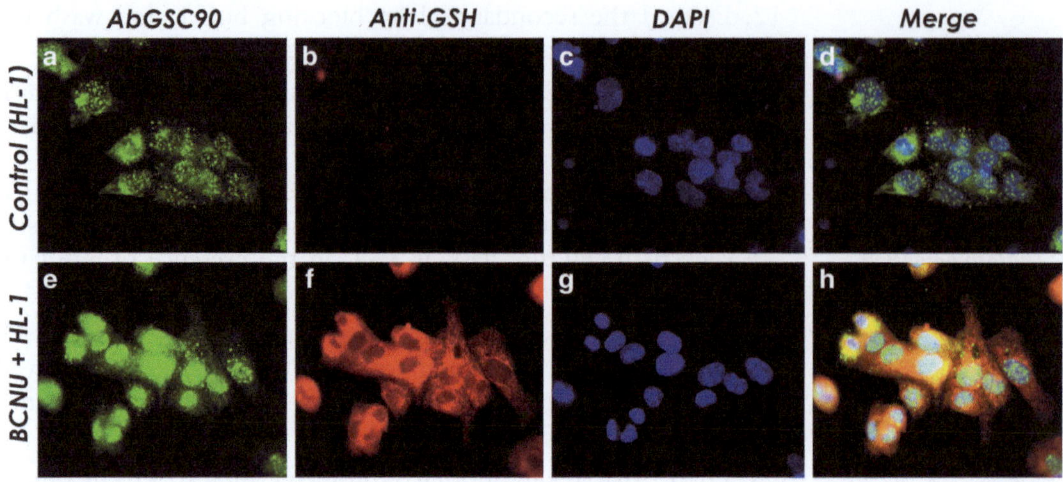

Fig. 3 Increased S-glutathionylation levels overlaid with anti-complex II antibody AbGSC90 from HL-1 myocytes under oxidative stress induced by BCNU treatment. *Upper row*, HL-1 without BCNU treatment (control); *lower row*, HL-1 with BCNU (80 μM) treatment. Images **a** and **e** are cellular proteins probed with AbGSC90. Images **b** and **f** are HL-1 without or with BCNU (80 μM) treatment staining with anti-GSH, demonstrating enhancement of cellular S-glutathionylation by BCNU. Images **c** and **g** are HL-1 nuclei counter stained with DAPI. The merged images **d** and **h** demonstrate marked cellular S-glutathionylation (*red*) induced by BCNU is partially localized (*yellow* image) with the 70 kDa subunit (*green*). Note that we could observe enhancement of overall glutathionylation levels in HL-1 cells after BCNU treatment. Nevertheless, complex II (probed with AbGSC90) is translocated at the perinucleus region and colocalized with the enhanced anti-GSH signals

2. Cells are placed in a Modular Incubator Chamber (Billups-Rothenberg, Inc., Del Mar, CA).
3. Hypoxic treatment: Nitrogen gas is flushed on the surface of the medium and incubating at 37°C for 1 h (see Note 9).
4. Reoxygenation: Aspirate glucose-free medium, and incubation of cell in medium with glucose for 1 h at 37°C in the presence of 4.5 % CO_2 under normoxic conditions.
5. Repeat the procedures steps 4–15 of Subheading 3.2.1, except that anti-3-nitrotyrosine monoclonal antibody and AbGSC90 are used as the primary Ab.

3.3 In Vitro Protein Oxidative Modifications of Complex II and Mass Spectrometry

3.3.1 In Vitro Protein S-Glutathionylation of Complex II

1. Purified complex II (or SQR, succinate-ubiquinone reductase) is prepared from bovine heart submitochondrial particles (SMP) (6, 10) (see Fig. 1b). Purified complex II (1 μM, based on heme *b*) is incubated with GSSG (1 mM) at R.T. for 1 h. The reaction is terminated by addition of sample buffer, and then heated at 70°C for 5 min (see Note 10).
2. Protein of complex II (50 pmol, based on heme *b*) are subjected to SDS-PAGE and followed by immunoblotting with anti-GSH monoclonal Ab under nonreducing conditions (see Fig. 2b). This step is done by repeating the procedure of **steps 1–12** in Subheading 3.1.2.

3.3.2 In Vitro Protein Tyrosine Nitration of Complex II

1. Purified complex II (1 μM, based on heme b) is incubated with $OONO^-$ (60 μM) at 37°C for 1 h. uric acid (1 mM) is then added to reaction mixture to remove excess $OONO^-$.
2. $OONO^-$-treated complex II (50 pmol, based on heme b) are subjected to SDS-PAGE and followed by immunoblotting with anti-3-nitrotyrosine polyclonal Ab under reducing conditions (in the presence of DTT or β-mercaptoethanol) (see Fig. 1). This step is done by repeating the procedure of steps 1–12 in Subheading 3.1.2.

3.3.3 Mass Spectrometry to Identify the Sites of Oxidative Modifications: In-Gel Digestion

1. Remove the gel upon completion of SDS-PAGE (Subheadings 3.3.1 and 3.3.2). The gel is rinsed in the fixing solution (methanol: acetic acid: water = 40:10:50) for 30 min at R.T.
2. The fixing solution is discarded and the gel is incubated with 20 mL of Coomassie blue staining solution (0.25 % Coomassie Brilliant Blue R-250 in solution containing 45.4 % (v/v) methanol and 9.2 % (v/v) acetic acid) for 1 h at R.T.
3. The staining solution is discarded and the gel is destained by incubation with destaining solution containing 0.5 % methanol and 0.75 % acetic acid for overnight at R.T.
4. The gel is equilibrated with 50 mL of water with three changes of water.
5. The band of 70 kDa FAD-binding subunit of complex II is trimmed as closely as possible to minimize background polyacrylamide material.
6. The trimmed gels are washed twice with washing buffer (50 % methanol/5 % acetic acid) for several hours, and the gels are dehydrated with acetonitrile.
7. The gels are reconstituted with dithiothreitol to reduce cysteinyl residues, and iodoacetamide is then added to alkylate cysteine sulfhydryls (see Note 11).
8. Gels are washed with cycles of acetonitrile and ammonium bicarbonate buffer. Gels are then dried by speed vac.
9. In order to obtain maximum sequential information, samples are digested with (i) trypsin, (ii) chymotrypsin and (iii) trypsin and chymotrypsin. For digestion (i) and (ii), 50 μL of sequencing grade trypsin (20 ng/μL) or chymotrypsin (25 ng/μL) are added to the dehydrated gel, respectively. In digestion (iii), both 50 μL of sequencing grade trypsin (20 ng/μL) or chymotrypsin (25 ng/μL) are added together to the dehydrated gel.
10. The gels are set on ice 10 min for rehydration, and 20 μL of 50 mM ammonium bicarbonate buffer is added. The mixture is incubated at R.T. for overnight.

11. The peptides in gels are extracted with 50 % acetonitrile with 5 % formic acid several times, and extracted peptides are pooled together and concentrated in a speed vac to ~25 μL for further analysis.

3.3.4 Nano-LC MS/MS

1. Capillary liquid chromatography–tandem mass spectrometry is performed on a Thermo Fisher LTQ mass spectrometer equipped with a nanospray source operated in positive ion mode. The LC system is Dionex UltiMate 3000 from Thermo Fisher. A 5 cm 75 μm ID BioBasic C18 column packed directly in the nanospray tip is used for chromatographic separation.
2. Mobile phase preparation for capillary LC system is as follows: solvent A is water containing 50 mM acetic acid, solvent B is acetonitrile. 5 μL aliquots of each sample are injected onto the column for the analysis. Peptides are eluted off the column into the LTQ system using a gradient of 2–80 % solvent B over 48 min with a flow rate of 300 nL/min. A total run time is 65 min.
3. A spray voltage of 3 KV and a capillary temperature of 200°C are used in LTQ.
4. The scan sequence of the mass spectrometer is based on the TopTen™ method: the analysis is programed for a full scan recorded between 350 and 2,000 Da and consecutive MS/MS scans to generate product ion spectra for the ten most abundant peaks in the full spectrum. The CID fragmentation energy is set to 35 %. Dynamic exclusion is enabled with a repeat count of 30 s, exclusion duration of 350 s and a low mass width of 0.50 and high mass width of 1.50 Da. Multiple MS/MS detection of the same peptide is excluded after detecting it three times.
5. The RAW data files collected on the mass spectrometer are converted to Mascot Generic Format (MGF) files using MassMatrix data conversion tools (http://www.massmatrix.net/download). Resulting .mgf files are searched on Mascot Daemon (Matrix Science, Boston, MA) and MassMatrix (7), or the identification of 70 kDa FAD-binding subunit of complex II and investigation of its modifications. The mass tolerance of precursor ions is set to 1.8 Da to accommodate accidental selection of the C13 ion and the fragment ion mass tolerance is set at 0.8 Da. Number of missed cleavages permitted in the search is set at 2 for tryptic or chymotryptic digestions. Number of missed cleavages is set at 4 for trypsin and chymotrypsin digestion. Modified peptides identified in the programs are all manually checked for validation (see Tables 1 and 2).
6. The identification of 70 kDa subunit and some of the modifications (glutathionylation and nitration) is done by

Table 1
Summary of the peptide sequence and corresponding OONO⁻-mediated oxidative modifications obtained from MS/MS analysis

Amino acid residue in the 70 kDa	Theoretical m/z	Observed m/z	Peptide sequence and OPTM	Remarks
Y_{56}	982.1527^{3+}	982.90^{3+}	^{48}VSDAISAQ**Y**$_{56(NO2)}$PVVDHEFDAVV VGAGGAGLR^{76} (*tyrosine nitration*)	(3)
Y_{142}	1286.9052^{3+}	1286.76^{3+}	^{130}GSDWLGDQDAIHYMTEQAPASVV ELEN**Y**$_{142(NO2)}$GMPFSR^{163} (*tyrosine nitration*)	(3)
C_{267}	1121.4730^{2+}	1121.56^{2+}	^{263}TYFS**C**$_{267(SO3)}$TSAHTSTGDGTAMV TR^{283} (*S-sulfonation*)	(2)
C_{476}	813.3665^{2+}	813.35^{2+}	^{467}AC$_{(CAM)}$ALSIAESC$_{476(SO3)}$RPGDK^{481} (*S-sulfonation*)	$O_2^{\bullet-}$–mediated S-sulfonation (2)
C_{537}	1101.0414^{2+}	1101.67^{2+}	^{529}VGSVLQEG**C**$_{537(SO3)}$EKISSLYGDLR^{548} (*S-sulfonation*)	(2)
C_{306}/C_{312}	802.3820^{2+}	802.42^{2+}	^{303}GAG**C**$_{306(SS)}$LITEG**C**$_{312(SS)}$ RGEGGIL^{319} (*protein thiyl radical*)	(2)
C_{439}/C_{444}	941.1017^{3+}	941.43^{3+}	^{425}HVNGQDQVVPGLYA**C**$_{439(SS)}$ GEAA**C**$_{444(SS)}$ASVHGANR^{452} (*disulfide formation*)	(2)
C_{288}/C_{575}	860.4094^{2+}	860.36^{2+}	^{287}P**C**$_{288(SS)}$QDL$^{291/568}$ELQNLMLC$_{575(SS)}$ AL^{577} (*disulfide formation*)	(2)
C_{288}	789.3871^{2+}	789.85^{2+}	^{284}AGLP**C**$_{288(DMPO)}$QDL EFVQF^{296} (*protein thiyl radical*)	(2)
C_{655}	856.4378^{2+}	856.48^{2+}	^{649}TLN ETD**C**$_{655(DMPO)}$ATVPPA IR^{663} (*protein thiyl radical*)	$O_2^{\bullet-}$–mediated (2, 6)

Table 2
Summary of the peptide sequence and corresponding GSSG-mediated S-glutathionylation obtained from MS/MS analysis

Amino acid residue in the 70 kDa	Theoretical m/z	Observed m/z	Peptide sequence and OPTM	Remarks
C_{90}	996.4455^{2+}	996.63^{2+}	^{77}AAFGLSEAGFNTA**C**$_{90(SG)}$VTK^{93}	Glutathionylation (6)
C_{537}	725.0136^{3+}	725.04^{3+}	^{528}RVGSVLQEG**C**$_{537(SG)}$EKISSLY^{544}	(6)
C_{655}	971.4195^{2+}	971.64^{2+}	^{651}NETD**C**$_{655(SG)}$ATVPPAIRSY^{665}	(6)

searching the data on MASCOT against SwissProt mammal database; considered modifications (variable) are oxidation (met), carbamidomethylation (Cys), glutathionylation (Cys) and nitration (Tyr).

7. Modifications on 70 kDa subunit are also investigated using MassMatrix. The data are searched against the protein sequence of 70 kDa subunit. Mass shifts caused by the modifications, chemical structures of the modification groups and the residues where the modifications occur are programed into the search engine. Glutathionylation ($C_{10}H_{15}N_3O_6S_1$, 305.0681 Da, cys), DMPO modification ($C_6H_9N_1O_1$, 111.0684 Da), nitration (N_1O_2H-1, 44.9851 Da, tyr), S-sulfonation (O_3, 47.7847 Da, cys), S-sulfination (O_2, 31.9898 Da, cys), and disulfide linkage formation (H-2, –2.0156 Da, cys) are included in the search as variable modifications in addition to oxidation (met) and carbamidomethylation (cys).

4 Notes

1. 3-nitrotyrosine is reduced to 3-aminotyrosine by dithionite. To prepare 1 M sodium dithionite dissolved in the phosphate buffer (pH 7.4), the phosphate buffer has to be bubbled with argon for 30 min to remove the oxygen in buffer.
2. Fibronectin pretreated 22 mm coverslips or 35 mm coverslip-bottom dish (BD BioCoat, Bedford, MA) is used to improve the attachment of HL-1 cells on the glass surface.
3. We dissolve high concentration pharmacological agents into fresh growth medium first, and then replace the old growth medium by aspiration. We do not directly add highly concentrated stock chemicals into the culturing plate with living myocytes.
4. Never directly pipette liquid onto adherent cells. Tilt the container and gently pipette liquid toward the bottom edge of the container. Set the container flat so that the liquid would slowly cover cells.
5. Expensive primary antibody in blocking solution can be recycled. Nevertheless, since the dilution factor of the antibody in the recycled solution is unknown, this recycled antibody solution would be only good for qualitative pioneering experiment.
6. Secondary antibody conjugated with fluorochrome could be light sensitive. Always cover sample with aluminum foil in black container.
7. Pipette about 9 μL of mounting medium on a glass slide with a wide open pipette tip. Use tweezers to retreat the coverslip

from the bottom of the Petri dish. Blot-dry the coverslip by touching the edge of the coverslip on Kimwipes. Slowly place the coverslip on the drop of the mounting medium.

8. H9c2 cardiac myoblasts are seeded at a density of 10^4 cells/35-mm Petri dish (3). Dependent on different research purposes, 1 % FBS could be included in the glucose-free medium to reduce cell death rate after the relatively harsh hypoxic experimental conditions.
9. Hypoxic treatment of myocytes can also be accomplished via formation of a layer of oil on the surface of the glucose-free medium. The reoxygenation time course (e.g., 1, 6, or 24 h of reoxygenation time) should be established if interesting in the longer term responses such as hypoxia induced factor (HIF) relative pathways.
10. Prior to sample digestion and SDS-PAGE, the mixture of isolated complex II and GSSG should be dialyzed against PBS (200× volume) at 4°C for 1 h to remove excess GSSG at the end of reaction.
11. This step should be skipped for glutathiolation and disulfide linkage formation studies since dithiothreitol will break the disulfide bond linkage between glutathione/cysteine or cysteine/cysteine.

Acknowledgments

*The author would like to thank Professor Pravin P.T. Kaumaya (The Ohio State University, College of Medicine, Department of Ob/Gyn, Columbus, OH) for development of polyclonal antibody against the 70 kDa subunit of complex II, namely, AbGSC90. This work is supported by RO1 HL83237.

References

1. Zhao X, He G, Chen YR, Pandian RP, Kuppusamy P, Zweier JL (2005) Endothelium-derived nitric oxide regulates postischemic myocardial oxygenation and oxygen consumption by modulation of mitochondrial electron transport. Circulation 111:2966–2972
2. Zhang L, Chen C, Kang PT, Garg V, Hu K, Green-Church K, Chen Y (2010) Peroxynitrite-mediated oxidative modifications of complex II: relevance in myocardial infarction. Biochemistry 49:2529–2539
3. Chen CL, Chen J, Rawale S, Varadharaj S, Kaumaya PP, Zweier JL, Chen YR (2008) Protein tyrosine nitration of the flavin subunit is associated with oxidative modification of mitochondrial complex II in the post-ischemic myocardium. J Biol Chem 283:27991–28003
4. Chen J, Chen C, Rawale S, Chen C, Zweier J, Kaumaya P, Chen Y (2010) Peptide-based antibodies against glutathione-binding domains suppress superoxide production mediated by mitochondrial complex I. J Biol Chem 285:3168–3180
5. Liu B, Tewari AK, Zhang L, Green-Church KB, Zweier JL, Chen YR, He G (2009) Proteomic analysis of protein tyrosine nitration after ischemia reperfusion injury: mitochondria as the major target. Biochim Biophys Acta 1794:476–485

6. Chen YR, Chen CL, Pfeiffer DR, Zweier JL (2007) Mitochondrial complex II in the post-ischemic heart: oxidative injury and the role of protein s-glutathionylation. J Biol Chem 282:32640–32654
7. Xu H, Freitas MA (2007) A mass accuracy sensitive probability based scoring algorithm for database searching of tandem mass spectrometry data. BMC Bioinformatics 8:133
8. Kang PT, Yun J, Kaumaya P, Chen Y (2011) Design and use of peptide-based antibodies decreasing superoxide production by mitochondrial complex I and complex II. Biopolymers (Peptide Science) 96:207–220
9. White S, Constantin P, Claycomb W (2004) Cardiac physiology at the cellular level: use of cultured HL-1 cardiomyocytes for studies of cardiac muscle cell structure and function. Am J Physiol Heart Circ Physiol 286: H823–H829
10. Yu L, Yu CA (1982) Quantitative resolution of succinate-cytochrome c reductase into succinate-ubiquinone and ubiquinol-cytochrome c reductases. J Biol Chem 257:2016–2021

Chapter 13

Detection of *O*-GlcNAc Modifications on Cardiac Myofilament Proteins

Genaro A. Ramirez-Correa, Isabel Martinez Ferrando, Gerald Hart, and Anne Murphy

Abstract

In this chapter it is described a general method that has been used successfully by more than one laboratory interested in detecting *O*-GlcNAc in myofilament proteins. Alternative reagents for chemo-enzymatic or metabolic labeling will be indicated, as well as references for more details in alternative methods. The outline is divided into (1) Enrichment of *O*-GlcNAc Stoichiometry, (2) Cardiac Myofilament Protein Isolation, (3) SDS-PAGE, (4) "Reduction and Alkylation," (5) In-Gel Protein Digestion, (6) Chemo-enzymatic Labeling of *O*-GlcNAc Moieties (Click Chemistry), (7) Biotin Alkyne Tagging, (8) Strong Cation Exchange (SCX) and Streptavidin, and (9) β-Elimination and Michael Addition (BEMAD) for *O*-GlcNAc Site-Mapping.

Key words O-linked-*β*-D-N-acetylglucosamine, *O*-GlcNAcylation, Cardiac myofilaments, Post-translational modification, Chemo-enzymatic labeling, Click chemistry, *β*-Elimination, Michael addition

1 Introduction

O-GlcNAcylation consists of the addition of single *O-linked* β-D-*N*-acetylglucosamine (*O*-GlcNAc) sugar to serine and threonine residues of nuclear and cytoplasmic proteins. This is a posttranslational modification (PTM) derived directly from glucose metabolism, and it is centrally involved in cellular homeostasis and viability (1, 2). The reaction is catalyzed by *O*-GlcNAc transferase (OGT), whereas *O*-GlcNAc removal is catalyzed by *O*-GlcNAcase (OGA) activity (1, 2). Changes in *O*-GlcNAcylation can result from a wide variety of stimuli, such as stress, nutrient deprivation, and also by glucose- or other nutrient-induced overload (3). The latter increases glucose shunt to hexosamine biosynthesis pathway (HBP), and this is emerging as major pathogenic factors for insulin resistance (2) and "glucose toxicity" during diabetes. Similar to phosphorylation, *O*-GlcNAc is widely distributed and highly dynamic (1, 2). However, phosphorylation depends upon multiple

Fernando Vivanco (ed.), *Heart Proteomics: Methods and Protocols*, Methods in Molecular Biology, vol. 1005, DOI 10.1007/978-1-62703-386-2_13,

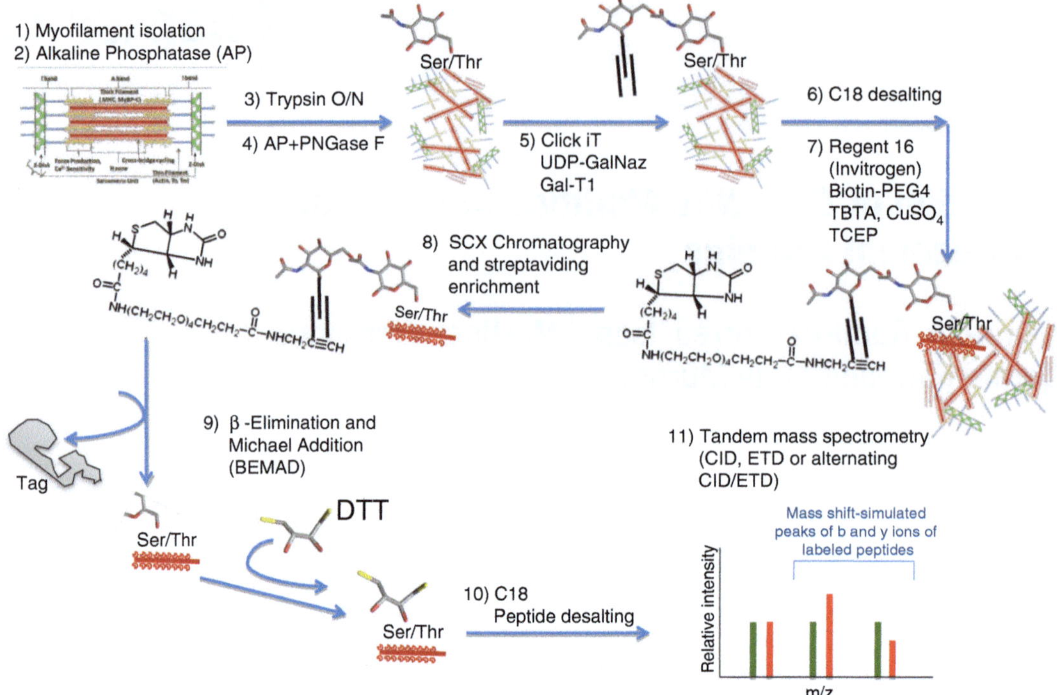

Fig. 1 Schematic representation of method used to detect and site-map *O*-GlcNAc in cardiac myofilaments. (1) Myofilament proteins were isolated as previously described (20, 24) and (2) dephosphorylated at whole protein level using Alkaline Phosphatase (Roche). (3) After 1D SDS-PAGE gel electrophoresis, prominent bands of myofilament proteins were excised, pooled, and trypsinized. (4) Dephosphorylation and removal of N-linked glycosylation, the peptides were treated with Alkaline Phosphatase and PNGaseF at 37 °C for 2 h. (5) Using Click Chemistry, GalNAz was attached to *O*-GlcNAc using a mutant (Y289L) of β-1, 4-galactosyltransferase (Gal-T1, Molecular Probes). (6) Peptides were desalted in a C18 column, and (7) a Biotin-PEG4 tag was attached by cycloaddition to GalNAz-O-GlcNA, and (8) peptides were passed through a SCX chromatography and enriched by streptavidin agarose beads. (9) To label the specific sites, mild β-elimination was performed on the streptavidin beads and DTT used to replace the GlcNAc-GalNAz-biotin (β-elimination and Michael addition or BEMAD). (10) The modified peptides were desalted again in C18 column and (11) subsequently identified by Finnigan LTQ mass spectrometer online coupled with a nano-LC, as reported by Wang et al. (11)

kinases and phosphatases, and, in contrast, *O*-GlcNAc is regulated only by two enzymes, i.e., OGT and OGA, which are highly conserved from multicellular eukaryotes to humans (4).

The presence of *O*-GlcNAc modifications in nucleocytoplasmic proteins was not discovered until 1984 (5). Since then *O*-GlcNAc has been identified in proteins from every cellular compartment and every functional classification. Site-specific mapping of *O*-GlcNAc modifications remains challenging mainly due to three factors: (1) The modification is easily loss during protein purification, (2) *O*-GlcNAc-modified peptides are suppressed or don't fly when ionized by matrix-assisted laser deionization (MALDI), and (3) enrichment of *O*-GlcNAcylated proteins and a more stable tagging is indispensable for consistent unambiguous detection and site-mapping (2). Fortunately, in the last decade new proteomic approaches and better instrumentation have advanced this field by providing tools for identification, enrichment, and site-mapping of *O*-GlcNAc-modified peptides (6–13). Currently *O*-GlcNAc modifications have been detected in more than 1,500 proteins, and the list grows continuously. Determination of in vivo *O*-GlcNAc stoichiometry in crude or enriched protein fraction was not possible until recently. This was achieved by attaching aminooxy-functionalized PEG (polyethylene glycol) derivatives to tag *O*-GlcNAc modifications by chemo-enzymatic (Click Chemistry) labeling with various well-defined molecular weights (2 kD and 5 kD)(14).

In this chapter we will focus on describing a general method that has been used successfully by more than one laboratory interested in detecting *O*-GlcNAc in myofilament proteins (15, 16). Alternative reagents for chemo-enzymatic or metabolic labeling will be indicated, as well as references for more details in alternative methods. The outline is divided into (1) "Enrichment of *O*-GlcNAc Stoichiometry," (2) "Cardiac Myofilament Protein Isolation," (3) "SDS-Polyacrylamide Gel Electrophoresis (SDS-PAGE)," (4) "Reduction and Alkylation," (5) "In-Gel Protein Digestion," (6) "Chemo-enzymatic Labeling of *O*-GlcNAc Moieties (Click Chemistry)," (7) "Biotin Alkyne Tagging," (8) "Strong Cation Exchange (SCX) and Streptavidin," and (9) "β-Elimination and Michael Addition (BEMAD) for *O*-GlcNAc Site-Mapping" (see Fig. 1 for work flow).

2 Materials

All solutions should be prepared with the purest H_2O available (double distillated ultrapure H_2O 18 MΩ or HPLC-grade H_2O).

2.1 Enrichment of O-GlcNAc Stoichiometry

1. NAG-thiazoline(1,2-dideoxy-2′-methyl-alpha-D-glucopyranoso (2, 1-d)-Delta2′-thiazoline) (17).
2. Thiamet-G (2-(ethylamino)-3aR, 6S, 7R, 7aR-tetrahydro-5R-(hydroxymethyl)-5H-pyrano (3, 2-d) thiazole-6, 7-diol from Cayman Chemicals Inc (18).
3. Adult or neonatal ventricular myocytes in cell culture (DMEM or M199).
4. Phosphate-buffered saline, PBS:137 mM NaCl, 2.7 mM KCl, 10 mM Na_2 HPO_4, 2 mM KH_2 PO_4.
5. Dimethyl sulfoxide (DMSO) or dimethylformamide (DMF).
6. 3 ml syringes and 27 gauge needles.

2.2 Cardiac Myofilament Protein Isolation

1. PUGNAc (*O*-(2-acetamido-2-deoxy-d-gluropyranosylidene) amino *N*-phenylcarbamate).
2. Protease inhibitor cocktail (Mini-EDTA free, Roche).
3. Sterile 2.0 ml microcentrifuge tube.
4. Two-dimensional lysis buffer (2D LB): 6 mM urea, 2 mM thiourea, 4 % CHAPS, 30 mM Tris–HCl pH 8.8.
5. Ammonium bicarbonate (NH_4HCO_3) 40 mM, pH 7.8.
6. 2DQuant Kit (GE Health Care).
7. BCA-RAC (Thermo Scientific).
8. Sonic Dismembrator 60 (Fisher Scientific).
9. Porcelain mortar and pestle.
10. Liquid nitrogen.
11. Protective cloth ware: goggles and/or face shield, cryogenic liquid handling gloves (Cryo-Gloves™).

2.3 SDS-PAGE, Reduction and Alkylation, and In-Gel Protein Digestion

1. $MgCl_2$ 1 mM.
2. Alkaline phosphatase (Roche).
3. SDS-PAGE loading buffer: 50 mM Tris-HCl pH 6.8, 100 mM DTT, 5 % SDS, 0.05 % bromophenol blue, 10 % glycerol.
4. NuPAGE 4–12 % Bis-Tris Mini-Gel (Invitrogen).
5. Coomassie blue (Safe Stain, Invitrogen).
6. Sterile cutter or scalpels.
7. Sterile 1.5 ml microcentrifuge tube.
8. Acetonitrile (HPLC grade).
9. Alkylation solution: 50 mM iodoacetamide in 50 mM NH_4HCO_3, pH 8.5.
10. Trypsin (Sequence grade, Promega).
11. Acetic acid solution 10 % (v/v).

2.4 Chemo-enzymatic Labeling of O-GlcNAc Moieties (Click Chemistry)

1. PNGase F (New England Biolabs).
2. SDS-HEPES buffer: 1 % SDS in 20 mM HEPES, pH 7.9.
3. HEPES buffer: 10 mM HEPES buffer, pH 7.9.
4. SDS-Tris buffer: 1 % SDS in 50 mM Tris–HCl pH 8.0.
5. Click-iT™ *O*-GlcNAc enzymatic labeling buffer 2.5×: 125 mM NaCl, 50 mM HEPES, 5 % NP-40, pH 7.9 (Invitrogen).
6. $MnCl_2$, 100 mM.
7. UDP-GalNAz (Invitrogen).
8. Gal-T1 (mutant enzyme Y289L) (Invitrogen) or 20 μg/ml of homemade recombinant enzyme.
9. α-crystallin.

10. MacroSpin Column (SCX) (the Nest Group, Inc., MA).
11. TCEP (Tris(2-carboxyethyl)phosphine hydrochloride) Sigma-Aldrich.
12. TBTA (Tris((1-benzyl-1H-1, 2,3-triazol-4-yl)methyl)amine) Sigma-Aldrich.
13. Cycloaddition reaction buffer: 2 mM of TCEP, 2 mM TBTA, 2 mM $CuSO_4$ in HPLC-grade water.
14. Biotin Alkyne (PEG4 carboxamide-propargyl biotin 1 mg) from Invitrogen.
15. *N*-(Aminooxyacetyl)-*N'*-biotinyl-hydrazine (30 mM aqueous stock) from Dojindo, MD.
16. Aminooxy-biotin (Biotium, CA).
17. $CuSO_4$, 40 mM.
18. Methanol HPLC grade.
19. Chloroform HPLC grade.

2.5 β-Elimination and Michael Addition for O-GlcNAc Site-Mapping

1. Synthetic peptide as controls, such as
   c-Myc peptide, KKFELLPT (*O*-GlcNAc) PPLSPSRR
   BPP peptide, PSVPVS(*O*-GlcNAc) GSAPGR (19).
2. C18 reversed-phase MacroSpin Columns (75–150 μl) (the Nest Group, Inc., MA).
3. TFA (trifluoroacetic acid 99 %), HPLC grade.
4. Mild BEMAD solution: 1 % triethylamine, 0.1 % NaOH, 10 mM DTT.
5. C18 capillary column: 5 μm C18, 6 cm × 75 μm ID × 375 μm OD.
6. Mobile phase A: 0.1 % formic acid.
7. Mobile phase B: 80 % acetonitrile in 0.1 % formic acid.

3 Methods

3.1 Enrichment of O-GlcNAc Stoichiometry

O-GlcNAcase (OGA) is the only mammalian known enzyme that removes *O*-GlcNAc. The stoichiometry of *O*-GlcNAc modification can be efficiently increased by inhibiting OGA. So far, the most potent and specific OGA inhibitor is Thiamet-G (18) (Cayman Chemicals Inc). Cellular and in vivo studies are highly recommended to inhibit OGA before analysis. Thiamet-G is water soluble and stable at a wide range of pH, including physiologic (7.4–7.6). The most important advantage of this inhibitor is that it can be added to cells in culture or administered to animals via tail vein or intraperitoneal injection and it crosses the blood–brain barrier.

1. Keep Thiamet-G (FW 248.3, Cayman Chemicals Inc) at −20 °C protected from humidity. To prepare a stock of 80 mM, dissolve 20 mg/ml in DMSO or DMF and keep at −20 °C. To prepare organic-free solutions, dissolve 3 mg/ml in PBS, pH 7.2–7.4 (12 mM Thiamet-G stock). Once diluted in physiological solution, do not use for more than 1 day.
2. Add to cell culture media 0.1–1 μM and culture for a minimum of 4–18 h.
3. For in vivo dilute Thiamet-G in drinking water (200 mg/kg/day) and harvest heart after 1 day. Alternatively, deliver Thiamet-G (50 mg/kg) by i.v. injection and harvest heart after 10–16 h.

3.2 Cardiac Myofilament Protein Isolation

1. Prechill porcelain mortar, pestle, and microcentrifuge tubes in liquid nitrogen. Place rat heart (~200 mg) or mouse heart in the mortar and grind the tissue (see Note 1).
2. Washout excess blood 2× in 40 ml of ice-cold PBS+ protease inhibitor cocktail (Mini-EDTA free, Roche) + 10 μM PUGNAc in 50 ml Falcon.
3. Decant into a pre-weighted ice-cold 2 ml Eppendorf tubes. Weigh again and calculate tissue mass. Add 10 volumes (volume/weight) of 2D lysis buffer (6 M urea, 2 M thiourea, 4 % CHAPS, 30 mM Tris–HCl pH 8.8) (20).
4. Homogenize the mixture by sonication (6–10 pulses at 2.5–3 W) using a Sonic Dismembrator 60 (Fisher Scientific).
5. Incubate on an end-to-end rocking platform at 4 °C for 1 h.
6. Repeat steps 4 and 5.
7. Repeat step 4 and centrifuge 16,200 × g × 30 min.
8. Transfer supernatant to new tubes, and if they are not to be used immediately, flash-freeze the samples in liquid nitrogen and store at −80 °C until needed.
9. Measure protein concentration with a 2DQuant Kit (Roche) or other reducing agent compatible kit (BCA-RAC Thermo Scientific), 2D lysis buffer compatible kit (see Note 2). Adjust concentration to 2–5 mg/ml in 40 mM ammonium bicarbonate (pH 7.8).

3.3 SDS-Polyacrylamide Gel Electrophoresis

1. Dephosphorylate samples (~200–250 μg) and a positive control (α-crystallin 20–40 μg) by dissolving in 40 mM NH_4HCO_3 and 1 mM $MgCl_2$ with alkaline phosphatase (1 unit/10 μl) for 3 h at 37 °C.
2. For each sample, mix 100–200 μg total protein and 20–40 μg of α-crystallin in a volume of 10–15 μl with 5 μl of 5 % SDS-PAGE loading buffer (see Note 3).

3. Heat the sample at 95–100 °C for 5 min.
4. Allow the samples to cool at room temperature.
5. Load the samples onto a mini-gel and perform electrophoresis at 200 V.
6. Upon completion of electrophoresis, transfer the gel into water in a covered container (see Note 4).
7. Rinse the gel in H_2O briefly twice.
8. Stain the gel with colloidal Coomassie blue.

3.4 Reduction and Alkylation

1. Cut the gel bands in small pieces (1 × 1 mm) with a sterile cutter or scalpel (see Note 5).
2. Transfer the excised gel bands to a sterile 1.5 ml microcentrifuge tube.
3. Destain the excised gel bands with a 1:1 solution of 50 mM NH_4HCO_3/acetonitrile, and incubate with occasional vortexing for a minimum period of 30 min (depends on the staining).
4. Dehydrate the excised gel bands by adding 500 μl of 100 % acetonitrile, and incubate with occasional vortexing, until gel bands get white and shrink (~10–15 min), then remove acetonitrile.
5. Swell the excised gel bands with 1 ml of reducing solution (10 mM DTT in 50 mM NH_4HCO_3, pH 8.5). Incubate for 60 min at 60 °C.
6. Cool the reaction at room temperature.
7. Discard the reducing solution.
8. Alkylate the sample by adding 1 ml of alkylation solution (50 mM iodoacetamide in 50 mM NH_4HCO_3, pH 8.5), and incubate at room temperature, protected from light for 30 min.

3.5 In-Gel Protein Digestion

1. Wash the excised gel bands with H_2O first followed by 50 % acetonitrile for 15 min each. Repeat this procedure once.
2. Dehydrate the excised gel bands with 100 % acetonitrile over 15 min.
3. Use prechilled 50 mM NH_4HCO_3 to dissolve one vial of trypsin at a concentration of 10–15 ng/μl (see Note 6).
4. Immerse the dehydrated excised gel bands in freshly dissolved trypsin and keep on ice. Use enough volume of trypsin to entirely cover the excised gel band.
5. Incubate the samples in each tube over night at 37 °C (see Note 7).
6. Remove excess liquid. Add another equal volume of 50 mM NH_4CO_3 (without trypsin).

7. Incubate the samples in each tube at 37 °C for 12–14 h.
8. Stop the digestion reaction with 1:5 (v/v) 10 % acetic acid.
9. Transfer the supernatant to a sterile 1.5 ml microcentrifuge tube.
10. Extract the peptides remaining in the gel matrix by incubating it first with 0.1 % acetic acid, followed by 0.1 % acetic acid in 50 % acetonitrile, and finally with 100 % acetonitrile.
11. Transfer the peptides from the above wash and dehydration steps to the same tube where the supernatant was transferred.
12. Dry samples by vacuum centrifugation.

3.6 Chemo-enzymatic Labeling of O-GlcNAc Moieties (Click Chemistry)

1. Resuspend peptides in 40 mM NH_4HCO_3 and 1 mM $MgCl_2$ and add alkaline phosphatase (1 unit/10 μl) and PNGase F (25 units/2 μl) for 4 h at 37 °C.
2. For enzymatic labeling make a 0.5 mM UDP-GalNAz solution by reconstituting a 50 μg UDP-GalNAz stock with 144 μl of 10 mM HEPES, pH 7.9.
3. Add 11 μl of $MnCl_2$ (100 mM) to the sample, vortex, and centrifuge briefly.
4. Add 30 μl of UDP-GalNAz (Invitrogen), and mix by pipeting up and down (see Note 8).
5. Remove 50 μl of the sample and place in a fresh tube (this is the negative control).
6. Add 15 μl of Gal-T1 (Y289L).
7. For a positive control of enzymatic labeling, make a 5 μg/μl α-crystallin solution (α-crystallin 250 μg is a positive control) by adding 50 μl of 1 % SDS in 20 mM HEPES, pH 7.9.
8. Set up at new 1.5 ml tube, in tube one add 4 μl of α-crystallin control, 4.5 μl of ultrapure water, 8 μl of Click-iT™ *O*-GlcNAc enzymatic labeling buffer (2.5×), 1.5 μl of $MnCl_2$ (100 mM), vortex briefly, and centrifuge briefly.
9. Add 1 μl of UDP-GalNAz (0.5 mM), and mix by pipeting up and down.
10. Add 1 μl of Gal-T1 (Y289L), and mix by pipeting up and down. (This is the negative control.)
11. Incubate the reactions, including your negative control, after covering the tube in aluminum foil at 4 °C overnight (12–14 h).
12. Store samples at −20 °C until analyzed or proceed to clean up and tag labeling.

3.7 Biotin Alkyne Tagging

1. Clean up samples by passing them through a C18 reversed-phase MacroSpin Columns (75–150 μl) following manufacturer's instructions (the Nest Group, Inc., MA).

2. Elute samples in 80 % acetonitrile and dry samples by vacuum centrifugation.
3. Prepare the cycloaddition reaction buffer (2 mM of TCEP, 2 mM TBTA, 2 mM $CuSO_4$ in HPLC-grade water) (see Note 9).
4. Reconstitute 1 mg of Biotin-PEG4 Alkyne (Invitrogen) in 63 μl of DMSO to make a 30 mM stock (see Note 10).
5. Reconstitute the dry samples in 18 μl of cycloaddition reaction buffer (2 mM of TCEP, 2 mM TBTA, 2 mM $CuSO_4$), and add 2 μl of Biotin-PEG4 Alkyne (3 mM final concentration) (see Note 11).
6. Set up GalNAz labeled α-crystallin as a positive control for Biotin Alkyne labeling reaction in a 1.5 ml Eppendorf as follows: 5 μl of peptide/protein mixture (maximum 20 μg) diluted in 1 % SDS, 50 mM Tris–HCl pH 8.0, 10 μl of 2× Click-iT Reaction Buffer, 2 μl of Biotin Alkyne, and 1 μl of H_2O.
7. Close the tube and vortex for 5 s.
8. Add 1 μl of $CuSO_4$ (40 mM) and vortex for 5 s.
9. Add 1 μl of Click-iT Reaction Buffer Additive 2 and vortex for 5 s. (The solution should turn bright orange.)
10. Incubate the cycloaddition reaction mixture for 12 h with gentle shaking at room temperature. (This is typically performed overnight and the tube is protected from light with aluminum foil.)

3.8 Strong Cation-Exchange Chromatography and Streptavidin Enrichment

1. Dilute the samples in up to 200 μl of strong cation-exchange (SCX) loading buffer (5 mM KH_2 PO_4, 10 % acetonitrile, pH 3.0) for MicroSpin Columns (the Nest Group Inc).
2. Condition and equilibrate the column following manufacturer's instructions.
3. Elute samples in one fraction by releasing peptides in high-salt SCX buffer (5 mM KH_2 PO_4, 10 % acetonitrile, and 300 mM KCl, pH 3.0) by spinning at 110×*g* for 1 min.
4. Incubate eluted samples with agarose-conjugate streptavidin (Thermo Scientific) for 2 h.
5. Wash the agarose-conjugate streptavidin beads extensively with low-salt buffer (1 mM Na_2HPO_4, 150 mM NaCl, 1 % Triton X-100, 0.5 % sodium deoxycholate, and 0.1 % SDS, pH 7.5). Use three times more volume than recommended by manufacturer's instructions.

3.9 β-Elimination and Michael Addition for O-GlcNAc Site-Mapping

1. Spike a sample with 10 or 20 pmoles of control peptide (c-Myc, BPP peptide, or commercially available *O*-GlcNAcylated peptide controls)
2. Add 200 μl of mild β-elimination and Michael addition (BEMAD) solution (1.5 % triethylamine, 20 mM DTT, pH

adjusted to 12.0–12.5 with NaOH). Add up to 0.1 % of NaOH or add triethylamine if necessary. For more details, consult reference (21).

3. Incubate the reaction at 54 °C for 4 h with shaking.
4. Add 40 μl of TFA (2 % final concentration) to neutralize the pH and stop the reaction.
5. The samples are ready to be desalted by C18 and analyzed by high-accuracy LC/MS/MS. A protocol is described below.

3.10 LTQ-MS/MS

The authors used a Finnigan LTQ mass spectrometer online coupled with a nano-LC to identify labeled peptides. The experimental conditions are as reported previously (16).

1. Clean up samples by removing free DTT and salts by use of a C18 capillary column such as a 5 μm C18, 6 cm × 75 μm ID × 375 μm OD at a flow rate of 300 nl/min. The mobile phase A: 0.1 % formic acid and mobile phase B: 80 % acetonitrile in 0.1 % formic acid were run in a linear gradient from 5 % A to 100 % B in 20 min.
2. One full MS scan (350–1,800 m/z) was followed by fragmentation of precursor ions. MS/MS scans were done only on the top eight most intensive ions. A dynamic exclusion was set (repeat count 2, repeat duration at 30 s, exclusion duration at 60 s; spray voltage, 2.1 kV).
3. MS data were queried by Mascot software against the nonredundant Swiss-Prot 50.4 Rattus database. The search parameters were set to allow a maximum of one missed cleavages, and variable modifications accounted were Met oxidation (+32 Da), DTT modifications on Ser/Thr (+136.2 Da), overall peptide tolerance (±1.5 Da), and MS/MS tolerance (±0.8 Da). Output peptides cutoff Mascot score is 32 (see Note 11).

4 Notes

1. It is best not to use more than 200 mg of the preparation; otherwise, it will be necessary to use several Eppendorf or a 15 ml Falcon tube to dissolve the tissue in 10 volumes of lysis buffer. Flash-freeze in liquid nitrogen and pulverize in mortar. Store at −80 °C until needed.
2. Typical myofilament protein yield is 10–15 mg/ml; if nonreducing agent compatible protein quantification reagents are at hand, dilute samples 1:10 or 1:100 in 40 mM ammonium bicarbonate (pH 7.8), and determine protein concentration.
3. SDS-PAGE loading buffer should contain at least 5 % SDS to ensure appropriate solubility of myofilament proteins.

4. Once the gel electrophoresis is completed, extreme care should be taken when handling the gel in order to diminish the possible contamination of the sample with traces of keratin. For this reason, work in a clean hood; use sterile materials and powder-free gloves.
5. For a nonbiased alternative approach, SDS-PAGE can be avoided, and myofilaments can be trypsinized directly in solution (keeping a protease: protein ratio of 1:10 to 1:100) before proceeding to "Reduction and Alkylation."
6. Use fresh trypsin dissolved in cold buffer. Keep this solution on ice at all times in order to prevent unwanted trypsin autoproteolysis.
7. You can shorten the digestion time down to 30 min (22). To optimize the digestion, it is best to perform a time-course trypsin incubation and analyze for missed cleavage on MS/MS.
8. This is about twofold in excess than recommended in manufacturer's instructions.
9. Prepare a fresh TCEP 50 mM stock in water. Prepare a fresh TBTA 10 mM stock in DMSO. Prepare a fresh $CuSO_4$ 50 mM stock in water ($CuSO_4 \cdot H_2O$). *Freshly prepared additives for cycloaddition reaction buffer are critical for the success of the Biotin Alkyne tagging.
10. Other biotin derivates that are less expensive and serve to tag the ketone derivate (GalNAz) Aminooxy-biotin (from Biotium, CA) and *N*-(Aminooxyacetyl)-*N'*-biotinyl-hydrazine (30 mM aqueous stock from Dojindo, MD) were utilized at 2.75–4 mM in the early studies using noncommercial Click-iT chemistry reagents (7, 8, 23).
11. This is also about threefold more than recommended by the manufacturer.
12. All MS/MS spectra from DTT modified peptides should be confirmed manually. A good control for specificity will include a sample pretreated with β-N-Acetyl-hexosaminidase$_f$ (NEB) to remove *O*-GlcNAc prior to enrichment.

References

1. Hart GW, Housley MP, Slawson C (2007) Cycling of O-linked beta-N-acetylglucosamine on nucleocytoplasmic proteins. Nature 446:1017–1022
2. Hart GW, Slawson C, Ramirez Correa GA, Lagerlof O (2011) Cross talk between O-GlcNAcylation and phosphorylation: roles in signaling, transcription, and chronic disease. Annu Rev Biochem 80:825–858
3. Zachara NE (2012) The roles of O-linked β-N-acetylglucosamine in cardiovascular physiology and disease. Am J Physiol Heart Circ Physiol 302:1905–1918
4. Rao FV, Dorfmueller HC, Villa F, Allwood M, Eggleston IM, van Aalten DM (2006) Structural insights into the mechanism and inhibition of eukaryotic O-GlcNAc hydrolysis. EMBO J 25:1569–1578

5. Torres CR, Hart GW (1984) Topography and polypeptide distribution of terminal N-acetylglucosamine residues on the surfaces of intact lymphocytes. Evidence for O-linked GlcNAc. J Biol Chem 259:3308–3317
6. Greis KD, Hart GW (1998) Analytical methods for the study of O-GlcNAc glycoproteins and glycopeptides. Methods Mol Biol 76:19–33
7. Hang HC, Bertozzi CR (2001) Ketone isosteres of 2-N-acetamidosugars as substrates for metabolic cell surface engineering. J Am Chem Soc 123:1242–1243
8. Khidekel N, Arndt S, Lamarre-Vincent N, Lippert A, Poulin-Kerstien KG, Ramakrishnan B et al (2003) A chemoenzymatic approach toward the rapid and sensitive detection of O-GlcNAc posttranslational modifications. J Am Chem Soc 125:16162–16163
9. Roquemore EP, Chou TY, Hart GW (1994) Detection of O-linked N-acetylglucosamine (O-GlcNAc) on cytoplasmic and nuclear proteins. Methods Enzymol 230:443–460
10. Vocadlo DJ, Hang HC, Kim EJ, Hanover JA, Bertozzi CR (2003) A chemical approach for identifying O-GlcNAc-modified proteins in cells. Proc Natl Acad Sci USA 100:9116–9121
11. Wang Z, Pandey A, Hart GW (2007) Dynamic interplay between O-linked N-acetylglucosaminylation and glycogen synthase kinase-3-dependent phosphorylation. Mol Cell Proteomics 6:1365–1379
12. Wells L, Vosseller K, Cole RN, Cronshaw JM, Matunis MJ, Hart GW (2002) Mapping sites of O-GlcNAc modification using affinity tags for serine and threonine post-translational modifications. Mol Cell Proteomics 1:791–804
13. Zachara NE, Cheung WD, Hart GW (2004) Nucleocytoplasmic glycosylation, O-GlcNAc: identification and site mapping. Methods Mol Biol 284:175–194
14. Rexach JE, Clark PM, Hsieh-Wilson LC (2008) Chemical approaches to understanding O-GlcNAc glycosylation in the brain. Nat Chem Biol 4:97–106
15. Hedou J, Bastide B, Page A, Michalski JC, Morelle W (2009) Mapping of O-linked beta-N-acetylglucosamine modification sites in key contractile proteins of rat skeletal muscle. Proteomics 9:2139–2148
16. Ramirez-Correa GA, Jin W, Wang Z, Zhong X, Gao WD, Dias WB et al (2008) O-linked GlcNAc modification of cardiac myofilament proteins: a novel regulator of myocardial contractile function. Circ Res 103:1354–1358
17. Whitworth GE, Macauley MS, Stubbs KA, Dennis RJ, Taylor EJ, Davies GJ, Greig IR, Vocadlo DJ (2007) Analysis of PUGNAc and NAG-thiazoline as transition state analogues for human O-GlcNAcase: mechanistic and structural insights into inhibitor selectivity and transition state poise. J Am Chem Soc 129:635–644
18. Yuzwa SA, Macauley MS, Heinonen JE, Shan X, Dennis RJ, He Y et al (2008) A potent mechanism-inspired O-GlcNAcase inhibitor that blocks phosphorylation of tau in vivo. Nat Chem Biol 4:483–490
19. Greis KD, Hayes BK, Comer FI, Kirk M, Barnes S, Lowary TL, Hart GW (1996) Selective detection and site-analysis of O-GlcNAc-modified glycopeptides by beta-elimination and tandem electrospray mass spectrometry. Anal Biochem 234:38–49
20. Yuan C, Guo Y, Ravi R, Przyklenk K, Shilkofski N, Diez R, Cole RN, Murphy AM (2006) Myosin binding protein C is differentially phosphorylated upon myocardial stunning in canine and rat hearts—evidence for novel phosphorylation sites. Proteomics 6:4176–4186
21. Whelan SA, Hart GW (2006) Identification of O-GlcNAc sites on proteins. Methods Enzymol 415:113–133
22. Shevchenko A, Tomas H, Havlis J, Olsen JV, Mann M (2006) In-gel digestion for mass spectrometric characterization of proteins and proteomes. Nat Protoc 1:2856–2860
23. Khidekel N, Ficarro SB, Peters EC, Hsieh-Wilson LC (2004) Exploring the O-GlcNAc proteome: direct identification of O-GlcNAc-modified proteins from the brain. Proc Natl Acad Sci USA 101:13132–13137
24. Murphy AM, Solaro RJ (1990) Developmental difference in the stimulation of cardiac myofibrillar Mg2(+)-ATPase activity by calmidazolium. Pediatr Res 28:46–49

Chapter 14

Quantification of Mitochondrial S-Nitrosylation by CysTMT^6 Switch Assay

Christopher I. Murray, Hea Seung Chung, Helge Uhrigshardt, and Jennifer E. Van Eyk

Abstract

S-nitrosylation (SNO) is an important oxidative posttranslational modification in the regulation of cardiac mitochondria. SNO modification of several mitochondrial proteins has been associated with cardiac preconditioning and improved cell survival following ischemia/reperfusion injury. Due to their labile nature, SNO modifications are challenging to study using traditional biochemical techniques; particularly, the identification of individual modified cysteine residues. Here, we describe the details of the cysTMT^6 switch assay, a variation of the classic biotin switch protocol. The cysTMT^6 reagent provides a simplified and powerful approach to SNO detection by combining unambiguous identification of the modified cysteine residue and relative quantification of up to six samples by mass spectrometry analysis.

Key words S-nitrosylation, Biotin switch assay, Cysteine-reactive tandem mass tags, Quantitation, Oxidative modification, Mitochondria

1 Introduction

S-nitrosylation (SNO), the reversible covalent addition of a nitrosonium (NO+) to a cysteine thiol, plays an important role in the regulation and dysfunction of the cardiac mitochondria (1, 2). SNO modifications have been found on proteins involved in a variety of key pathways including oxidative phosphorylation (3–7), the tricarboxylic cycle (4), and fatty acid oxidation (4, 7). In addition, SNO modifications have been implicated as a protective mechanism in cardiac preconditioning and ischemia/reperfusion injury (7–10). Cardioprotection has been significantly linked to mitochondria and the preservation of its function during periods of ischemia (11, 12), suggesting that SNO modifications of critical proteins in the mitochondria serve to resist the oxidative damage sustained during ischemia, preserving function (8).

Fernando Vivanco (ed.), *Heart Proteomics: Methods and Protocols*, Methods in Molecular Biology, vol. 1005, DOI 10.1007/978-1-62703-386-2_14,

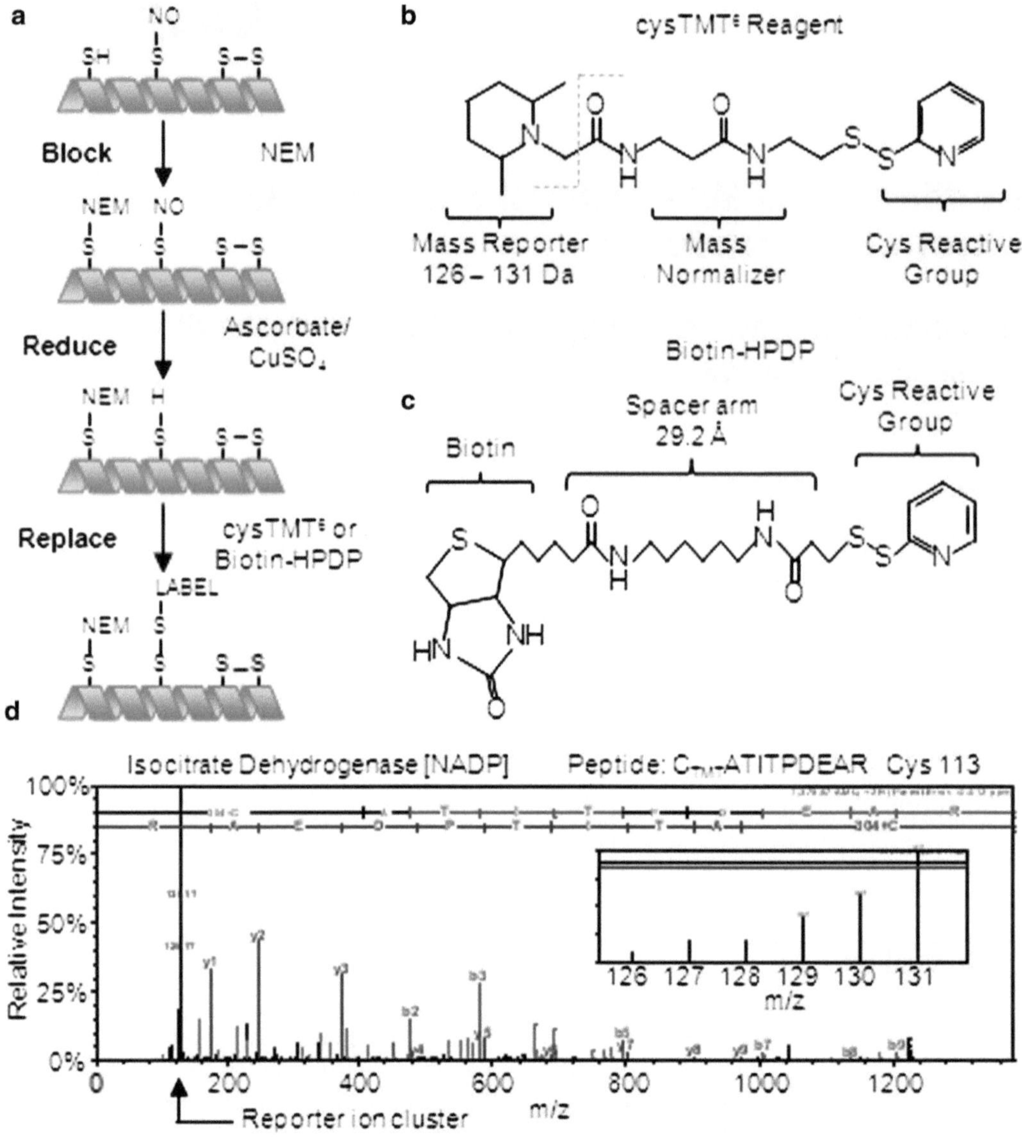

Fig. 1 Multiplex analysis of SNO-modified cysteine residues using the cysTMT[6] switch assay. (**a**) Reaction schema is presented outlining the basic steps in the biotin/cysTMT[6] switch assays including blocking, reduction, and labeling of the once modified cysteine residue with a stable moiety for enrichment or detection (14, 19). Structure of the cysTMT (**b**) and biotin-HPDP (**c**) reagents indicating both reagents have the same pyridyldithiol thiol-reactive group. Note the mass reporter region in the cysTMT molecule, this region fragments releasing a reporter ion of mass 126–131 Da. (**d**) Representative MS/MS spectrum identified as a peptide in isocitrate dehydrogenase (NADP), mitochondrial indicating SNO modification at Cys 113 as demonstrated by the presence of cysTMT label. The *inset* is a blow-up of the *m*/*z* region containing TMT reporter ions indicating the extent of labeling for each sample. In this case, cardiac rat mitochondrial lysates were treated as follows: 100 μM GSH (126 Da), 50 μM GSSG (127 Da), 2 μM GSNO (128 Da), 20 μM GSNO (129 Da), 50 μM GSNO (130 Da), and 100 μM GSNO (131 Da). The pattern demonstrates the specific and increasing labeling of Cys113 with increasing GSNO treatments

The labile nature of these modifications makes detection by traditional biochemical techniques challenging. In 2001, Jaffrey and Snyder introduced the biotin switch technique which utilizes a replacement strategy to add a stable biotin moiety in place of the SNO modification (see Fig. 1) (13, 14). To achieve this, free cysteines are blocked using a thiol-reactive agent like NEM, and then the SNO modifications are specifically reduced by ascorbate. Newly exposed thiols are then labeled with biotin-HPDP which allows for easy detection and enrichment with streptavidin. Since the initial introduction of this method, several variations have emerged which have improved and expanded on the original approach (15–20). Recently, a multiplex version of the biotin switch assay was presented, using the cysteine-reactive tandem mass tag (cysTMT) reagent in place of biotin (19). CysTMT provides an enrichable permanent mass tag that can be detected in the MS analysis. This eliminated any ambiguity in the assignment of SNO-modified sites that had previously existed due to nonspecific binding of peptides in the capture process or for peptides with more than one cysteine. Additionally, because of the multiplex nature of the cysTMT^6 label, it was possible to perform relative quantification on the extent of SNO at each modified cysteine residue. Samples can be labeled with one of the six isotopically balanced tags that will fragment the MS2 analysis releasing a reporter ion (126–131 Da) alongside the fragmented peptide ions. When the differently labeled samples are combined, the reporter ion intensities indicate the relative amount of each sample in the analysis. This approach allowed the direct comparison of NO donor and control treated samples to assess the unintended ascorbate reduction of disulfide bonds or other cysteine modifications (19).

This protocol chapter outlines the steps and points of concern involved for multiplex analysis of SNO modification in cardiac mitochondria. The protocol focuses on the treatment of isolated mitochondria; however, it can be adapted for the detection and comparison of endogenous SNO modifications between control and stimulated conditions.

2 Materials

2.1 Stock Solutions

Prepare all stock solutions using Chelex 100-treated ultrapure water (unless otherwise noted) and analytical grade reagents (see Note 1). Store all reagents at room temperature unless otherwise indicated.

1. 250 mM HEPES pH 7.4, Chelex 100 treated (see Note 2), store light protected at 4 °C.
2. 25 % SDS (w/v).
3. 10 % (w/v) 3-((3-cholamidopropyl) dimethylammonio]-1-propanesulfonate (CHAPS), store at −20 °C.

4. 100 mM EDTA, store at 4 °C.
5. 240 mM neocuproine dissolved in MeOH, store at −20 °C.
6. 1 M mannitol.
7. 2 M sucrose.
8. 100 mM EGTA.
9. 0.5 M KCl.
10. 1.5 M NaCl.
11. 50 mM Na-pyruvate.
12. 20 mM ADP.
13. 0.5 M KH_2PO_4.
14. 0.1 M $MgCl_2$.
15. 50 mM $CaCl_2$.
16. 10 % (w/v) RapiGest (Waters) dissolved in ultrapure water, store at 4 °C.

2.2 Mitochondrial Treatment Components

1. Intact mitochondria isolated from cardiac tissue, prepared fresh, or stored in cryo-medium (200 mM mannitol, 50 mM sucrose, 1 mM EGTA, 1 % (w/v) BSA, 10 % (v/v) DMSO, 10 mM HEPES, pH 7.4) (21–23) (see Note 3).
2. Isolation buffer (IB): 200 mM mannitol, 50 mM sucrose, 1 mM EGTA, 10 mM HEPES, pH 7.4.
3. Incubation buffer (ICB): 135 mM KCl, 10 mM NaCl, 5 mM Na-pyruvate, 2 mM ADP, 0.5 mM KH_2PO_4, 0.5 mM $MgCl_2$, 2 mM $CaCl_2$, 20 mM HEPES, pH 7.4.
4. NO-donor stock (see Notes 4 and 5).
5. Zeba spin desalting column (Thermo Fisher Scientific).

2.3 CysTMT[6] Switch Assay Components (Prepare from Stock Solutions Immediately Before Use)

1. HEN buffer: 250 mM HEPES pH 7.4, 1 mM EDTA, and 0.1 mM neocuproine.
2. 400 mM N-ethylmaleimide (NEM) dissolved in EtOH, prepare fresh immediately before use.
3. Blocking buffer: HEN including 2.5 % (w/v) SDS and 20 mM NEM.
4. Acetone stored at −20 °C.
5. Labeling buffer: HEN including 1.0 % (w/v) SDS.
6. 4 mM cysTMT isotopic labels 126–131 (Thermo Fisher Scientific), resuspend each 0.2 mg tube with 20 μL of acetonitrile (ACN), vortex, and collect liquid to the bottom of tube using bench top centrifuge. Dilute volume up to 120 μL with HENS. Resuspend immediately before use and keep light protected during handling.
7. 50 mM ascorbate dissolved in ultrapure water.
8. 50 mM $CuSO_4$ dissolved in ultrapure water.

9. Sequencing grade trypsin (or other protease) dissolved at 0.5 μg/μL in TBS: 50 mM Tris–HCl pH 7.4, 150 mM NaCl.
10. 50 mM phenylmethanesulfonylfluoride (PMSF) dissolved in isopropanol.

2.4 CysTMT⁶ Switch Peptide Enrichment Components

1. Immobilized anti-TMT resin (Thermo Fisher Scientific).
2. TBS: 50 mM Tris–HCl pH 7.4, 150 mM NaCl.
3. TBS containing 4 M urea.
4. Empty spin column tubes (10 mL).
5. End-over-end shaker.
6. 50 % (v/v) ACN/0.4 % (v/v) trifluoroacetic acid (TFA) in ultrapure water.
7. Detergent removal columns (Thermo Fisher Scientific).
8. C_{18} UltraMicroSpin™ column (Nest Group).

3 Methods

3.1 Treatment of Cardiac Mitochondria

1. To prepare intact mitochondria, take or thaw one aliquot and centrifuge for 7 min at 8,000 × *g*, 4 °C. The supernatant which should be colorless can be discarded.
2. To remove residual BSA and DMSO from the cryo-medium and any potentially ruptured mitochondria, carefully resuspend the reddish-brown pellet with 1 mL of ice-cold IB and centrifuge for 7 min at 8,000 × *g*, 4 °C and discard the supernatant.
3. Repeat step 2 with 1 mL of ice-cold ICB.
4. Carefully resuspend the pellet in 1 mL of ICB, and put on ice.
5. Remove 10 μL for protein determination, and then adjust the sample to 1 mg/mL protein with ICB (see Note 6).
6. For NO-donor treatment, mitochondria can be left intact or lysed. For lysis, remove an aliquot of 1 mg/mL mitochondria, and adjust to 1.0 % (v/v) Triton X-100, and incubate on ice for 5 min.
7. For each treatment remove 200–300 μg to a separate Eppendorf tube.
8. Add NO donor or control conditions to samples, and incubate for 20–30 min protected from light at room temp or 37 °C (see Note 7).
9. Following incubation, remove NO donor or control compounds. For intact mitochondria centrifuge for 3 min at 8,000 × *g*, and then wash the pellet in 1 mL of ICB containing 0.1 mM neocuproine (7 min, 8,000 × *g*, 4 °C). Lysed mitochondria can be passed through a Zeba spin desalting column equilibrated with HEN.

3.2 CysTMT[6] Switch Protocol

Cautionary note: SNO modifications are sensitive to light and need to be kept in the dark as much as possible until after labeling step (step 6).

1. For intact mitochondria, resuspend the pellet to 0.5 μg/μL in blocking buffer (400 μL for 200 μg of starting material). The pellet should completely dissolve in the buffer. For lysed mitochondria, prepare a blocking buffer with 2× the SDS and NEM, and dilute the sample 1:1 so the final concentration is 2.5 % (w/v) SDS and 20 mM NEM.
2. Incubate samples for 20 min at 50 °C, vortexing occasionally taking care not to foam the solution.
3. Remove any unreacted NEM by acetone precipitation. Add at least 5× volumes of cold acetone to each sample and vortex. Incubate sample for 20 min at −20 °C. A light colored suspension may be visible. Centrifuge for 10 min at 18,000 × *g*, 4 °C to pellet precipitate.
4. Aspirate acetone and carefully wash the pellet and walls of the tube with an additional 0.75 mL of cold acetone taking care not to disrupt the pellet.
5. Resuspend the pellet for labeling. The final concentration of all the labeling solution should be ~2–5 μg/μL protein, 1 mM cysTMT, ascorbate, and $CuSO_4$ in HEN containing 1 % (w/v) SDS. For a sample starting with 200 μg, the pellet can be resuspended in 71 μL of labeling buffer followed by the addition of 25 μL of 4 mM cysTMT, then 2 μL each of 50 mM ascorbate and 50 mM $CuSO_4$. Be sure to clearly indicate which sample is to be labeled with which of the cysTMT isotopic labels (126–131) (see Note 8).
6. Incubate samples for 2 h at 37 °C protected from light. After the completion of this incubation, samples are no longer light sensitive.
7. To remove excess cysTMT label, add 2 volumes of ultrapure water followed by 2 volumes (adjusted to the new volume) of cold acetone. For example, for a 100 μL labeling volume, add 200 μL of water followed by 600 μL of acetone. Incubate for 20 min at −20 °C and centrifuge for 10 min at 8,000 × *g*, 4 °C.
8. Aspirate acetone and carefully wash the pellet and walls of the tube with 2 × 0.75 mL of cold acetone.
9. Resuspend the pellet to 2–5 μg/μL in labeling buffer.
10. To be confidant of the equal loading and to compensate for any unequal losses that may have occurred during the procedure, it is recommended to determine the concentration of protein in each in the 6-plex (see Note 6).

11. Before proceeding to MS analysis, it is recommended to check the success of labeling by western blot. For this, remove 10 μg of each sample, and store at −20 °C for analysis. Refer to Subheading 3.4 for details on western blot analysis.
12. Combine equal amounts of protein from the 6-plex, and dilute up to 1.8 mL with TBS.
13. Pass samples through a Zeba spin desalting column equilibrated with TBS.
14. Adjust to 0.02 % RapiGest using stock solution.
15. Add 1 μg of trypsin for every 50 μg of total sample. For example, add 24 μg of trypsin to digest 1,200 μg for sample. Other proteases, like chymotrypsin, can be used. Be sure to consider the optimal conditions for each protease.
16. Digest overnight at 37 °C. If during the digest, protein begins to precipitate, increase the concentration of RapiGest, or add a minimum amount of another surfactant such, as SDS, to ensure solubility (see Note 9).
17. To halt digestion, add PMSF to a final concentration of 0.25 mM, and incubate for 20 min at room temp (see Note 10). Once digestion has been halted, samples can be stored at −80 °C until capture.

3.3 Capture of CysTMT-Labeled Peptides

1. Add 600 μL of TMT affinity beads to an empty spin column and equilibrate with 3 × 5 mL of TBS.
2. Dilute 6-plex cysTMT switch sample up to 5 mL with TBS, combine with beads, and seal column.
3. Incubate for 2 h at room temp on an end-over-end shaker.
4. Drain flow through, and wash beads 3 × 5 mL of TBS followed by 3 × 5 mL of TBS containing 4 M urea and then another 3 × 5 mL of TBS.
5. Add 5 mL of TBS and seal column.
6. Incubate for 2 h at room temp as in step 3 of Subheading 3.3.
7. Drain flow through, and wash with 3 × 5 mL TBS followed by 2 × 5 mL ultrapure water to reduce buffering capacity and salts.
8. Elute the bound cysTMT peptides with 2 × 600 μL of 50 % ACN, 0.4 % (v/v) TFA combining the elution fractions in an Eppendorf tube.
9. Dry the samples in a vacuum centrifuge (see Note 11).
10. Resuspend peptides in 50 mM ammonium acetate, and apply them to a detergent removal spin column following the manufacture's protocol.
11. Dry peptides in a vacuum centrifuge (see Note 11).
12. Resuspend peptides in 0.1 % (v/v) TFA, and apply to a C_{18} micro-column according to the manufacture's protocol. Elute with 50 % (v/v) ACN, 0.1 % (v/v) formic acid.

13. The samples are now ready for MS/MS analysis in an instrument with a mass range suitable for detecting peptide fragments and the reporter ions (126–131 Da) such as the Orbitrap Velos, Velos plus, Q-TOF, TOF-TOF, or triple quadrupole instruments.
14. After spectra have been acquired, database searching can be performed using cysTMT[6] and NEM as variable modifications. Reporter ion intensities can be harvested using a variety of software applications including Libra which is part of the trans-proteomic pipeline.

3.4 Analysis of CysTMT Switch Assay by Western Blot

1. Dilute 10 μg of cysTMT switch sample to 0.25 μg/μL of 1× LDS sample buffer without DTT (see Note 12).
2. Load 5 μg/lane on to a 4–12 % MES NuPAGE gel, and separate proteins according to the manufacture's protocol.
3. Transfer proteins to a nitrocellulose membrane for 1 h at 100 V in NuPAGE transfer buffer.
4. Stain membrane with Direct Blue 71; the detailed protocol can be found here (24) to observe equal loading of lanes.
5. Block membrane with 10 mL of blocking solution (5 % milk in TBS containing 0.1 % (v/v) Tween-20 (TBS-t)) for 1 h.
6. Incubate with anti-TMT antibody diluted 1:1,000 in blocking solution for 1 h.
7. Rinse membrane in TBS-t to remove excess primary antibody.
8. Incubate with anti-mouse secondary antibody diluted 1:40,000 in blocking solution for 1 h.
9. Wash membrane overnight in blocking solution at 4 °C. Perform at least four washing volume changes, the final two in TBS-t.
10. Expose membrane to chemiluminescent substrate and develop signal on X-ray film.
11. Differences in S-nitrosylation can be observed by differences in cysTMT labeling. If no difference is observed between experimental and control samples, try rerunning the gel with a reduced protein load.

4 Notes

1. To prepare Chelex 100 treated water, add 50 g/L of Chelex 100 resin to ultrapure water and stir for 1 h at room temp. Remove Chelex 100 resin with a filter. Store at 4 °C. Chelex 100 is used to remove bivalent transition metal ions which can reduce SNO modifications.

2. Dissolve HEPES powder in 0.5 L of water, and adjust pH to 7.4 by adding NaOH. Treat with Chelex 100 as in Note 1. Store protected from light at 4 °C.
3. This chapter describes a protocol for the analysis of SNO modifications induced in vitro in either intact or lysed mitochondria. It is also possible to extend this method to endogenous detection of SNO-modified proteins. To achieve this, mitochondria from a stimulated (or unstimulated) tissue source can be isolated and processed directly. Care must be taken to ensure samples are protected from light and that the buffers used in the purification procedure are free of free bivalent transition metals. Yields of endogenous SNO-modified proteins have been found to be less than in vitro stimulations (19).
4. A variety of NO donors and controls can be prepared to treat intact mitochondria or lysates depending on the goals of the experiment. S-nitrosoglutathione (GSNO), diethylamine NONOate (DEANO), *S*-nitroso-*N*-acetylpenicillamine (SNAP), and SNO-Cys are all common choices. Preparation depends on the individual properties of each reagent. GSNO, DEANO, and SNAP can all be obtained commercially and dissolved in ultrapure water, 10 mM NaOH, and DMSO, respectively, to make stock solutions that can be stored at −80 °C for 2–10 weeks. SNO-Cys can be prepared fresh by incubating 10 mM Cys, 10 mM $NaNO_2$, 10 mM HCl, and 1 mM EDTA for 1 h at RT protected from light. This results in a 5–7 mM SNO-Cys solution which can be quantified at 336 nm using a light spectrometer with an extinction coefficient of 900 M^{-1} cm^{-1} (25). Common control reagents include donor controls like glutathione in its reduced (GSH) and oxidized (GSSG) form or reduced Cys as well as decomposed versions of the NO donor (DEANO left at pH 7.4 and room temp for several days).
5. NO donors can be used at a range of concentrations. Physiological NO concentrations have been reported to be 1 μM or below (26, 27). Many biochemical and proteomic studies have used concentrations ranging from 1 to 100 μM NO donor; however, cysteines modified in the presence of excessively high NO-donor concentrations should be considered SNO modifiable and not assumed to be true physiological posttranslational modifications. Working stock solutions should be prepared (910–100×) immediately prior to use in HEN buffer.
6. Protein concentration can be determined using the bicinchoninic acid (BCA) protein assay (Pierce) or other compatible proteins assays.
7. SNO modifications are light sensitive and should be protected from light whenever possible during sample processing. Common strategies to limit light exposure include wrapping

samples or sample racks in tin foil, truing off/dimming laboratory lights, closing window shades, and working out of direct sunlight. Opaque Eppendorf tubes are also available; however, several steps require the careful washing of protein pellets from acetone precipitation. Failure to observe these pellets during washing may result in sample loss and reduced sensitivity.

8. Samples can be labeled with up to six different isotopic tags for multiplex analysis (126–131). Labeling a series of samples in reverse order for one or more of the replicates can be beneficial to assess the reproducibility or any bias in labeling.
9. RapiGest is an MS compatible detergent and is a preferred option to maintain protein solubility in the tryptic digest. If RapiGest is insufficient for solubility, the use of other detergents may be necessary. It is important to add the minimum required to achieve solubility to avoid contamination in the MS analysis. This amount will depend on the composition of the sample, but 0.05 % (w/v) SDS has been effective. The using of the detergent removal column (Subheading 3.3, step 10) will aid in cleaning the sample and should reduce contamination of any detergent added during tryptic digestion.
10. Tryptic digestion needs to be halted with PMSF or an equivalent inhibitor prior to the capture step to prevent proteolysis of the anti-TMT antibodies present on the affinity resin. Failure to inhibit proteolytic activity may reduce yield in the affinity capture.
11. When drying samples in a vacuum centrifuge, it is optimal to evaporate nearly but not all liquid from the sample tube. Overdrying a sample can result in peptides adhering to the tube walls making it difficult to resuspend them. It is preferable to leave approximately 5 μL of liquid in the tube to aid in resuspension.
12. The cysTMT reagent forms a mixed disulfide bond with the thiol side chain of the cysteine residue. In order to observe TMT labeling by western blot, samples must be run under nonreducing conditions. Inclusion of a reducing agent will remove the TMT label producing a negative result. Samples can be boiled for approximately 5 min in LDS sample buffer to promote solubilization.

Acknowledgement

The authors would like to thank John Rogers at Thermo Fisher Scientific for the generous gift of cysTMT materials. This work was supported by American Heart Association Pre-Doctoral Fellowship 0815145E to CIM and NIH grants P01 HL77180-0, N01-HV-28180, and P50 HL 084946-01 to JVE.

References

1. Jones SP, Bolli R (2006) The ubiquitous role of nitric oxide in cardioprotection. J Mol Cell Cardiol 40:16–23
2. Foster DB et al (2009) Redox signaling and protein phosphorylation in mitochondria: progress and prospects. J Bioenerg Biomembr 41:159–168
3. Borutaite V, Brown GC (2006) S-nitrosothiol inhibition of mitochondrial complex I causes a reversible increase in mitochondrial hydrogen peroxide production. Biochim Biophys Acta 1757:562–566
4. Chouchani ET et al (2010) Identification of S-nitrosated mitochondrial proteins by S-nitrosothiol difference in gel electrophoresis (SNO-DIGE): implications for the regulation of mitochondrial function by reversible S-nitrosation. Biochem J 430:49–59
5. Dahm CC et al (2006) Persistent S-nitrosation of complex I and other mitochondrial membrane proteins by S-nitrosothiols but not nitric oxide or peroxynitrite: implications for the interaction of nitric oxide with mitochondria. J Biol Chem 281:10056–10065
6. Burwell LS et al (2006) Direct evidence for S-nitrosation of mitochondrial complex I. Biochem J 394:627–634
7. Sun J et al (2007) Preconditioning results in S-nitrosylation of proteins involved in regulation of mitochondrial energetics and calcium transport. Circ Res 101:1155–1163
8. Burwell LS, Brookes PS (2008) Mitochondria as a target for the cardioprotective effects of nitric oxide in ischemia-reperfusion injury. Antioxid Redox Signal 10:579–599
9. Prime TA et al (2009) A mitochondria-targeted S-nitrosothiol modulates respiration, nitrosates thiols, and protects against ischemia-reperfusion injury. Proc Natl Acad Sci USA 106:10764–10769
10. Nadtochiy SM et al (2009) In vivo cardioprotection by S-nitroso-2-mercaptopropionyl glycine. J Mol Cell Cardiol 46:960–968
11. Halestrap AP et al (2007) The role of mitochondria in protection of the heart by preconditioning. Biochim Biophys Acta 1767: 1007–1031
12. Murphy E, Steenbergen C (2007) Preconditioning: the mitochondrial connection. Annu Rev Physiol 69:51–67
13. Jaffrey SR et al (2001) Protein S-nitrosylation: a physiological signal for neuronal nitric oxide. Nat Cell Biol 3:193–197
14. Jaffrey SR, Snyder SH (2001) The biotin switch method for the detection of S-nitrosylated proteins. Sci STKE, pl1
15. Hao G et al (2006) SNOSID, a proteomic method for identification of cysteine S-nitrosylation sites in complex protein mixtures. Proc Natl Acad Sci USA 103: 1012–1017
16. Forrester MT et al (2009) Proteomic analysis of S-nitrosylation and denitrosylation by resin-assisted capture. Nat Biotechnol 27: 557–559
17. Huang B, Chen C (2010) Detection of protein S-nitrosation using irreversible biotinylation procedures (IBP). Free Radic Biol Med 49:447–456
18. Paige JS et al (2008) Nitrosothiol reactivity profiling identifies S-nitrosylated proteins with unexpected stability. Chem Biol 15: 1307–1316
19. Murray CI et al (2011) Identification and quantification of S-nitrosylation by cysteine reactive tandem mass tag switch assay. Mol Cell Proteomics 11:M111.013441
20. Zhou X et al (2010) ESNOQ, proteomic quantification of endogenous S-nitrosation. PLoS One 5:e10015
21. Cavadini P et al (2002) Protein import and processing reconstituted with isolated rat liver mitochondria and recombinant mitochondrial processing peptidase. Methods 26:298–306
22. Storrie B, Madden EA (1990) Isolation of subcellular organelles. Methods Enzymol 182:203–225
23. Abadir PM et al (2011) Identification and characterization of a functional mitochondrial angiotensin system. Proc Natl Acad Sci USA 108:14849–14854
24. Hong HY et al (2000) Direct Blue 71 staining of proteins bound to blotting membranes. Electrophoresis 21:841–845
25. Park JK, Kostka P (1997) Fluorometric detection of biological S-nitrosothiols. Anal Biochem 249:61–66
26. Kanai AJ et al (1997) Beta-adrenergic regulation of constitutive nitric oxide synthase in cardiac myocytes. Am J Physiol 273: C1371–C1377
27. Pinsky DJ et al (1997) Mechanical transduction of nitric oxide synthesis in the beating heart. Circ Res 81:372–379

Chapter 15

Identification of Thioredoxin Target Protein Networks in Cardiac Tissues of a Transgenic Mouse

Cexiong Fu, Tong Liu, Andrew M. Parrott, and Hong Li

Abstract

The advent of sensitive and robust quantitative proteomics techniques has been emerging as a vital tool for deciphering complex biological puzzles that would have been challenging to conventional molecular biology methods. The method here describes the use of two isotope labeling techniques—isobaric tags for relative and absolute quantification (iTRAQ) and redox isotope-coded affinity tags (ICAT)—to elucidate the cardiovascular redox-proteome changes and thioredoxin 1 (Trx1)-regulated protein network in cardiac-specific Trx1 transgenic mouse models. The strategy involves the use of an amine-labeling iTRAQ technique, gauging the global proteome changes in Trx1 transgenic mice at the protein level, while ICAT, labeling redox-sensitive cysteines, reveals the redox status of cysteine residues. Collectively, these two quantitative proteomics techniques can not only quantify global changes of the cardiovascular proteome but also pinpoint specific redox-sensitive cysteine sites that are subjected to Trx1-catalyzed reduction.

Key words Quantitative proteomics, Liquid chromatography, Tandem mass spectrometry, iTRAQ, Redox ICAT, Hypertrophy, Thioredoxin 1

1 Introduction

Chemical labeling of peptides/proteins with isotope-coded reagents (1), rendering peptide/proteins with mass differences that are readily discernible in mass spectrometers, enables the comparative proteome quantitation from multiple biological samples. One advantage of the chemical-labeling technique is its versatile applicability to all sources of proteins (cells, tissues, serum, bone, hair, etc.). Unlike stable isotope labeling with amino acid in cell culture (SILAC) technique (2), which incorporates stable isotopic amino acids during cell culture, but is limited to proteins that can be retrieved from cultured cells that undergo rapid protein turnover, here we will introduce the application of two distinct chemical-labeling approaches—iTRAQ (1, 3) and ICAT (4, 5)—to quantify

Fernando Vivanco (ed.), *Heart Proteomics: Methods and Protocols*, Methods in Molecular Biology, vol. 1005,
DOI 10.1007/978-1-62703-386-2_15,

the proteome changes in left ventricular tissues of wild-type and a cardiac-specific Trx1 transgenic mouse model (6, 7).

iTRAQ reagents label the primary amines on the N-termini and lysine residues of peptides and can accommodate the quantification of up to eight different samples simultaneously (8-plex iTRAQ (8)). The isobaric nature of iTRAQ reagents does not add to the complexity of chromatography and the mass spectrum (MS) and only releases signature fragments (*m/z* 114–117 for 4-plex and 113–121 for 8-plex) of individual tags upon Collision-Induced Dissociation or Higher Energy Collision Dissociation that can be observed in tandem mass (MS/MS) spectra for peptide identification and quantification (Fig. 1a). On the other hand, ICAT reagents, available in light and heavy versions, label free thiol groups of cysteine

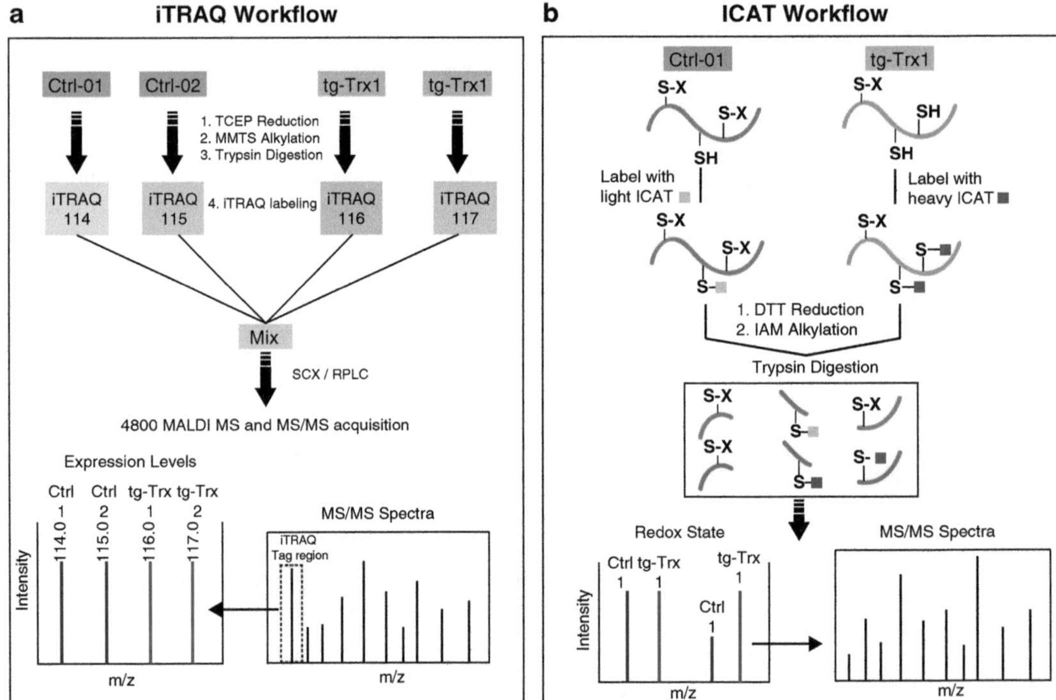

Fig. 1 Typical workflow for ICAT and iTRAQ quantitation. (**a**) In the iTRAQ workflow, protein samples are first subjected to TCEP reduction, MMTS alkylation, and trypsin digestion. The resulting tryptic peptides are then labeled by designated iTRAQ reagents separately. After quenching the reaction, labeled peptides are mixed and separated by multidimensional chromatography. Finally, MS data is acquired on a 4800 MALDI TOF/TOF in a data-dependent acquisition mode. (**b**) In the ICAT workflow, free protein thiols are first labeled by either the light ICAT (*control*) or heavy ICAT (Trx1-overexpressing tissue) reagents. Protein disulfide bonds are then reduced by DTT and alkylated with IAM, sequentially. The labeled proteins are mixed, digested with trypsin and separated sequentially using SCX, avidin affinity, and RPLC separations. ICAT-labeled peptides are identified and quantified by a 4800 MALDI-TOF/TOF mass spectrometer. Peptides containing Trx1-reduced cysteines had an ICAT H/L ratio larger than one and can be quantified by the precursor peak intensity and identified by the MS/MS spectrum. Modified from Molecular & Cellular Proteomics, 2009 (8), 1674–1687 with permission

residues. ICAT reagents incorporate a biotin tag to enable selective avidin-based enrichment of ICAT-labeled peptides from non-cysteine-containing peptides, therefore reducing sample complexity. The light and heavy ICAT-labeled peptides appear as doublets in MS spectra, within which the peak intensity/integrated chromatographic peak area of the doublet are used for peptide quantitation. Peptide sequence is obtained from the MS/MS spectra of either the light or heavy ICAT-labeled peptide (Fig. 1b). Many derivatives of the ICAT technique were created to gauge the redox status of cysteines in peptides by introducing different reduction agents and workflows (5, 9, 10).

A general shotgun proteomics approach commonly deals with a massive number of tryptic peptides (20,000–100,000) in a single liquid chromatography coupled with a tandem mass spectrometer (LC/MS/MS) experiment (11). To maximize proteome coverage and discovery of low-abundant proteins, multiple chromatographic separations are routinely applied in conjunction with these chemical-labeling techniques for peptide fractionation and enrichment. Some of the most popular multidimensional chromatographic methods include multidimensional protein identification technology (MUDPIT) (12), OFFGEL (13, 14), strong cation exchange coupled with reversed phase liquid chromatography (SCX-RPLC) (3, 15, 16), and SCX-affinity chromatography-RPLC (5, 7). Here we will describe the application of the latter two techniques for the preparation of iTRAQ and ICAT-labeled peptides for LC/MS/MS identification and quantification of peptides and their reduction by Trx1.

Many lines of evidence (17, 18) have established Trx1, an 11 kDa antioxidant protein, as a negative regulator of oxidative stress-induced hypertrophy. Here we demonstrate a detailed proteomics method involving the use of two complementary stable isotope labeling proteomics techniques to identify the cardiac Trx1-targeted protein network in a Trx1 transgenic mouse model. By use of this protocol, we were able to identify 78 putative Trx1 reductive sites in 55 proteins (7), including many metabolic enzymes within the protein networks regulating the tricarboxylic acid (TCA) cycle and oxidative phosphorylation pathways that have been shown previously to be regulated by Trx1 (17). Some novel target protein networks, including the creatine–phosphocreatine shuttle, the mitochondrial permeability transition pore complex, and the cardiac contractile apparatus, were observed for the first time. By using the two comparative proteomics methods including iTRAQ and redox ICAT, we were able to find that Trx1 plays not only a conventional role as an antioxidant but also a role in remodeling the cardiovascular system to regulate cardiac energy dynamics and muscle contraction.

2 Materials

2.1 Tissue Homogenization and Protein Extraction

1. The left ventricular heart tissues from control and Trx1-overexpressed mice (three in each group).
2. ICAT Lysis Buffer: 6 M urea, 2 % CHAPS, 1 % Triton X-100, and 30 mM Tris–HCl at pH 7.5 and 0.1 % (v/v) of protease inhibitor cocktail (Sigma, cat no. P8340, St Louis, MO, USA) (see Note 1).
3. Omni Tissue Homogenizer: (Omni International Inc., Marietta, GA, USA).
4. Bicinchoninic Acid (BCA) Protein Assay Kit: (Pierce, cat #. 23225, Rockford, IL, USA).
5. Spectra MAX 190 Microplate Spectrophotometer (Molecular Devices, Sunnyvale, CA, USA).
6. iTRAQ Lysis Buffer: 150 mM TEAB, 1.0 % Igepal CA630 (NP-40), 1.0 % Triton X-100, 0.1 % v/v protease inhibitor cocktail (see Note 1).

2.2 ICAT Labeling

1. Cleavable ICAT® Reagent—10 Assay kit (Sciex Cat# 4339036, Forster City, CA, USA).
2. Accessory for ICAT: Cartridge holder (4326688), needleport adaptor (4326689), outlet connector (4326690), avidin affinity cartridges (4326694), cation exchange cartridges (4326695).
3. Cysteine Reducing Reagent: 50 mM Dithiothreitol (DTT, BioRad Cat #161-0611, Hercules, CA, USA). Cysteine.
4. Alkylation Reagent: 50 mM Iodoacetamide (IAM, BioRad Cat # 163-2109, Hercules, CA, USA).
5. Eppendorf Vacufuge concentrator 5301 (Eppendorf North America, Inc. Westbury, NY, USA).

2.3 iTRAQ Labeling

1. Reducing Reagent: 50 mM Tris-(2-carboxyethyl) phosphine (TCEP).
2. Cysteine-Blocking Reagent: 200 mM methyl methanethiosulfonate (MMTS).
3. HPLC grade ethanol.
4. HPLC grade water.
5. Trypsin (20 μg/vial, Promega, cat no. V5111, Madison, WI, USA).
6. 4-plex iTRAQ™ reagents: 114, 115, 116, 117, (Applied Biosystems Inc., ABI, Forster City, CA, USA).

7. Eppendorf Vacufuge concentrator 5301 (Eppendorf North America, Inc. Westbury, NY, USA).

2.4 Liquid Chromatography Systems

2.4.1 Strong Cation Exchange Liquid Chromatography

1. Mobile Phase A: 10 mM KH_2PO_4 and 20 % acetonitrile (ACN), pH 3.0.
2. Mobile Phase B: 600 mM KCl, 10 mM KH_2PO_4 and 20 % ACN, pH 3.0.
3. BioCAD Sprint™ Perfusion Chromatography System (PerSeptive BioSystems).
4. Column: Polysulfoethyl-A column (4.6×200 mm, 5 μm, 300 Å, Poly LC Inc., Columbia, MD, USA).

2.4.2 Peptide Desalting

1. PepClean C_{18} spin columns (Pierce, cat #. 89870, Rockford, IL, USA).
2. Loading Solution: 5 % ACN containing 0.5 % trifluoroacetic acid (TFA, Pierce, cat # 28904, Rockford, IL, USA).
3. Activation Solution: 50 % ACN containing 0.5 % TFA.
4. Elution Solvent: 70 % ACN.
5. Eppendorf Vacufuge concentrator 5301 (Eppendorf North America, Inc. Westbury, NY, USA).

2.4.3 Reversed-Phase Liquid Chromatography

1. Mobile Phase A: 5 % ACN containing 0.1 % TFA.
2. Mobile Phase B: 95 % ACN containing 0.1 % TFA.
3. LC-Packings Ultimate Chromatography System equipped with a Probot MALDI spotting device (Dionex, Sunnyvale, CA, USA).
4. C_{18} PepMap trapping column (0.3×5 mm, 5 μm, 100 Å, Dionex, P/N 160454).
5. C_{18} PepMap capillary column (0.1×150 mm, 3 μm, 100 Å, Dionex, P/N 160321).
6. Matrix-Assisted Laser Desorption Ionization (MALDI) Matrix Solution: 7 mg/ml α-cyano-4-hydroxycinnamic acid (Sigma, cat #. 476870, St Louis, MO, USA) in 60 % ACN, 5 mM ammonium monobasic phosphate and internal peptide calibrants (50 fmol/ml each of (Glu1)-fibrinopeptide B (GFP, *m/z* 1,570.677, Sigma, cat #. F3261) and adrenocorticotropic hormone 18–39 (ACTH 18–39, *m/z* 2,465.199, Sigma, cat #. A8346)).

2.5 Mass Spectrometry

1. 4800 Proteomics Analyzer (ABI).
2. MALDI plates (ABI).
3. Mass Standards Kit containing a six-peptide mixture (ABI, cat# 4333604).

2.6 Data Analysis Software

1. 4000 Series Explorer (ABI).
2. GPS Data Explorer v3.5 (ABI).
3. Mascot Search Engine v1.9 (Matrix Science Ltd. London, UK).

3 Methods

3.1 Protein Extraction

1. Mouse left ventricular tissues (~100 mg) are diced into 2×2 mm cubes and wash thoroughly by ice-cold PBS (repeat twice) to remove blood content in a 2 ml Eppendorf tube.
2. Spin down heart tissues and remove supernatant.
3. Add 500 μl of either ICAT or iTRAQ lysis buffer to each sample tube (see Note 2).
4. Perform heart tissue homogenization on an Omni Tissue Homogenizer at 4 °C. Six strike cycles (15 s each) were carried out with 2 min cooling intervals to avoid overheating (see Note 3).
5. Remove tissue debris in the homogenates by centrifugation for 30 min at 14,000×*g* at 4 °C in a bench-top centrifuge. Transfer supernatants into a fresh 1.5 ml Eppendorf tube and keep it on ice.
6. Measure protein concentrations for all six samples using the BCA protein assay with bovine serum albumin (BSA) diluted in the lysis buffer as standards. Protein yield will be in the range of 4–10 mg/ml depending on the lysis buffer of choice.
7. Adjust protein concentration of each sample to the same level with either ICAT or iTRAQ lysis buffer.

3.2 ICAT Labeling

1. Pipette 120 μg proteins from each sample into separate tubes (see Note 4).
2. Precipitate proteins in cold acetone (5:1 ratio at −20 °C) overnight (see Note 5).
3. Pellet the protein contents by high-speed centrifugation for 15 min at 14,000×*g* at 4 °C.
4. Remove supernatant and wash the pellets three times with cold acetone (−20 °C).
5. Solubilize the proteins with 80 μl of ICAT-labeling buffer: 6 M urea, 2 % CHAPS, 0.01 % SDS, and 30 mM Tris–HCl at pH 8.3 (see Note 6).
6. Bring the ICAT reagent tubes to room temperature and briefly spin down the powder to the bottom of the tubes.
7. Add 20 μl of ACN to each ICAT tube and vortex the solution.
8. Spin down the solution to the bottom of the tubes and transfer the entire content to designated sample tubes for protein labeling.

9. Incubate the mixture for 2 h at 37 °C.
10. Briefly centrifuge to bring all the solution to the bottom of each tube.
11. Quench excess ICAT reagents at room temperature by adding 10 mM DTT (final concentration) and incubate for 15 min (see Note 7).
12. Alkylate newly generated sulfhydryls by 15 mM IAM (final concentration) and incubate for 15 min at room temperature in dark.
13. Mix light and heavy ICAT-labeled sample pairs (see Note 8).
14. Dilute the sample volumes at least 6 times with 20 mM ammonium bicarbonate buffer (see Note 9).
15. Add trypsin solution to a final 1:50 ratio (enzyme: protein) and digest overnight at 37 °C (see Note 8).
16. Dry the digested peptide samples in a SpeedVac.

3.3 SCX-LC

The combined peptide mixture is separated by strong cation exchange liquid chromatography (SCX-LC) on a polysulfoethyl-A column to remove excess ICAT reagents and unwanted detergent (SDS and CHAPS), prior to fractionation of the peptides.

1. Reconstitute the ICAT-labeled peptides by adding ~500 μl of SCX Mobile Phase A. Adjust pH to 2.5–3.0 with phosphoric acid if necessary.
2. Centrifuge the sample at 20,000 × *g* for 10 min to remove any particulates.
3. Equilibrate the SCX column with Mobile Phase A, and then inject the ICAT-labeled peptides onto the SCX column through a 500 μl sample loading loop.
4. The gradient profile of SCX consisted of 10 min of 100 % Mobile Phase A followed by 30 min of 0–25 % Mobile Phase B and 20 min of 25–100 % B at 1 ml/min. Collect peptide fractions at 2 min/fraction after the elution of both neutral and anionic interferences.
5. Dry all the SCX fractions in a SpeedVac for subsequent desalting steps.

3.4 Peptide Desalting Using C_{18} Spin Columns

1. Reconstitute each dried SCX fraction in 150 μl of the Loading Solution (see Note 10).
2. Add 200 μl of the Activation Solution into a C_{18} spin column and centrifuge at 1,500 × *g* for 1 min. Repeat this step once.
3. Equilibrate the spin column with 200 μl of the Loading Solution and centrifuge the column at 1,500 × *g* for 1 min. Repeat this step twice.
4. For each SCX fraction, load all 150 μl of the peptides in the Loading Solution onto the spin column and centrifuge at

1,000 × *g* for 1 min and collect the flow through. Reload the flow through materials onto the spin column.

5. Wash the bound peptides with 200 μl of the Loading Solution and centrifuge at 1,500 × *g* for 1 min to remove salts. Repeat this step twice.
6. Elute the peptides using 100 μl of the Elution Solution by centrifugation at 1,500 × *g* for 1 min and repeat twice. Collect all three eluants into the same Eppendorf tube.
7. Dry the peptides solution in a SpeedVac.

3.5 Enrichment of ICAT-Labeled Peptides by Avidin Affinity Chromatography

1. Reconstitute each dried peptide fraction in 500 μl of the Affinity-Load Buffer, vortex to mix the solution. Confirm the pH of solution is ~7.0 (see Note 11).
2. Briefly centrifuge to bring all the solution to the bottom of the tubes.
3. Assemble avidin cartridge system.
4. Load 2 ml Affinity-Elution Buffer to the cartridge and discard the eluate.
5. Load 2 ml Affinity-Load Buffer to the cartridge and discard the eluate.
6. Slowly load (drop by drop) the peptide samples in 500 μl of the Affinity-Load Buffer and collect the flow through (see Note 12).
7. Reload the flow through on the avidin cartridge and save the subsequent flow through.
8. Wash the avidin cartridge with 1 ml of Wash1 and divert the eluate to waste.
9. Wash the avidin cartridge with 1 ml of Wash2 and divert the eluate to waste.
10. Wash the avidin cartridge with 1 ml of HPLC grade water and divert the eluate to waste.
11. Load 800 μl of Affinity-Elution Buffer into syringe and inject slowly to the cartridge (~1 drop/5 s) and discard the first 50 μl of eluate. Collect the remaining 750 μl of eluate into a glass vial.
12. Repeat steps 1–11 for the remaining peptide fractions.

3.6 TFA Cleavage of Biotin Moiety from ICAT Peptides

1. SpeedVac the affinity eluates to complete dryness (see Note 13).
2. Prepare cleavage mixture of 95 μl of cleavage reagent A with 5 μl of cleavage reagent B and mix them with dry peptide samples.
3. Vortex the reaction mixture and incubate at 37 °C for 2 h.
4. Centrifuge and dry the reaction mixtures.

3.7 RPLC and Spotting MALDI Plate

1. Reconstitute cleaved ICAT peptide samples in 20 μl of RPLC Mobile Phase A (MPA), vortex vigorously, then centrifuge at 10,000 × *g* for 5 min.
2. Equilibrate the RPLC column with 5 % RPLC Mobile Phase B for at least 15 min for stable column pressure.
3. Each peptide sample (6.4 μl) is loaded onto a C_{18} trapping column using a Microliter Pickup method at a flow rate of 20 μl/min. An online desalting step is carried out by a 5-min MPA wash.
4. Peptides bound to the trapping column are subsequently resolved on a C_{18} capillary PepMap column with the following gradient profiles at a flow rate of 400 nl/min (Table 1).
5. Mix the RPLC eluants in line with MALDI matrix in a 1:2 volume ratio through a 30 nl mixing tee, and deposited onto a MALDI plate using the Probot, at 12 s per spot.
6. Repeat the RPLC steps for each SCX fraction.

3.8 Mass Spectrometry

1. Mix 50 fmol of 6-peptide calibrants at a ratio of 1:1 with MALDI matrix solution. Deposit the freshly prepared calibrant solution on the calibration spots on the MALDI plate.
2. Create a new spot set and load and align the ICAT sample plates. Load the sample plates into the plate loader of the 4800 Proteomics Analyzer.
3. Acquire and update the MS calibration file using the six-peptide mixture in the Mass Standards Kit. The MS/MS calibration file needs to be updated using GFP MS/MS ions (*m/z* 1,570.677).
4. Acquire MS spectra for each spot in the positive ion mode with a laser intensity of 3,200 and a mass range of 850–3,500, and sum of the first 1,500 laser shots. In the MS processing method,

Table 1
Gradient profile for C_{18} PepMap column separation of ICAT-labeled peptides (flow rate: 400 nl/min)

Time (min)	Mobile Phase A	Mobile Phase B
0	95	5
2	95	5
75	70	30
90	10	90
100	10	90
105	95	5
115	95	5

use internal calibration standards (GFP (m/z 1,570.677) and ACTH 18–39 (m/z 2,465.199)) to achieve a mass accuracy better than 50 ppm.

5. After MS analysis, identify and extract ICAT ion pairs (Δ m/z 9.03 ± 0.03 Da) by GPS Explorer software (v3.5 ABI). Compute the relative ICAT ratios with the integrated chromatographic areas of each ICAT ion pair. Submit ICAT pairs with over 20 % change ratios and signal/noise ratios (S/N) > 50 to MS/MS spectra acquisition (see Note 14).
6. In the MS/MS acquisition method, spectra are accumulated with 2,000 laser shots at a laser intensity of 3,200 using 2-keV collision energy and 5×10^{-7} Torr collision gas pressure. Generate peak lists with 4000 Series Explorer with the following settings: set S/N threshold to 10, local noise window width at 250 m/z, and minimum peak width bin size was 2.9; set target resolution at 22,000 at m/z 2,400 for MS and 8,000 at m/z 2,000 for MS/MS. Smooth MS/MS spectra with the Savitzky–Golay algorithm (FWHM = 9, polynomial order = 4).

3.9 Database Search

1. Perform peptide identification on a Mascot Search Engine (v1.9) integrated in the GPS Explorer software with the following search parameters: one missed tryptic cleavage, 50 ppm for MS mass error tolerance, and 0.3 Da for MS/MS mass error tolerance, variable modifications included ICAT L/H modifications and carbamidomethylation of cysteines, and methionine oxidation.
2. Unique peptides identified with confidence interval (C.I.) values at or above 95 % from the MS/MS database search are considered significant (see Note 15).

3.10 Trypsin Digestion and iTRAQ Labeling

1. Pipette 100 µg of proteins in iTRAQ Lysis Buffer from the two controls and two transgenic samples into four separate tubes (see Note 16). To each sample, add 2 µl of Reducing Reagent and vortex. Bring down the contents with a brief centrifugation. Incubate the sample tubes at 60 °C for 1 h. Spin briefly to settle the liquid to the bottom of each tube.
2. To each sample, carefully add 1 µl of the Cysteine-Blocking Reagent. Mix by vortexing and centrifuge briefly to collect the solutions at the bottom of the tube. Incubate at room temperature for 10 min.
3. Reconstitute two vials of trypsin (20 µg/vial) with 25 µl each of HPLC grade water. Vortex briefly.
4. To each sample tube, add 10 µl of the trypsin solution, vortex, and centrifuge briefly to collect the solution at the bottom of the tube. Incubate at 37 °C for 12–16 h. Spin briefly to bring the sample solution to the bottom of the tubes (see Note 17).

5. Bring the iTRAQ reagents to room temperature. Add 70 μl of ethanol into each reagent vial, cap the vial and vortex vigorously, and then centrifuge briefly to settle the iTRAQ reagents to the bottoms of the vials (see Note 18).
6. Transfer the entire content of one iTRAQ reagent vial into each of the four sample tubes, and vortex to mix thoroughly. Spin briefly to collect the liquid at the bottom of the tubes. Peptides derived from the two control samples are labeled with iTRAQ Reagents 114 and 115, whereas peptides obtained from the two Trx1-overexpressed samples are labeled with iTRAQ Reagents 116 and 117. Incubate the reaction vials at room temperature for 1 h.
7. Carefully combine the entire contents of all four iTRAQ-labeled samples into one tube, mix thoroughly by vortexing, and then centrifuge briefly.

3.11 2D-LC Separation and MS Analysis

1. The combined peptide mixture is first separated by SCX-LC to remove excess iTRAQ reagents. In order to remove both TEAB and the organic solvent from the sample, dry the combined sample completely in a SpeedVac (see Note 19). Follow the same steps in Subheading 3.3 to fractionate the peptides.
2. Follow the same steps in Subheading 3.4 to desalt the peptides. Dry the resulting peptides by SpeedVac and reconstitute the peptides with 10 μl Mobile Phase A. Use Nano-RPLC for peptide separations following the same steps in Subheading 3.7, except using the following gradient: (Table 2).
3. Mix the RPLC eluants in line with MALDI matrix and deposited onto a MALDI plate. Follow steps 1–3 in Subheading 3.8 to calibrate the 4800 MALDI TOF/TOF analyzer.

Table 2
Gradient profile for Nano-RPLC C_{18} column separation of iTRAQ-labeled peptide mixture

Time (min)	Mobile Phase A	Mobile Phase B
0	95	5
4	92	8
34	82	18
57	62	38
64	5	95
69	5	95
70	95	5
85	95	5

4. Create an acquisition, a processing, and a job-wide interpretation method for both MS and MS/MS analyses. Use the MS acquisition method in a positive MS reflector with a mass range of 850–3,000 (in Da) and a focus mass of 1,950 Da. Set the laser intensity to 3,000 and the detector voltage multiplier at 0.90. MS spectrum is an average of 1,000 laser shots. In the processing method, GluFib (m/z 1,570.677) and ACTH 18–39 (m/z 2,465.199) masses are used as the internal calibrants.
5. For the interpretation method, the precursor selection is based on a minimum S/N filter of 50, a precursor mass tolerance of 200 ppm, with an MS/MS acquisition order from weakest to strongest ions.
6. Use a 2 kV positive MS/MS method to acquire the tandem spectra. Set the laser intensity to 4,000 and detector voltage multiplier at 0.90. Specify the metastable suppression as "on" and the precursor mass window at relative 400 resolution (FWHM). Each MS/MS spectrum is accumulated over 4,000 laser shots. In the MS/MS processing method, each spectrum is smoothed using the Savitzky–Golay algorithm with points across the peak set at 3 and polynomial order set at 4. Set the medium CID gas recharge pressure to medium with a threshold of 5.0×10^{7} Torr.

3.12 Bioinformatics Analysis

1. Peptide identification is performed by searching the MS/MS spectra against Swiss-Prot mouse database (see Note 20), using a local Mascot Search Engine (v. 1.9) on a GPS (v. 3.5, ABI) server. The following search parameters are used: trypsin with one missed cleavage, mass tolerance of 50 ppm for the precursor ions, and 0.3 Da for the MS/MS fragment. iTRAQ-labeled N-termini and lysines, and cysteine methanethiolation are selected as fixed modifications, while methionine oxidation and iTRAQ-labeled tyrosine are considered as variable modifications. Only peptides identified with confidence interval (C.I.) values at or greater than 95 % should be used for protein identification and quantitation.
2. Extract the iTRAQ reporter ions cluster areas using GPS Explorer. Only ion counts greater than 5,000 are used for quantification analysis to reduce interference from analytical noise. The individual reporter ion peak areas for each iTRAQ channel are normalized by the population medians.
3. For each peptide, the ratio of normalized reporter ion peak areas at 115, 116, and 117 are divided by the normalized reporter ion peak areas at 114. Such ratios are then transformed into $\log_2$ values. In cases of multiple MS/MS spectra matched to the same

peptide sequence, the peptide ratio is calculated and weighted based on the relative proportion of each spectrum.

4. The mean of all peptide ratios from the same protein are averaged to obtain protein ratios.
5. The *p*-values in Student's t-tests are calculated by comparing each protein $\log_2$ ratio in the control group (P_{114} and P_{115}) to those in the Tg-Trx group (P_{116} and P_{117}) using Microsoft Excel. Anti-$\log_2$ of P_i values is calculated to produce the exact protein fold change values.

4 Notes

1. It is recommended to use freshly prepared lysis buffer. Selection of proper detergents is discretional upon protein of interest (e.g., membrane proteins). Be sure to confirm the compatibility between the detergent and the iTRAQ chemistry (http://www.absciex.com/Documents/Downloads/Literature/mass-spectrometry-4375249C.pdf).
2. To preserve the native redox states of protein cysteines, it is highly recommended to minimize sample exposure to air, keep samples on ice during sample preparation if compatible, and purge with high-purity nitrogen for extended incubation (e.g., trypsin digestion step).
3. Depending on the detergents of choice, excessive bubbles may be formed in the lysis process. A quick spin-down of sample tubes at 4 °C will help to remove bubbles.
4. Estimate free protein thiol contents in the samples. For example, for 100 μg protein with an average mass of 50 kDa and 6 cysteines per protein, the total cysteine content can be estimated as 100×10^{-6} g/50,000 g/mol $\times 6 = 12$ nmol. Each ICAT tube contains 175 nmol of labeling reagent to maintain excessive reagent/free cysteines ratio >10 times for complete labeling.
5. Whole cell lysates contain many small molecules that could have adverse effects on protein ICAT labeling. For example, glutathione and other cysteine-containing antioxidants are observed at high levels (mM) and will consume ICAT reagents at much faster reaction rates than protein thiols. Alternative precipitation methods (such as TCA precipitation and methanol/chloroform precipitation) and buffer exchange methods (membrane ultrafiltration) can be implemented to remove interfering molecules.

6. To avoid protein loss and sample variation, complete solubilization of proteins is a key step. Mild agitation with Eppendorf pipette tips and sonication in a water bath will facilitate protein solubilization. Increasing SDS concentration to 0.05 % can also enhance the degree of protein solubilization.
7. In addition to scavenging excess ICAT reagents, DTT also reduces disulfide bonds and other reversible cysteine modifications. This is important in the forward redox ICAT-labeling scheme, since the reduction of disulfide bonds after ICAT labeling will facilitate tryptic digestions and improve protein and peptide identifications.
8. For quality control (QC) of ICAT labeling and protease digestion efficiency, preserve 1 μl of individual labeling solution (section 3.2, ICAT labeling step 13) and 1 μl of solution before and after tryptic digestion. Load the samples in separate lanes of 1D-SDS PAGE. After electrophoresis and protein staining, evaluate the initial sample concentration and digestion efficiency. In the second QC test, mix 1 μl each of tryptic heavy and light ICAT-labeled peptides after desalting with ZipTips. Spot the peptides at 1:1 ratio with the MALDI matrix solution on a MALDI plate. Acquire a MS spectrum of the peptides and evaluate the abundance and relative ratio of ICAT peptide pairs with 9 Da (or multiplier) mass differences.
9. Reducing urea concentration to <1 M is important for effective trypsin digestion. It is also important to adjust the pH of the solution to a range of 8.0–8.5 for optimum trypsin activity.
10. Combining adjacent fractions with less peptide abundance according to the SCX chromatogram greatly reduces the LC/MS/MS processing time without much loss of total protein identification and quantification. Late-eluted fractions might have more salts and require additional RPLC loading buffer for full dissolution.
11. The loading capacity of the avidin cartridge is ~10 μg of peptide. A new avidin cartridge can be used up to 50 times with proper usage, cleaning and storage.
12. Using a syringe pump (e.g., Standard Infusion Only Pump 11 Elite Syringe Pumps, Harvard Apparatus, Holliston, MA, USA) for solvent delivery yields more consistent and robust results.
13. Residual water content will have an adverse effect on the cleavage reaction. It is critical to dry the peptide samples completely before TFA cleavage. Perform the reaction in a hood with proper ventilation. Since TFA is reactive and corrosive to a broad range of materials, it is suggested to use glass vials and glass syringe with metal plunger to transfer and hold the reaction mixtures.

14. Peptides containing multiple cysteines can result in multiple ICAT labeling; thus, the mass difference for these peptides may be observed in multipliers of (Δ m/z 9.03 ± 0.03 Da). This can be addressed by setting allowance of multiple ICAT labeling in the GPS Explorer software.
15. One-hit-wonder is one of the caveats in using ICAT-labeled peptides for protein identification and quantifications, given the low observation frequency of cysteines (< 3.3 %) in vertebrate proteomes. It is not uncommon to find only one ICAT-labeled peptide for a given protein. Careful inspection of MS and MS/MS spectra for positive identification and removal of potential interference is important to reduce false-positive results. One confirmative hallmark for ICAT-labeled peptide is the signature mass differences of 339.1 for heavy ICAT labeled and 330.1 for light ICAT labeled, between y^n and y^{n-1} ions, where n indicates the location of ICAT-labeled cysteine (Fig. 2).
16. Based on iTRAQ instructions from ABI, each protein sample should be between 5 and 100 μg for each iTRAQ-labeling reaction. To ensure maximum labeling efficiency, sample volumes should be less than 50 μl each. If the sample volume is larger than 50 μl, a SpeedVac can be used to reduce the sample volumes before iTRAQ labeling.

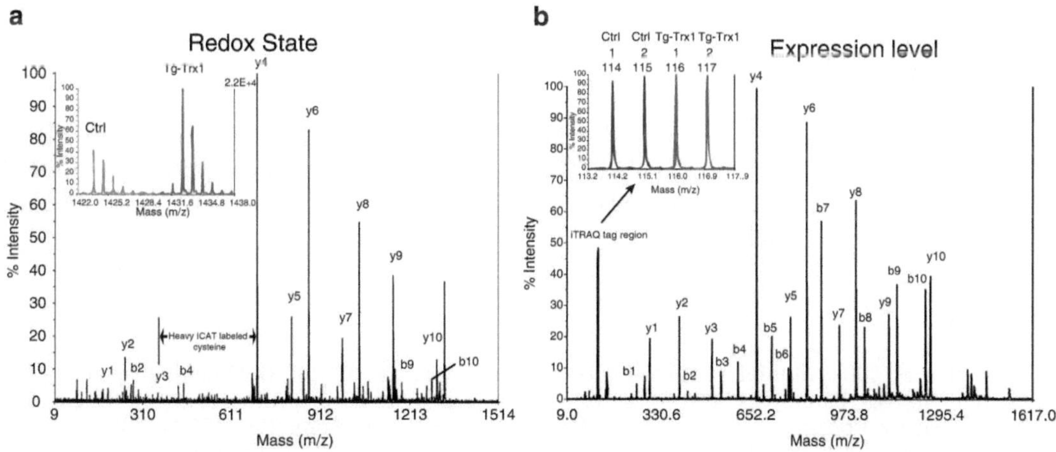

Fig. 2 Peptide identification and quantitation by ICAT and iTRAQ. Panel (**a**) sequence of a cysteine-containing peptide [153]EFNGLGDCLTK[163] from ADP/ATP translocase 1 is identified by matching MS^2 fragment ion mass to the predicted theoretical fragments. The addition of 227 Da (236 Da for the heavy ICAT tag) to a cysteine residue is a hallmark of an ICAT-labeled peptide (mass difference between y3 and y4 ions). Quantitation of light and heavy labeled peptides (redox states of cysteines in this case) is computed from the precursor intensities acquired from the MS^1 scan (*left inset*) or from the integrated extracted ion chromatogram. (**b**) iTRAQ quantitation is carried out on the reporter region (mass range 114–117 for 4-plex iTRAQ) in the MS^2 spectra (*see inset of panel b*). The individual peak intensity of each mass reporter ion reflects the corresponding peptide/protein level in each sample. The peptide sequence is identified in a similar manner as ICAT technology. Modified from Molecular & Cellular Proteomics, 2009 (8), 1674–1687 with permission

17. It is important to check the protein digestion efficiency before iTRAQ labeling. Take 1 μl of each digested sample and desalt it using a C_{18} ZipTip (Millipore, Billerica, MA). Mix the eluted peptides with the MALDI matrix solution in a 1:1 ratio and spot them onto a MALDI plate. Acquire MS spectra to check if the peptide ion signals are comparable.
18. To maximize labeling efficiency, the concentration of organic reagents (ethanol and iTRAQ reagents) in iTRAQ-labeling reactions should be larger than 60 % (v/v).
19. To remove all of the TEAB, reconstitute the combined iTRAQ-labeled samples in 100 μl of HPLC grade water and dry the sample in a SpeedVac. Repeat this step twice to ensure all the TEAB is evaporated.
20. It is important to use the latest version of the protein database to ensure comprehensive peptide identification. Swiss-Prot, IPI, NCBI protein database, or EST (6 frame translation into protein sequences) can be used, with an increasing number of entries and database size. Generally speaking, using bigger databases will likely increase one's chance to match a spectrum to a peptide sequence. However, it will also increase the odds for random matching. We chose the Swiss-Prot database for our study because of its high protein sequence accuracy and low redundancy.

Acknowledgements

The authors are grateful for the funding support from NIH/ NINDS grant P30 NS046593 to H.L for the support of a NINDS NeuroProteomics Core Facility at UMDNJ-New Jersey Medical School.

References

1. Ross PL et al (2004) Multiplexed protein quantitation in Saccharomyces cerevisiae using amine-reactive isobaric tagging reagents. Mol Cell Proteomics 3:1154–1169
2. Ong SE et al (2002) Stable isotope labeling by amino acids in cell culture, SILAC, as a simple and accurate approach to expression proteomics. Mol Cell Proteomics 1:376–386
3. Liu T et al (2007) Identification of differentially expressed proteins in experimental autoimmune encephalomyelitis (EAE) by proteomic analysis of the spinal cord. J Proteome Res 6:2565–2575
4. Gygi SP et al (1999) Quantitative analysis of complex protein mixtures using isotope-coded affinity tags. Nat Biotechnol 17:994–999
5. Fu C et al (2008) Quantitative analysis of redox-sensitive proteome with DIGE and ICAT. J Proteome Res 7:3789–3802
6. Yamamoto M et al (2003) Inhibition of endogenous thioredoxin in the heart increases oxidative stress and cardiac hypertrophy. J Clin Invest 112:1395–1406
7. Fu C et al (2009) Elucidation of thioredoxin target protein networks in mouse. Mol Cell Proteomics 8:1674–1687
8. Choe L et al (2007) 8-Plex quantitation of changes in cerebrospinal fluid protein expression in subjects undergoing intravenous immunoglobulin treatment for Alzheimer's disease. Proteomics 7:3651–3660

9. Sethuraman M et al (2004) Isotope-coded affinity tag (ICAT) approach to redox proteomics: identification and quantitation of oxidant-sensitive cysteine thiols in complex protein mixtures. J Proteome Res 3:1228–1233
10. Leichert LI et al (2008) Quantifying changes in the thiol redox proteome upon oxidative stress in vivo. Proc Natl Acad Sci USA 105:8197–8202
11. Michalski A, Cox J, Mann M (2011) More than 100,000 detectable peptide species elute in single shotgun proteomics runs but the majority is inaccessible to data-dependent LC–MS/MS. J Proteome Res 10:1785–1793
12. Washburn MP et al (2001) Large-scale analysis of the yeast proteome by multidimensional protein identification technology. Nat Biotechnol 19:242–247
13. Chenau J et al (2008) Peptides OFFGEL electrophoresis: a suitable pre-analytical step for complex eukaryotic samples fractionation compatible with quantitative iTRAQ labeling. Proteome Science 6:9
14. Obad S, dos Santos N et al (2011) Silencing of microRNA families by seed-targeting tiny LNAs. Nat Genet 43:371–378
15. Ralhan R et al (2008) Discovery and verification of head-and-neck cancer biomarkers by differential protein expression analysis using iTRAQ labeling, multidimensional liquid chromatography, and tandem mass spectrometry. Mol Cell Proteomics 7:1162–1173
16. Keshamouni VG et al (2006) Differential protein expression profiling by iTRAQ–2DLC–MS/MS of lung cancer cells undergoing epithelial-mesenchymal transition reveals a migratory/invasive phenotype. J Proteome Res 5: 1143–1154
17. Ago T et al (2006) Thioredoxin1 upregulates mitochondrial proteins related to oxidative phosphorylation and TCA cycle in the heart. Antioxid Redox Signal 8:1635–1650
18. Yang Y et al (2011) Thioredoxin 1 negatively regulates angiotensin II-induced cardiac hypertrophy through upregulation of miR-98/let-7. Circ Res 108:305–313

Chapter 16

Using Pure Protein to Build a Multiple Reaction Monitoring Mass Spectrometry Assay for Targeted Detection and Quantitation

Eric Grote, Qin Fu, Weihua Ji, Xiaoqian Liu, and Jennifer E. Van Eyk

Abstract

Multiple reaction monitoring (MRM) is an increasingly popular mass spectrometry-based method to simultaneously detect and quantify multiple proteins. MRM is particularly useful for validating biomarkers discovered with a mass spectrometer and any analite discovered by MS can be monitored by MR because an MRM assay can be developed without the need to generate specific antibodies. In this chapter, we present a robust and systematic procedure to rapidly build a high-sensitivity MRM assay using purified protein as the starting material. Theoretical digestion of the protein with trypsin is used to identify mass spectrometry-compatible peptides and to generate preliminary MRM transitions to detect these peptides. Peptides generated by trypsin cleavage of the actual protein are then run on a liquid chromatography column coupled to a triple quadrupole mass spectrometer, which is programmed with the preliminary transitions. Whenever a transition is detected, it triggers dissociation of the corresponding peptide and collection of a full mass range scan of the resulting fragment ions. From this scan, fragment ions yielding the strongest and most reproducible signals are utilized to design empirical MRM transitions. The assay is further refined by optimizing the collision energy and creating a standard curve to measure sensitivity. Once MRM transitions have been established for a particular protein, they can be combined with transitions for other target proteins to create multiplex assays and used to quantify proteins in samples arising from serum, urine, subcellular fractions, or any other specemen of interest.

Key words Quantitation, Multiple reaction monitoring, Transitions, Assay development, Collision energy, Lower limit of quantification, Biomarker

1 Introduction

Multiple reaction monitoring (MRM), also known as selective reaction monitoring (SRM), is a targeted form of mass spectrometry that can be used to detect and quantify specific peptides representing a protein of interest (1, 2). In comparison to ELISAs and

Eric Grote and Qin Fu have equally contributed to this work.

Fernando Vivanco (ed.), *Heart Proteomics: Methods and Protocols*, Methods in Molecular Biology, vol. 1005,
DOI 10.1007/978-1-62703-386-2_16, © Springer Science+Business Media New York 2013

other quantitative immunoassays, MRM provides high specificity uncompromised by antibody cross-reactivity and rapid assay development without the time- and labor-intensive steps of antibody generation and screening. A further advantage of MRM assays is that multiplexing to allow simultaneous quantification of 25 or even 100 proteins can be readily accomplished without an excessive amount of optimization (3). Thus, MRM assays are increasingly being used to verify and validate biomarkers discovered via mass spectrometry and to quantify posttranslational modifications including phosphorylation, ubiquitination, and glycosylation (4–8). MRM can also be used to detect target proteins in fractionated or immune-purified samples if low amounts of protein are present or if the assay is not sufficiently sensitive to detect its target protein in a more complex biological samples (9).

In an MRM assay, a protein sample is cleaved into peptides with a protease such as trypsin (which cleaves after lysine and arginine residues). The peptides are then separated by liquid chromatography and eluted into the ionization source of a triple quadrupole mass spectrometer, which analyzes the mass/charge ratio (m/z) of intact peptide ions and of fragment ions generated by colliding selected peptides with an inert gas (see Note 1). In MRM, the mass spectrometer is programmed to detect predetermined pairs of m/z values called transitions: each corresponding to a precursor ion (intact peptide) and a product ion (fragment). Targeting predetermined transitions provides for more robust detection by ensuring that target ions will be monitored even if other ions are present in greater abundance. In addition, MRM provides higher sensitivity by targeting a limited number of peptides with specific m/z values and monitoring each associated transition for a relatively long time instead of rapidly scanning across the entire available mass/charge range, as is typically done when the goal is to identify every peptide in a complex mixture. Specific detection is enhanced by tracking multiple transitions from the same peptide and confirming that they elute at the sometime from the LC column. The amount (quantity or concentration) of peptide can be accurately measured by spiking the sample with a known amount of a stable isotope-labeled peptide of the same sequence and then comparing the signal of the heavy and light peaks for each transition. If peptide standards are not available, MRM can still be used to compare the relative amounts of multiple targets.

In this chapter, we describe a method for building MRM assays by starting with a tryptic digest of purified protein and then systematically identifying MRM transitions with high sensitivity, specificity, and reproducibility (Fig. 1). MRM assays are commonly built using a limited number of synthetic peptides selected based upon the amino acid sequence of the protein of interest (3). Unfortunately, it is difficult to predict which peptides will have ionization and fragmentation characteristics suitable for sensitive and reproducible detection by mass spectrometry, and it is not economically practical

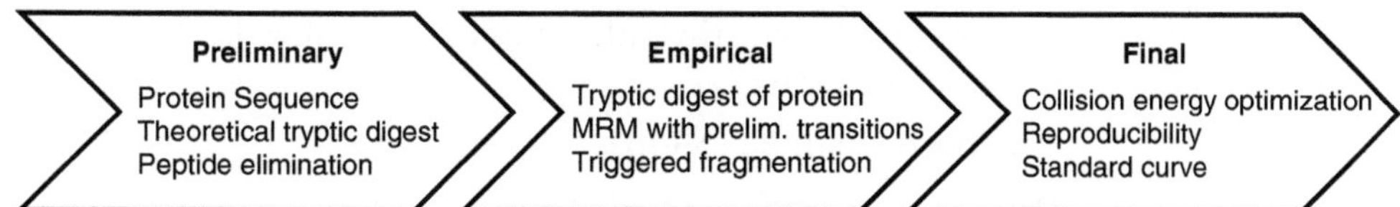

Fig. 1 Workflow for developing an MRM assay from purified protein

to synthesize every possible tryptic peptide from a given protein. An alternative approach identifies potential MRM transitions using existing mass spectrometry data from discovery proteomics or online databases (10). A disadvantage of this approach is that each type of mass spectrometer has unique ionization and fragmentation characteristics, so transitions with strong signals in discovery proteomics may not be suitable for quantitative MRM, and vice versa. In contrast, when a tryptic digest of pure protein is used to build an MRM method, every tryptic peptide is evaluated using the same mass spectrometer that will be used in the final assay, and the method can often be built in less time that it would take to prepare custom peptides. To demonstrate our method, we describe the development of MRM assays for ST2 protein.

2 Materials

2.1 Instrumentation

For best results, MRM assays should be performed on a triple quadrupole LC-MS/MS system. We used a Prominence UFLCXR high-performance liquid chromatography system (Shimadzu Scientific Instruments, Columbia, MD) with an XBridge BEH30 C18 reverse-phase chromatography column (2.1 mm × 100 mm, 3.5 μm, Waters, Milford, MA) coupled to a 5500 QTRAP triple quadrupole/linear ion trap mass spectrometer (AB SCIEX, Framingham, MA) (see Note 2).

2.2 Software

1. MRMPilotTM, Version 2.1 (AB SCIEX, Framingham, MA) (see Note 3).
2. Analyst, Version 1.5.2 (AB SCIEX, Framingham, MA).
3. MultiQuant, Version 2.1 (AB SCIEX, Framingham, MA).

2.3 Reagents

1. ST2 protein (a gift from Critical Diagnostics, San Diego, CA).
2. Water, HPLC grade (Thermo Fisher Scientific, San Jose, CA).
3. Acetonitrile (ACN), HPLC grade.
4. Formic acid.
5. Sample vials (VWR International, Radnor, PA).
6. Oasis HLB Extraction Cartridge, 1 cc/30 mg WAT094225 (Waters, Milford, MA).

7. Sequencing Grade Modified Trypsin (Promega, Madison, WI).
8. Tris(2-carboxyethyl) phosphine hydrochloride (TCEP) (Pierce, Rockford, IL).
9. Iodoacetamide (Sigma, St Louis, MO).
10. 37 °C and 55 °C incubators.
11. Speed/Vac concentrator, Savant SPD 20/D (Thermo Fisher Scientific, San Jose, CA).
12. Low-retention 1.5-ml microfuge tubes (Thermo Fisher Scientific, San Jose, CA).

3 Methods

3.1 In Silico Identification of Proteotypic Peptide Candidates

The identification of MRM transitions begins with an examination of the amino acid sequence of the protein of interest. Peptides for evaluation are identified by performing a theoretical digest with trypsin (see Note 4), selecting peptides with masses within the preferred detection range of the mass spectrometer (see Note 5), and then eliminating peptides with amino acids that would complicate analysis (see Note 6). For each of these peptides, commercial software is used to design preliminary transitions by predicting high-probability fragments and then calculating *m/z* values for the parental and fragment ions. Experimental data from online databases or prior mass spectrometry analysis conducted in the discovery phase of a project can also be used to define preliminary transitions, but this data often does not extend to all potentially useful tryptic fragments (11).

The method is illustrated using the ST2 protein, also called interleukin-1 receptor-like 1 (UniProt Q01638). ST2 has membrane bound and soluble isoforms. The membrane form is a receptor for IL-33. The soluble form is a decoy receptor that interferes with IL-33 signaling. ST2 is under consideration as a serum biomarker for asthma, heart failure, myeloma, and other diseases (12–14). It has 22 tryptic peptides (assuming no miscleavage) with lengths between 5 and 25 amino acids. After eliminating the N-terminal signal sequence and peptides containing methionine, cysteine, or a potential N-glycosylation site (Fig. 2). We designed preliminary.

1. Download the amino acid sequence of the protein of interest from the UniProtKB Protein knowledgebase: http://www.uniprot.org/ (see Note 7).
2. Import the sequence into MRMPilot (see Note 3). Open MRMPilot. Select "Choose Proteins." Select "Sequence or Accession #." Paste the sequence.
3. Set the Digest Settings to Enzyme = Trypsin, No missed cleavages or modifications. Under Peptide Selection, set the

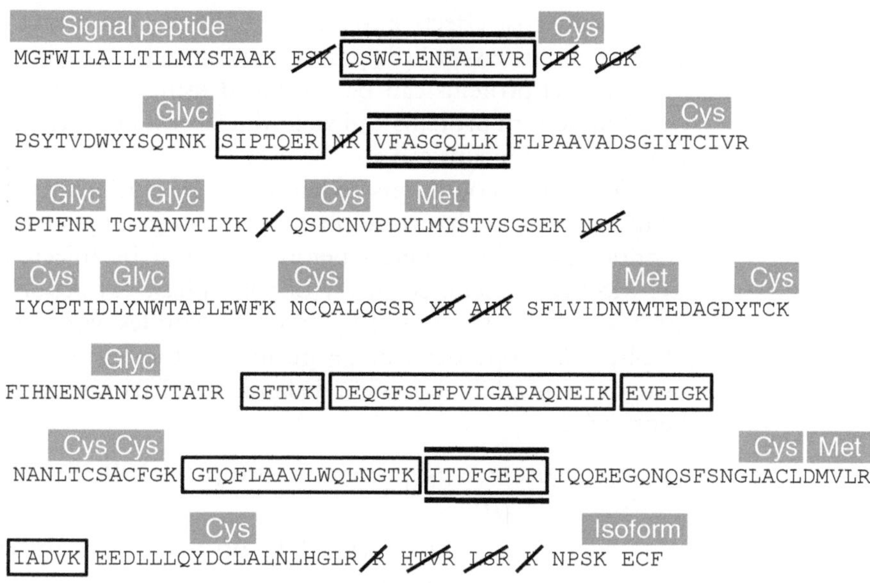

Fig. 2 Annotated sequence of ST2. The amino acid sequence of ST2 is segmented into tryptic peptides cleaved C-terminal to lysine (K) or arginine (R) residues. Preliminary MRM transitions were designed for the *boxed* peptides. *Bars* indicate the three peptides with the highest MRM signal. Short peptides (/) and peptides with methionine (Met), cysteine (Cys), and predicted N-linked glycosylation (Glyc) were excluded from further consideration. The *signal peptide* is removed in the secretory pathway, and the C-terminus is not conserved in other ST2 *Isoforms*

number of amino acids to range from 5 to 30 and reject peptides containing Met, Cys, and all posttranslational modifications. Select Perform Digest.

4. From the UniProt page, identify peptides with amino acid modifications or natural sequence variations. Deselect these peptides in MRMPilot (Fig. 2).
5. Run a BLASTP search (http://blast.ncbi.nlm.nih.gov/Blast.cgi?PAGE=Proteins) on each of the remaining peptides and eliminate any peptides that are not unique to the protein of interest to insure that the MRM results are specific (see Note 8).
6. Select "Add Selected Peptides" and then "Build Method" to build an MRM-IDA method using parameters appropriate for your triple quadrupole mass spectrometer.
7. Save this acquisition method.

3.2 Tryptic Digestion of Pure Target Protein

A tryptic digest of purified protein produces an approximately equimolar mix of peptides, facilitating a comparison of their MRM signals. Peptides that are unstable, insoluble, or have ends which are poorly cleaved by trypsin will be present in reduced amounts, reducing the likelihood that these undesirable peptides will be included in the final assay.

It is usually cost-effective to purchase the protein from a commercial source, since very little protein is required to develop an MRM assay. Novel proteins can be produced using conventional methods. Full-length proteins will yield the largest number of peptides, but isolated domains can also be used.

The buffer conditions required for an efficient tryptic digest would interfere with reverse-phase chromatography, so the digested peptides must be desalted before they can be injected into an LC-MS/MS system. A variety of solid-phase extraction products are available for this task (see Note 9). Peptides with a wide range of physical properties can be purified without substantial losses using a hydrophilic-lipophilic-balanced (HLB) reverse-phase sorbent (15).

1. Reconstitute 2 μg protein in 100 μl of 100 mM NH_4HCO_3 (see Note 10).
2. Add 1 μl of 100 mM TCEP. Incubate at 55 °C for 30 min to reduce of disulfide bonds.
3. Add 1 μl of 200 mM iodoacetamide. Incubate in the dark at 37 °C for 30 min to alkylate cysteines.
4. Add 100 ng trypsin (1 μg trypsin: 20 μg protein). Incubate at 37 °C for 4–6 h or overnight (see Note 11).
5. Add 5 μl of 10 % formic acid to stop the reaction.
6. Add 400 μl dd H_2O and 500 μl 4 % phosphoric acid (4 % H_3PO_4) to the trypsin digest.
7. Insert an HLB cartridge into the vacuum manifold.
8. Condition with 1 ml 100 % methanol.
9. Equilibrate three times with 1 ml 0.1 % formic acid.
10. Load the trypsin digest. Flow slowly/dropwise.
11. Wash three times with 1 ml 0.1 % formic acid.
12. Elute with 1 ml 0.1 % formic acid in 80 % acetonitrile (ACN). Flow slowly.
13. Transfer the eluate to a low-retention 1.5-ml microfuge tube and evaporate the solvent in a speed vac.
14. Reconstitute the peptides in 100 μl 20 % ACN/0.1 % formic acid and store at −20 °C.
15. Calculate the expected peptide concentration in pmol/μl (2/5 pmol/μl for ST2).

3.3 Experimental Identification of the Best Peptides and Transitions

The preliminary transitions from Subheading 3.1 are used to identify peptides from the trypsin digest of Subheading 3.2 as they elute into the mass spectrometer. A relatively high concentration of peptides is injected because the preliminary transitions may have low signals. A total ion chromatograph is collected for each transition,

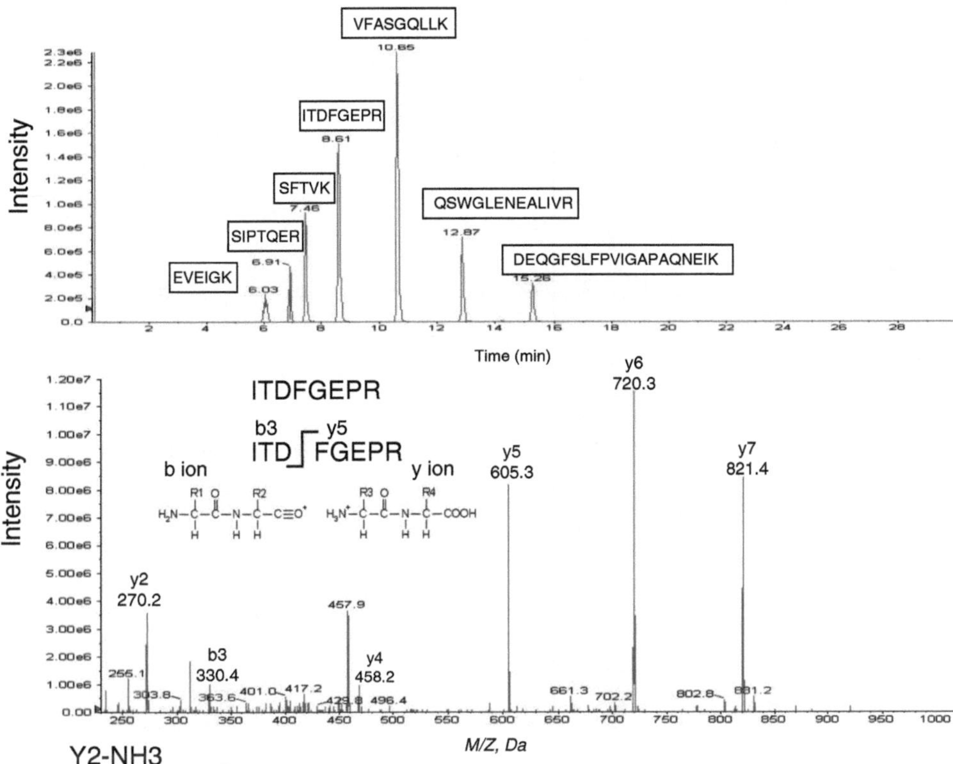

Fig. 3 MRM-triggered information-dependent analysis (MRM-IDA) of ST2 peptides. *Top panel*. Total ion chromatograph from an MRM analysis with preliminary transitions for 9 ST2 peptides. Two of the peptides were not detected. *Bottom panel*. MRM-triggered MS/MS spectra for the peptide ITDFGEPR. Product ion peaks are labeled with *m/z* values. *Inset*. Fragmentation of the ITDFGEPR peptide into b3 and y5 product ions

and whenever an MRM signal is detected above a predefined threshold, it triggers the acquisition of a full spectrum MS/MS fragmentation scan (Fig. 3). Peaks in the MS/MS data are matched to the mass/charge ratios of b and y ions in MRMPilot and can also be matched to a wider range of product ions using MS-Product http://prospector.ucsf.edu/prospector/cgi-bin/msform.cgi?form=msproduct. Low-energy collision-induced dissociation (CID) spectra exhibit mostly y and b ions (16). The three to five most abundant product ions are used to design new MRM transitions (Tables 1 and 2).

1. Prepare 50 μl of peptide diluted to 500 fmol/μl in 0.1 % formic acid, 20 % ACN.
2. Inject 5 μl of the peptides three times using the MRM-IDA method built in Subheading 3.1. Set "EPI Scan to Sum" at 2 to average product ion scans. Five preliminary transitions are monitored for each targeted peptide. Whenever one of these transitions is detected, a full MS/MS scan is collected. The MRM-IDA method also collects MRM data from the preliminary transitions.

Table 1
MRM-IDA results for ST2. Preliminary transitions were designed for 9 ST2 peptides. A tryptic digest of ST2 was injected into the LC-MS/MS system. When a peptide eluted into the mass spectrometer, it was detected by MRM with one or more of the preliminary transitions. Transitions that did not yield a robust and reproducible signal were eliminated. For the information-dependent analysis (IDA), detection above a set threshold triggered the collection of a complete product ion (MS2) scan. The most prominent peaks in the MS2 scan were used to generate new transitions

Peptide sequence	Highest peak intensity	Lowest peak intensity	Lowest CV (%)	Highest CV (%)	Preliminary transitions	Transitions eliminated	Transitions added	Post-IDA transitions
VFASGQLLK	300,000	500	5	60	5	1	1	5
SIPTQER	190,000	80	38	44	5	3	0	2
SFTVK	700,000	3,500	7	43	4	2	1	3
QSWGLENEALIVR	190,000	5,000	4	16	5	1	1	5
ITDFGEPR	570,000	500	7	16	5	2	2	5
IADVK	900	90	8	40	4	4	0	0
GTQFLAAVLWQLNGTK	70	60	20	100	5	5	0	0
EVEIGK	400,000	300	14	43	5	2	2	5
DEQGFSLFPVIGAPAQNEIK	150,000	20,000	4	16	5	0	0	5

Table 2
MRM-IDA analysis of the ITDFGEPR peptide of ST2. MRMPilot was used to design five preliminary transitions, three of which yielded robust peaks with a mean peak height greater than 100,000 and a %CV below 20 %. The other two peaks were removed. Two new transitions were identified from an MRM-triggered MS2 scan. Peak intensity and %CV were not determined for these untargeted transitions

Fragment type	Precursor *m/z*	Fragment *m/z*	Retention time (min)	Collision energy (V)	Mean peak height	% CV	Action
2+/y7	467.73	821.38	8.6	26	364,466	14	Keep
2+/b7	467.73	760.35	8.6	26	550	13	Remove
2+/y6	467.73	720.33	8.6	26	566,418	14	Keep
2+/b6	467.73	663.30	8.6	26	1,590	7	Remove
2+/y5	467.73	605.30	8.6	26	296,858	16	Keep
2+/y4	467.73	458.24	8.6	26			Add
2+/y2	467.73	272.17	8.6	26			Add

3. Add the MRM-IDA data to MRMPilot. In MRMPilot, select "Add MRM data." Import the *.wiff data files from the three replicate injections and review the data in MultiQuant to check that peaks are integrated correctly. Data from the three replicates will be averaged.
4. In the MRMPilot Table view, deselect any transitions with a %CV > 100.
5. Add new transitions from the triggered MS/MS scans. In the table, the results are grouped by peptide, with the highest intensity transition listed first. Select the highest intensity transition and view the spectral graph. Identify five product ions with the highest intensity. The software will indicate how these peaks were derived from the parent ion. If any of these transitions are different from the five preliminary transitions, add a new transition to the method. Deselect any preliminary transitions with intensity less than the top five.

3.4 Collision Energy Optimization

A variety of MS parameters can be adjusted to increase the MRM signal (see Note 12). The most important parameter is usually the collision energy (CE) because the CE affects the abundance and proportion of product ions.

1. For each peptide, calculate a theoretical CE optimum (3):
   $CE = 0.043 \times (\text{precursor ion } m/z) + 2.25$.
2. Modify the MRM method to run five scans for each peptide with collision energies ranging from 6 V above to 6 V below

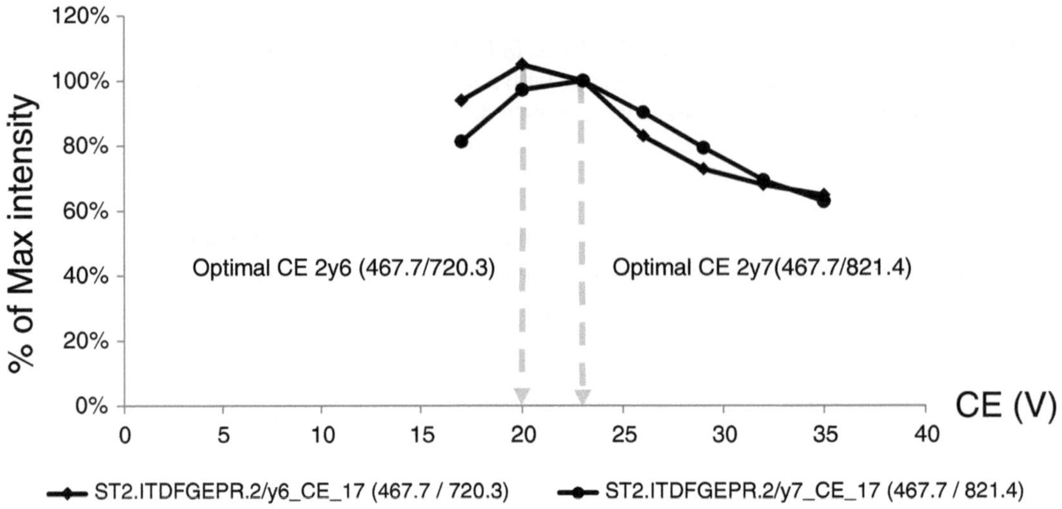

Fig. 4 CE optimization. MRM transitions for products of the ITDFGEPR peptide of ST2 were measured at the indicated collision energy voltages

the theoretical optimum. In MRMPilot, select "find best CE" and set the CE increment to "3" and the # of steps to "5".

3. Inject 5 μl of the peptides three times using this modified MRM method.
4. Use MultiQuant to obtain integrated peak areas and CVs for all transitions and CE settings.
5. For each transition, identify the CE producing the highest peak area with a CV < 20 % and delete the other collision energies from the method (Fig. 4).

3.5 Evaluate Assay Performance Using a Calibration Curve to Identify Transitions with a Linear Response

A standard curve is prepared to corrolate MRM signal intensity to digested protein concentration for each transition. This data is used to calculate the lower limit of quantification (LLOQ), providing a measure of the sensitivity for each transition. Assay performance can be further evaluated using methods commonly used for the evaluation of quantitative immunoassays (17). For the final MRM method, three peptides with four transitions for each are normally selected based upon LLOQ, linearity, recovery, and reproducibility (3, 18).

1. Prepare five-fold serial dilutions of the trypsin-digested peptide solution from Subheading 3.2 at concentrations ranging from 0.1 to 1,000 fmol/μl in 0.1 % formic acid, 20 % ACN.
2. Inject each dilution three times using the CE-optimized MRM method generated in Subheading 3.4.
3. Plot mean peak area vs. concentration to establish a standard curve for each transition (Fig. 5).
4. Fit the data with a linear curve (see Note 13).

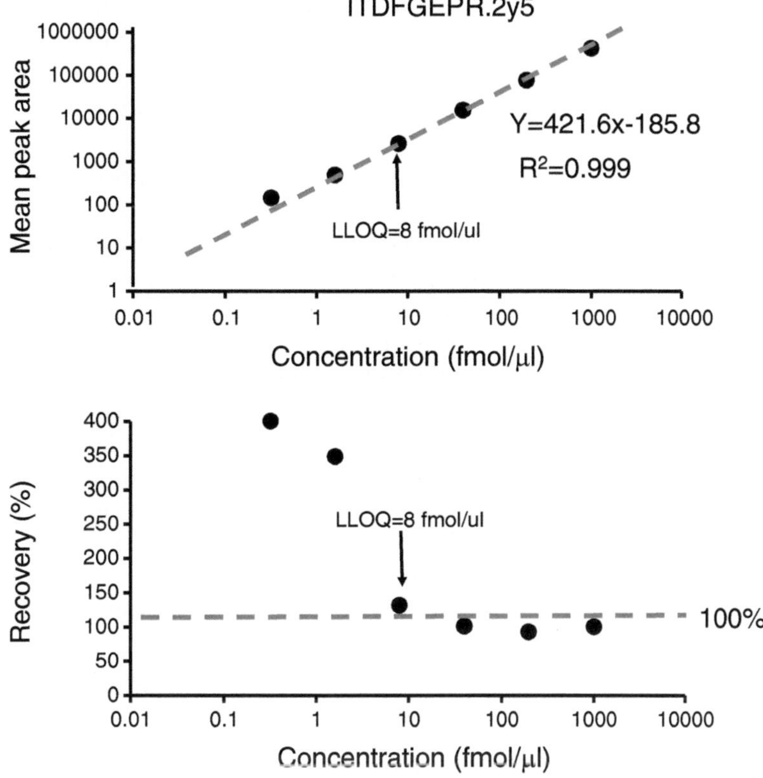

Fig. 5 Measuring the sensitivity of an MRM assay for the ITDFGEPR.2y5 transition. *Upper panel*. Titration of the ITDFGEPR peptide. The data was fit by linear regression. *Lower panel*. %Recovery calculated as measured/expected concentration. The %CV was >20 % and the % recovery was outside the range of 80–120 % at concentrations below the LLOQ

5. Calculate the percentage of coefficient of variation (%CV) and expected concentration using peak intensity data using the formula derived from fitting the linear curve.
6. Determine the LLOQ as the lowest concentration exhibiting a signal to noise ratio (S/N) above 5, accuracy between 80 % and 120 % of the true value, and a %CV below 20 %.
7. Finally, the peptides themselves are compared to determine which ones will be best for the final assay. Factors to evaluate include the intensity and sensitivity (LLOQ) of the best transition, the number of measurable transitions, and whether the measurable transitions provide the desired specificity (see Note 14).

3.6 Conclusion

The use of pure protein as a source of peptides to build an MRM method makes it possible to iteratively select the best of all possible peptides and transitions based upon sensitivity (LLOQ) and reproducibility. Once a method is built, it can be adapted to measure the amount of protein in a complex biological sample.

Heavy-isotope-labeled peptides are required for absolute quantitation (see Note 15). Tryptic peptides have a C-terminal lysine or arginine. Replacing these natural amino acids with ($^{15}N^{13}C$)-lysine or ($^{15}N^{13}C$)-arginine increases the mass of a peptide (and its y ion fragments) by 8 or 10 Da, respectively (see Note 16). Nevertheless, natural and heavy-isotope peptides are chemically identical—they will co-elute from an HPLC column and produce the same proportion of fragment ions. Thus, when a known amount of a heavy-isotope-labeled peptide is spiked into a tryptic digest prepared from a complex biological sample, it can serve as an internal standard to measure the absolute concentration of the native protein.

Complex biological samples may contain components that interfere with MRM assays, causing a "matrix" effect. Matrix effects depend upon the type of sample (plasma, urine, cell lysate, etc.) and preparation method (abundant protein depletion, immunoprecipitation, LC prefractionation, etc.). The matrix effect can be evaluated by spiking heavy-isotope peptides in control samples prepared using the same method.

MRM assays provide a powerful and versatile tool for the quantitative detection of targeted proteins. Once an MRM assay is built, it can be used for a wide range of analytical purposes limited only by the imagination of the investigator.

4 Notes

1. Triple quadrupole mass spectrometers currently have the best resolution, sensitivity, and multiplexing capabilities, but other mass spectrometers can be used as well.
2. Triple quadrupole mass spectrometers from other manufactures, such as Thermo Scientific (Thermo Fisher Scientific, Waltham, MA), Waters (Waters, Milford, MA), Agilent (Agilent Technologies, Santa Clara, CA), and Shimadzu (Columbia, MD), can also be used. Upstream separation can be performed by microflow liquid chromatography, nanoflow liquid chromatography, ultra performance liquid chromatography, or capillary electrophoresis.
3. MRMPilot is designed to work with AB SCIEX mass spectrometers. Other instrument suppliers provide their own MRM assay design software, such as MassHunter from Agilent. MRMaid, Skyline, and PChopper are freely available software packages that can be used to design MRM assays (10, 19–21).
4. Trypsin is the preferred protease for many mass spectrometry applications because it yields peptides with a positive charge at both ends: an amino group at the N-terminus and a lysine or arginine at the C-terminus. After collision-induced dissociation,

both fragments are charged and can be detected by MS/MS. Other specific proteases can be used, such as chymotrypsin, LysC, and AspN, if it is essential to detect a specific sequence within a protein (perhaps because of a posttranslational modification), and that sequence cannot be detected as a tryptic peptide. Another alternative is to use a chemical agent, such as cyanogen bromide, to cleave a protein into specific peptides.

5. A 5500 QTRAP triple quadrupole/linear ion trap mass spectrometer detects peptides with m/z values ranging from 300 to 1,200, corresponding to approximately 5–30 amino acids. The cost and available purity of extremely long or short synthetic peptides also places practical limits on peptide lengths for quantitative MRM assays.
6. Peptides containing methionine, cysteine, and glycosylated asparagine are generally unsuitable for MRM assays. Methionine is subject to an indeterminate degree of in vitro oxidation. Cysteines are typically alkylated and reduced during sample preparation. Alkylation with iodoacetamide yields carbamidomethyl cysteine, with a mass shift of +57.021465 amu. Peptides with cysteines can be used, if necessary, but the corresponding peptide standard for quantitation should be synthesized with carbamidomethyl cysteine. N-linked glycans are large and heterogeneous and are not incorporated into recombinant proteins expressed in *E. coli*. Peptides with glycosylated asparagines can be used for MRM assays if the carbohydrate is removed with PNGase F, which also converts the asparagine to aspartic acid.
7. Protein sequences are available from many other sources, including Entrez protein (http://www.ncbi.nlm.nih.gov/protein).
8. Peptides that are conserved between isoforms and members of a protein family can be used with caution.
9. C-18 reverse-phase media may also be used for peptide cleanup. Popular products sold for this purpose include ZipTips (Millipore, Billerica, MA), Stage tips (Thermo Fisher Scientific, San Jose, CA), and TopTips (PolyLC Inc, Columbia, MD).
10. It is convenient to prepare one large-scale digest rather than repeating the digest multiple times, but the scale of the digest can be reduced for proteins that are expensive or sold in smaller packages. Very little peptide is required per assay, but multiple injections are used to identify transitions and optimize detection.
11. Conditions for the trypsin digest may need to be optimized to ensure that the reaction proceeds to completion. Progress of the reaction can be monitored by running aliquots collected at

various time points on an SDS-PAGE gel and staining the gel with silver (22). Denaturants such as urea can be added if necessary. TX-100 and SDS should be avoided because they interfere with many solid-phase extraction procedures.

12. In addition to CE, other parameters that can be adjusted to optimize detection include the depolarization potential (DP) and collision cell exit potential (CXP).
13. The calibration curve can be calculated using either weighted or non-weighted linear regression models. The unweighted approach assumes that the curve is equally accurate over the entire concentration range. For a large dynamic range, a weighted approach is often preferred to minimize the relative error at low concentrations (23).
14. It may be essential to include some low-intensity transitions to ensure specificity. For example, phosphopeptides often have more than one amino acid that might be phosphorylated. Transitions arising from fragmentation between potentially phosphorylated amino acids are required to localize the phosphorylation site.
15. The concentration of peptide standards should be measured by an amino acid analysis.
16. Heavy-isotope tags can be incorporated at any position in a synthetic peptide standard. ($^{15}N^{13}C$)-Lysine and ($^{15}N^{13}C$)-arginine are convenient but more expensive than other heavy-isotope-labeled amino acids. Heavy-isotope-labeled peptides can also be produced by specific cleavage of labeled proteins (24). Care should be taken to ensure that heavy-isotope protein standards are 100 % labeled because any peptides containing contaminant natural amino acids would be indistinguishable from the peptides to be quantified in the biological sample.

References

1. Anderson L, Hunter CL (2006) Quantitative mass spectrometric multiple reaction monitoring assays for major plasma proteins. Mol Cell Proteomics 5:573–588
2. Lange V, Malmstrom JA, Didion J, King NL, Johansson BP, Schafer J et al (2008) Targeted quantitative analysis of Streptococcus Pyogenes virulence factors by multiple reaction monitoring. Mol Cell Proteomics 7: 1489–1500
3. Kuzyk MA, Smith D, Yang J, Cross TJ, Jackson AM, Hardie DB, Anderson NL, Borchers CH (2009) Multiple reaction monitoring-based, multiplexed, absolute quantitation of 45 proteins in human plasma. Mol Cell Proteomics 8:1860–1877
4. Addona TA, Abbatiello SE, Schilling B, Skates SJ, Mani DR, Bunk DM et al (2009) Multi-site assessment of the precision and reproducibility of multiple reaction monitoring-based measurements of proteins in plasma. Nat Biotechnol 27:633–641
5. Gerber SA, Rush J, Stemman O, Kirschner MW, Gygi SP (2003) Absolute quantification of proteins and phosphoproteins from cell lysates by tandem MS. Proc Natl Acad Sci USA 100:6940–6945
6. Stahl-Zeng J, Lange V, Ossola R, Eckhardt K, Krek W, Aebersold R, Domon B (2007) High sensitivity detection of plasma proteins by multiple reaction monitoring of N-glycosites. Mol Cell Proteomics 6:1809–1817

7. Unwin RD, Griffiths JR, Leverentz MK, Grallert A, Hagan IM, Whetton AD (2005) Multiple reaction monitoring to identify sites of protein phosphorylation with high sensitivity. Mol Cell Proteomics 4:1134–1144
8. Xu P, Peng J (2006) Dissecting the ubiquitin pathway by mass spectrometry. Biochim Biophys Acta 1764:1940–1947
9. Nicol GR, Han M, Kim J, Birse CE, Brand E, Nguyen A et al (2008) Use of an immunoaffinity-mass spectrometry-based approach for the quantification of protein biomarkers from serum samples of lung cancer patients. Mol Cell Proteomics 7:1974–1982
10. Mead JA, Bianco L, Ottone V, Barton C, Kay RG, Lilley KS, Bond NJ, Bessant C (2009) MRMaid, the web-based tool for designing multiple reaction monitoring (MRM) transitions. Mol Cell Proteomics 8:696–705
11. Chem Mead JA, Bianco L, Bessant C (2010) Mining proteomic MS/MS data for MRM transitions. Methods Mol Biol 604:187–199
12. Dhillon OS, Narayan HK, Quinn PA, Squire IB, Davies JE, Ng LL (2011) Interleukin 33 and ST2 in non-ST-elevation myocardial infarction: comparison with Global Registry of Acute Coronary Events Risk Scoring and NT-proBNP. Am Heart J 161:1163–1170
13. Hsu CL, Neilsen CV, Bryce PJ (2010) IL-33 is produced by mast cells and regulates IgE-dependent inflammation. PLoS One 5: e11944
14. Moffatt MF, Gut IG, Demenais F, Strachan DP, Bouzigon E, Heath S, von Mutius E, Farrall M, Lathrop M, Cookson WO, GABRIEL Consortium (2010) A large-scale, consortium-based genomewide association study of asthma. N Engl J Med 363:1211–1221
15. Kim MK, Lee TH, Suh JH, Eom HY, Min JW, Yeom H, Kim HJ et al (2010) Development and validation of a liquid chromatography-tandem mass spectrometry method for the determination of goserelin in rabbit plasma. J Chromatogr B Analyt Technol Biomed Life Sci 878:2235–2242
16. Sleno L, Volmer DA (2004) Ion activation methods for tandem mass spectrometry. J Mass Spectrom 39:1091–1112
17. Fu Q, Zhu J, Van Eyk JE (2010) Comparison of multiplex immunoassay platforms. Clin Chem 56:314–318
18. Almeida AM, Castel-Branco MM, Falcao AC (2002) Linear regression for calibration lines revisited: weighting schemes for bioanalytical methods. J Chromatogr B Analyt Technol Biomed Life Sci 774:215–222
19. Afzal V, Huang JT, Atrih A, Crowther DJ (2011) PChopper: high throughput peptide prediction for MRM/SRM transition design. BMC Bioinformatics 12:338
20. Cham Mead JA, Bianco L, Bessant C (2010) Free computational resources for designing selected reaction monitoring transitions. Proteomics 10:1106–1126
21. Stergachis AB, MacLean B, Lee K, Stamatoyannopoulos JA, MacCoss MJ (2011) Rapid empirical discovery of optimal peptides for targeted proteomics. Nat Methods 8: 1041–1043
22. Winkler C, Denker K, Wortelkamp S, Sickmann A (2007) Silver- and Coomassie-staining protocols: detection limits and compatibility with ESI MS. Electrophoresis 28:2095–2099
23. Johnson EL, Reynolds DL, Wright DS, Pachla LA (1988) Biological sample preparation and data reduction concepts in pharmaceutical analysis. J Chromatogr Sci 26:372–379
24. Lebert D, Dupuis A, Garin J, Bruley C, Brun V (2011) Production and use of stable isotope-labeled proteins for absolute quantitative proteomics. Methods Mol Biol 753:93–115

Chapter 17

A Sequential Extraction Methodology for Cardiac Extracellular Matrix Prior to Proteomics Analysis

Javier Barallobre-Barreiro, Athanasios Didangelos, Xiaoke Yin, Nieves Doménech, and Manuel Mayr

Abstract

Cardiac fibrosis is characterized by excessive deposition of extracellular matrix (ECM) and is a common complication of various cardiovascular diseases. However, little is known about proteins in the cardiac extracellular space. Proteomics analysis of cardiac ECM can be challenging due to the presence of more abundant intracellular proteins, the low degree of solubility of integral ECM proteins, and the presence of abundant posttranslational modifications. Here we describe an extraction methodology based on tissue decellularization, which allows the biochemical subfractionation of extracellular proteins in cardiac tissue. These relatively low-complexity protein fractions are suitable for analysis by gel-LC-MS/MS and other proteomics techniques.

Key words Extracellular matrix, Cardiac fibrosis, Fractionation techniques

1 Introduction

The proteomic analysis of cardiac tissues is technically challenging. Cardiac tissue is highly enriched in myofilament proteins, and a few high-abundant proteins, i.e., myosin, actin, and titin, hamper the identification of less-abundant proteins (1). Most proteomics studies performed so far have been biased towards cytosolic proteins or focused on cardiac mitochondria. Less is known about proteins in the cardiac extracellular space. Cardiac fibrosis is a common complication of various cardiovascular diseases, such as cardiomyopathy and myocardial infarction, and the characterization of this subproteome will be important in disease. At the tissue level, cardiac fibrosis is characterized by excessive deposition of extracellular matrix (ECM) (2). The ECM constitutes the main scaffold in cardiovascular tissues and is responsible for many tissue properties. It is composed of aggregating proteoglycans, collagens, and

Fernando Vivanco (ed.), *Heart Proteomics: Methods and Protocols*, Methods in Molecular Biology, vol. 1005,
DOI 10.1007/978-1-62703-386-2_17,

glycoproteins. Besides, the extracellular space is filled with interstitial fluid, which contains apolipoproteins, growth factors, cytokines, and proteases that associate with components of the ECM (3) and are essential to understand disease mechanisms.

Analysis of the ECM can be challenging for the following reasons: (1) More abundant intracellular proteins hamper the identification of scarce ECM components. (2) Integral ECM proteins are heavily cross-linked and difficult to solubilize. (3) The presence of abundant posttranslational modifications (i.e., glycosylation) alters their molecular mass, charge, and electrophoretic properties, affecting both separation and identification by mass spectrometry (4). We have recently developed an extraction methodology for cardiac tissues, which allows the biochemical subfractionation of extracellular proteins (5). This three-step procedure is an adaptation from a previous method used in our lab to extract extracellular proteins in vascular tissues (6): First, 0.5 M NaCl is used to extract ECM associated and loosely bound ECM proteins as well as proteins in the interstitial fluid, including apolipoproteins, proteases, and non-cross-linked glycoproteins, as well as newly synthesized proteins of the ECM. Then, a decellularization step is performed with SDS. Finally, ECM proteins are extracted with 4 M guanidine hydrochloride (GuHCl). The latter extract contains abundant aggregating proteoglycans (i.e., versican, aggrecan) and cross-linked collagens (i.e., collagens I, IV), integral components of the ECM (Fig. 1). Both, the NaCl and GuHCl fraction are of relatively low complexity and suitable for proteomic analysis by gel-LC-MS/MS or other techniques such as difference in gel electrophoresis (7). Based on the results from the proteomics analysis, samples can then be grouped based on their EMC profiles (Fig. 2).

2 Materials

2.1 General Solutions

1. 250 mM EDTA, pH 8.0, stock solution. Adjust pH using 12 M NaOH.
2. 1 % Sodium dodecyl sulfate (SDS).
3. 100 mM Tris–HCl stock solution, pH 7.5. Adjust pH with 12 M HCl.
4. 1× Phosphate-buffered saline (PBS): 150 mM NaCl, 1.7 mM KH_2PO_4, and 5 mM Na_2HPO_4. Adjust carefully to pH 7.4 using 6 M NaOH.

2.2 Extraction Buffers

Extraction buffers can be stored at room temperature. Before use, add 1/100 (v/v) of a broad-spectrum cocktail of protease inhibitors (104 mM AEBSF, 80 μM aprotinin, 4 mM bestatin, 1.4 mM E-64, 2 mM leupeptin, pepstatin A, 1.5 mM) (Sigma-Aldrich).

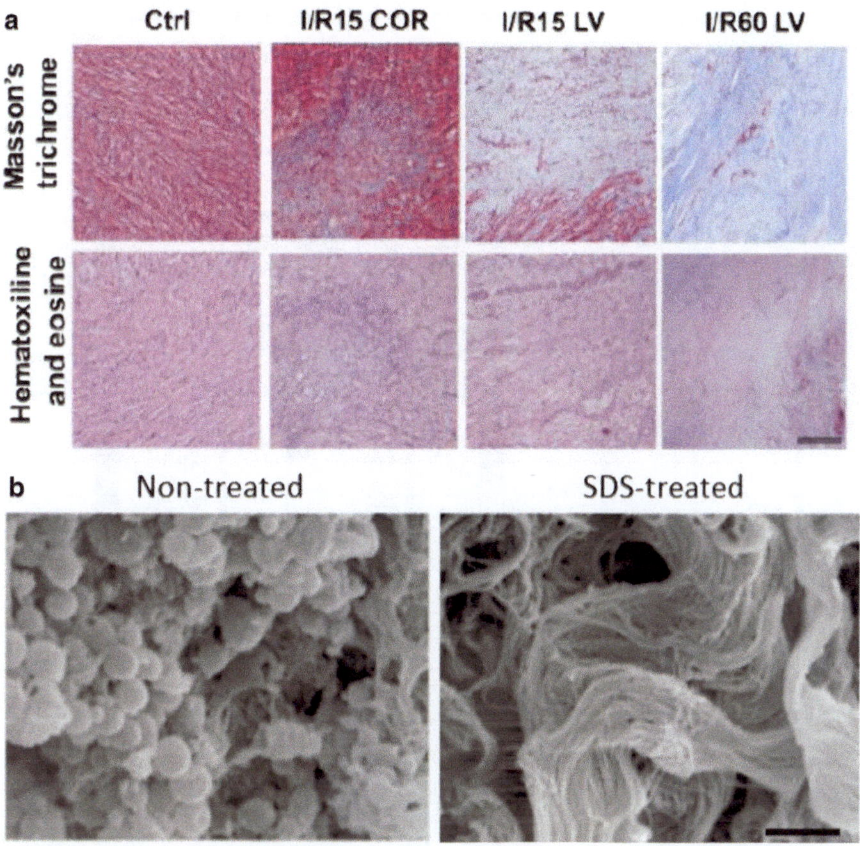

Fig. 1 (**a**) Masson's trichrome and hematoxylin and eosin staining in control and infarcted porcine hearts used for the extraction procedure. Different stages of fibrosis were considered: Non-fibrotic healthy controls (Ctrl), tissue from the border region 15 days after induced ischemia and reperfusion injury (I/R15 COR), focal injury in the left ventricle at 15 days (I/R15 LV), and focal injury in the left ventricle at 60 days (I/R60 LV). Scale bar: 200 μm. (**b**) High-resolution transmission electron microscopic (TEM) analyses were performed using a JSM-6400 scanning electron microscopy (Jeol) to visualize the effective decellularization and remaining ECM fibers after the SDS treatment. Scale bar: 2 μm. Adapted from Barallobre-Barreiro et al. (5)

1. Extraction buffer 1 (NaCl buffer): 0.5 M NaCl, 10 mM Tris–HCl, pH 7.5 (dilute the 100 mM Tris–HCl stock solution), and 25 mM EDTA.
2. Extraction buffer 2 (SDS buffer): 0.1 % (35 mM) sodium dodecyl sulfate (SDS), 25 mM EDTA (see Note 1).
3. Extraction buffer 3 (GuHCl buffer): 4 M guanidine hydrochloride, 50 mM sodium acetate, and 25 mM EDTA. pH should be adjusted to 5.8 using 12 M NaOH.

2.3 Deglycosylation Buffer (4×)

600 mM NaCl and 200 mM Na_2HPO_4 in ddH_2O, pH 6.8. Prior to deglycosylation, add 0.05 U of the following enzymes to the 1× deglycosylation buffers:

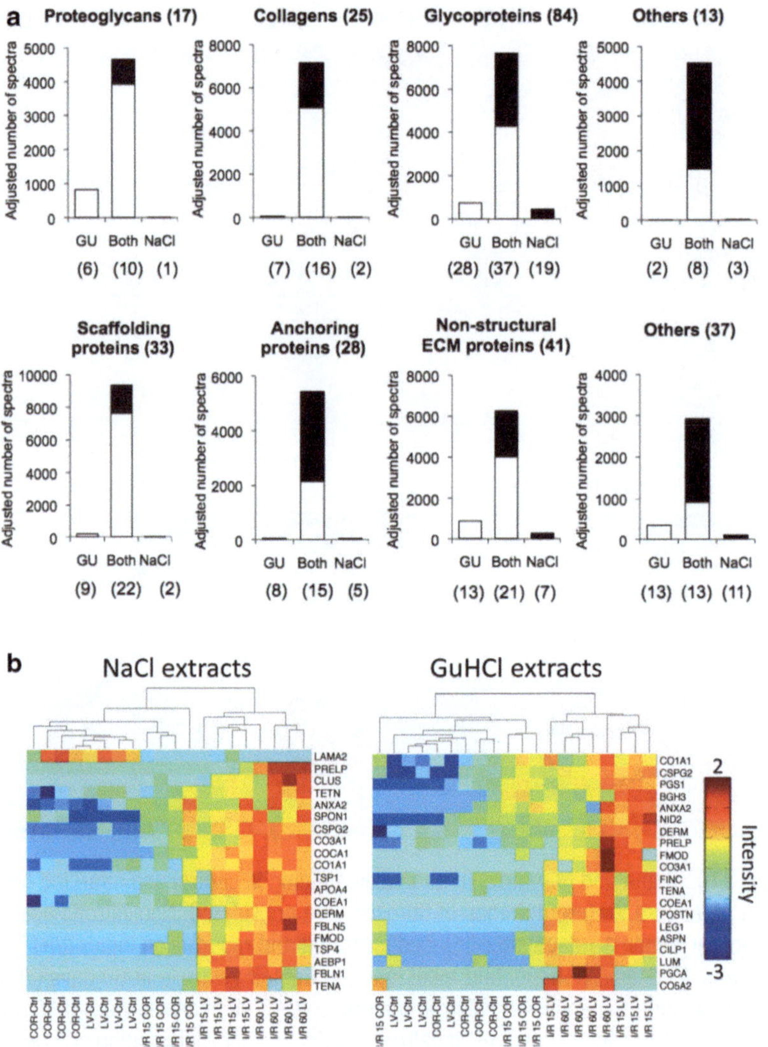

Fig. 2 (**a**) The distribution of ECM proteins was compared in the NaCl and GuHCl extracts (GU) according to their biochemical (*upper panels*) and functional classification (*lower panels*). For each category, the number of identified proteins (*n*) is shown while the total spectral counts are depicted in the charts. (**b**) Hierarchical clustering of the top 20 most differentially expressed proteins across experimental groups (one-way ANOVA) in the NaCl and GuHCl extracts. *Blue-red* heat map values correspond to low–high protein expression levels. Adapted from Barallobre-Barreiro et al. (5)

1. Chondroitinase ABC from Proteus vulgaris (Sigma-Aldrich): It catalyzes the removal of polysaccharides containing 1,4-β-D-hexosaminyl and 1,3-β-D-glucuronosyl or 1,3-α-L-iduronosyl linkages to disaccharides containing 4-deoxy-β-D-gluc-4-enuronosyl groups. It acts on chondroitin 4-sulfate, chondroitin 6-sulfate, and dermatan sulfate glycosaminoglycan side chains.

2. Keratanase (endo β-galactosidase) from Bacteroides fragilis (Sigma-Aldrich): it cleaves internal 1,4-β-galactose linkages in unbranched, repeating poly-*N*-acetyllactosamine and keratan sulfate.

2.4 Coomassie Blue

0.1 % Coomassie brilliant blue R-250, 5 % acetic acid, 25 % methanol.

3 Methods

3.1 Tissue Sampling and Preparation

1. Cardiac tissue samples are frozen in liquid nitrogen and stored at −80 °C (see Note 2).
2. Weigh the tissue. We tend to use 30–50 mg of tissue.
3. Dice the cardiac tissue into smaller pieces.
4. Perform five washes with PBS plus EDTA and proteinase inhibitors at 4 °C to minimize blood contamination.

3.2 Sequential Protein Extraction

A schematic workflow of the extraction procedure is shown in Fig. 3.

3.2.1 Step 1: Incubation with NaCl Buffer

This ionic buffer facilitates the extraction of loosely bound and newly synthesized extracellular matrix proteins. It is detergent-free, non-denaturing, and does not disrupt cell membranes (8).

1. After washing, put tissue samples into 1.5 ml tubes with screw caps and add NaCl buffer, exactly ten times (v/w) the tissue weight (see Note 3).
2. Vortex the tubes at room temperature for 1 h at 1000rpm (see Notes 4 and 5).
3. Collect the supernatants and centrifuge at maximum speed for 10 min at 4 °C, store at −20 °C until use.
4. Briefly wash the remaining tissue with fresh NaCl buffer followed by ddH_2O.

3.2.2 Step 2: Incubation with SDS Buffer

1. Add SDS buffer exactly ten times (v/w) the tissue weight.
2. Vortex the tubes at room temperature for 16 h at 1000 rpm. Particular care must be taken to ensure a low vortex speed to minimize mechanical disruption of the ECM.
3. Collect the supernatants, centrifuge at maximum speed for 10 min at 4 °C, and store at −20 °C until use.
4. Briefly wash the remaining tissue with ddH_2O to remove SDS.

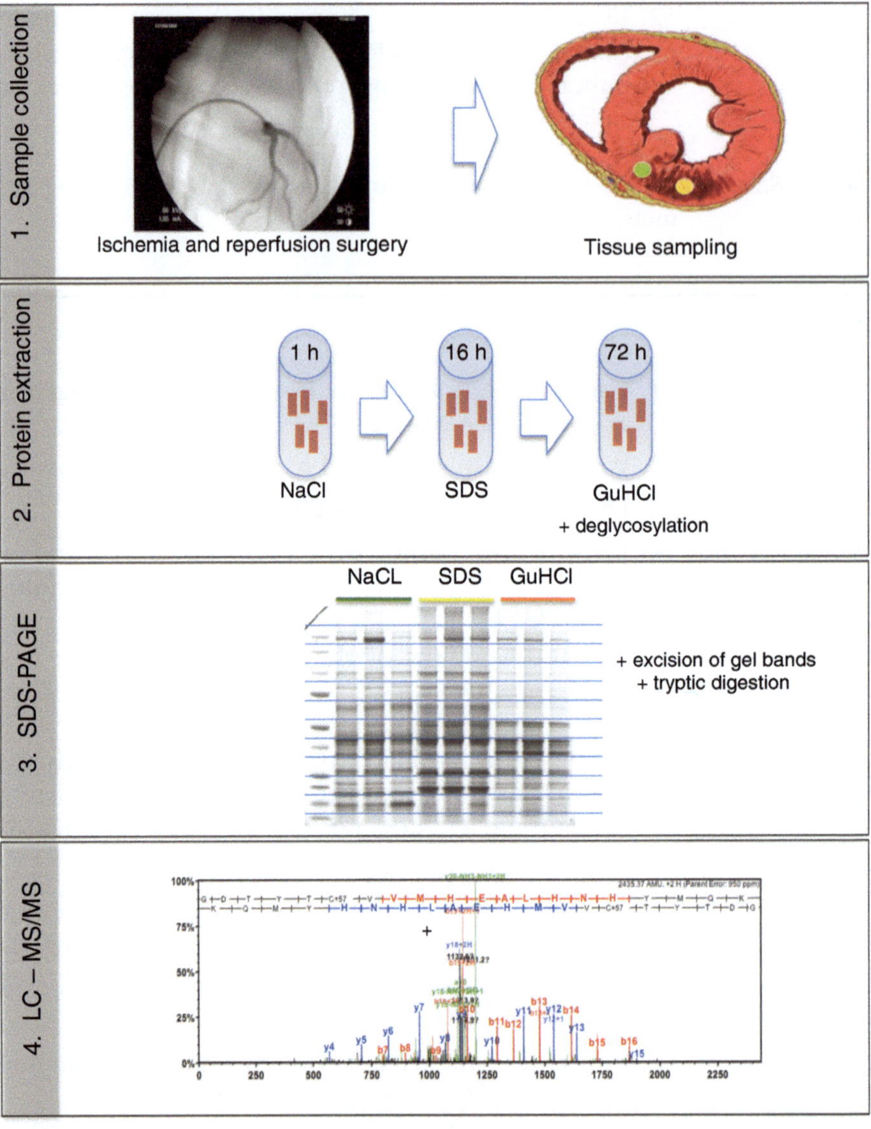

Fig. 3 Schematic workflow of the protocol followed for the proteomic analysis of cardiac extracellular space. Contaminant proteins, which are not extracellular, but will be present and identified in the NaCl or GuHCl extracts, can be readily identified based on gene ontology annotations. Myofilament proteins, in particular, are often not completely solubilized after SDS treatment and are detected in the GuHCl extracts, which can be of interest for a range of cardiac pathologies

3.2.3 Step 3: Incubation with GuHCl Buffer

1. Add GuHCl buffer five times (v/w) the tissue weight.
2. Vortex the tubes at room temperature for 48–72 h at maximum speed. Vigorous vortexing is recommended to facilitate mechanical disruption of the ECM.
3. Collect the supernatant, centrifuge at 16,000 × *g* for 10 min at 4 °C, and store at −20 °C until use.

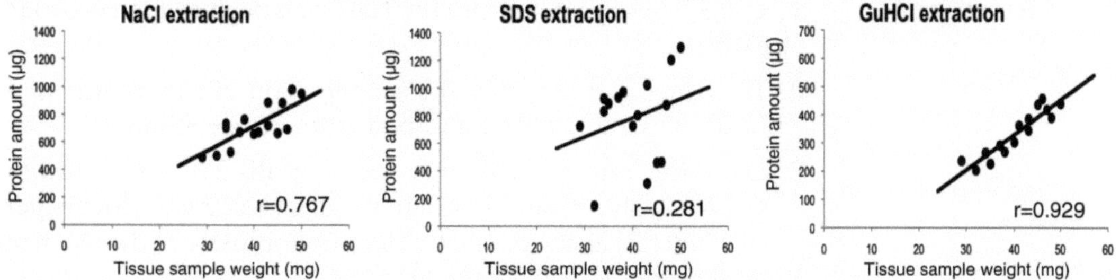

Fig. 4 Protein extraction yields for each of the steps including 0.5 M NaCl, 0.1 % SDS buffer, and 4 M GuHCl incubations. A strong correlation exists between the initial amount of tissue and the final amount of protein extracted after incubation with NaCl and GuHCl buffers. In contrast, tissue cellularity determines the yield in the SDS extracts

3.3 Measurements of Protein Concentration

The protein concentration for all extracts can be determined by UV absorbance at 280 nm using an extinction coefficient of 1.1 of a 0.1 % mg/ml solution calculated on the basis of the frequency of tyrosine and tryptophan, which are the main UV-light-absorbing amino acids at 280 nm in mammalian proteins (9). The constant tissue weight to buffer volume ratio should result in a linear relationship of the protein concentrations determined in NaCl and GuHCl extracts. The total protein amount obtained from the SDS extracts largely depends on the cellularity of the tissue (Fig. 4).

3.4 Precipitation of the GuHCl Extracts

GuHCl is not compatible with standard SDS PAGE gels. It precipitates in the presence of SDS. A complete removal is essential.

1. Precipitate GuHCl extracts using 8× the volume of absolute ethanol.
2. Incubate at −20 °C for at least 6 h (ideally overnight).
3. Centrifuge at 16,000 × g for 30 min at 4 °C.
4. Carefully remove the supernatant and completely dry the samples using a vacuum centrifuge.

3.5 Deglycosylation of the GuHCl Extracts

The 4 M GuHCl buffer is effective to extract proteoglycans from matrix-rich tissues such as tendons (4), cartilage (10), and vasculature (11). However, unless the lateral glycosaminoglycan chains are cleaved, large aggregating proteoglycans (i.e., versican and aggrecan) will not enter the SDS-PAGE gel.

1. Solubilize the dried GuHCl extracts using deglycosylation buffer containing the enzymes specified in the Subheading 2 (see Note 6).
2. Incubate the samples in agitation at 37 °C for 12 h.
3. Clean the extracts by centrifugation at 16,000 × g for 10 min.

3.6 Electrophoresis

1. Use 35 μg of protein from the NaCl and deglycosylated GuHCl extracts.
2. To reduce and denature the samples, add of the volume of 4× Laemmli sample buffer and boil at 96 °C for 5 min.
3. Load the samples (35 μg), i.e., on Bis-Tris discontinuous 4–12 % polyacrylamide gradient gels (NuPAGE, Invitrogen) (see Note 7) and separate by electrophoresis until the dye front reaches the bottom of the gel.

3.7 Gel Staining and Band Cutting

1. Submerse gels in fixation buffer (10 % acetic acid, 50 % methanol) for 30 min at room temperature.
2. Rehydrate the gels for at least 1 h in ddH_2O.
3. Gels can be stained for 2 h at room temperature with Coomassie Blue.
4. Destain the gels to visualize protein bands with several washes in 10 % acetic acid.
5. Excise gel bands across lanes, leaving no "empty" gel pieces behind.

3.8 Tryptic Digestion

1. Subject the gel bands to in-gel tryptic digestion using a robotic digestion system (12).
2. Digests are frozen at −80 °C and freeze-dried for 3 h.
3. Tryptic peptides are reconstituted in 0.1 % formic acid prior to analysis by LC-MS/MS (13).

4 Notes

1. In order to facilitate solubilization of 1 % SDS, warm the buffer under hot tap water. SDS readily crystallizes at <20 °C. Redissolve before use.
2. OCT (optimal cutting temperature) can be used for freezing. It will readily dissolve in the five PBS washes before tissue extraction.
3. A constant ratio between tissue weight and buffer volume is important for reproducible extraction. In our experience, 10:1 (v/w) for the NaCl and SDS extractions and 5:1 for GuHCl step provide sufficient protein amounts.
4. Adjusting vortex speed is critical. Excessive speed during the NaCl and SDS steps might result in mechanical disruption of the tissue. In contrast, maximum speed is recommended for the GuHCl step to induce disruption of the ECM.
5. A temperature-controlled environment is preferable for incubations at room temperature (at around 20 °C) to ensure consistency.

6. After addition of 4× Laemmli sample buffer, the final protein concentration is 1 μg/μl. This should be kept in mind when adjusting protein concentrations.
7. The use of ready-made gels ensures reproducibility and minimizes keratin contamination. A 4–12 % gradient provides a good separation of protein bands before LC-MS/MS analysis.

References

1. Agnetti G, Husberg C, Van Eyk JE (2011) Divide and conquer: the application of organelle proteomics to heart failure. Circ Res 108:512–526
2. Matsui Y, Morimoto J, Uede T (2010) Role of matricellular proteins in cardiac tissue remodeling after myocardial infarction. World J Biol Chem 1:69–80
3. Bowers SL, Banerjee I, Baudino TA (2010) The extracellular matrix: at the center of it all. J Mol Cell Cardiol 48:474–482
4. Vogel KG, Peters JA (2001) Isolation of proteoglycans from tendon. Methods Mol Biol 171:9–17
5. Barallobre-Barreiro J, Didangelos A, Schoendube FA, Drozdov I, Yin X, Fernández Caggiano M, Willeit P, Puntmann VO, Aldama-López G, Shah AM, Doménech N, Mayr M (2012) Proteomics analysis of cardiac extracellular matrix remodeling in a porcine model of ischemia-reperfusion injury. Circulation 125:789–802
6. Didangelos A, Yin X, Mandal K, Baumert M, Jahangiri M, Mayr M (2010) Proteomics characterization of extracellular space components in the human aorta. Mol Cell Proteomics 9:2048–2062
7. Didangelos A, Yin X, Mayr M (2012) Method for protein subfractionation of cardiovascular tissues before DIGE analysis. Methods Mol Biol 854:287–297
8. Mason RM, Mayes RW (1973) Extraction of cartilage protein-polysaccharides with inorganic salt solutions. Biochem J 131:535–540
9. Stoscheck CM (1990) Quantitation of protein. Methods Enzymol 182:50–68
10. Stanton H, Fosang AJ (2002) Matrix metalloproteinase are active following guanidine hydrochloride extraction of cartilage: generation of DIPEN neoepitope during dialysis. Matrix Biol 21:425–428
11. Didangelos A, Yin X, Mandal K, Saje A, Smith A, Xu Q, Jahangiri M, Mayr M (2011) Extracellular matrix composition and remodeling in human abdominal aortic aneurysms: a proteomics approach. Mol Cell Proteomics 10:M111.008128
12. Shevchenko A, Tomas H, Havlis J, Olsen JV, Mann M (2006) In-gel digestion for mass spectrometric characterization of proteins and proteomes. Nat Protoc 1:2856–2860
13. Yin X, Cuello F, Mayr U, Hao Z, Hornshaw M, Ehler E, Avkiran M, Mayr M (2010) Proteomics analysis of the cardiac myofilament subproteome reveals dynamic alterations in phosphatase subunit distribution. Mol Cell Proteomics 9:497–509

Chapter 18

Optimized Method for Identification of the Proteomes Secreted by Cardiac Cells

Miroslava Stastna and Jennifer E. Van Eyk

Abstract

In the past, various studies using different methods have been carried out to analyze proteins secreted by cells. There are several crucial steps that have to be followed to ensure successful secreted proteome detection and identification. Simultaneously with the optimization of the experimental conditions for various cell type culturing and subsequent cell conditioning to obtain conditioned medium with secreted proteins in vitro, the analytical separation methods for fractionation of complex protein mixture and mass spectrometry for protein identification are of high importance. The separation methods primarily used are either gel-based (e.g., 1-DE and 2-DE) or gel-free methods (e.g., liquid chromatography and capillary electrophoresis). Here we outline an optimized protocol for the preparation and analysis of conditioned medium containing proteins secreted by neonatal cardiac myocytes by using reversed-phase liquid chromatography (RPLC) followed by tandem mass spectrometry (LC MS/MS). Although optimized for neonatal cardiac myocytes, the general steps described in the following chapter can be adapted to other cell types as well.

Key words Cardiac cells, Conditioned medium, Secreted proteomes, Reversed-phase liquid chromatography, Tandem mass spectrometry, Proteomics

1 Introduction

The analysis and identification of proteins secreted by various cells are of great interest since they could provide valuable information on physiological and pathological conditions and they are an incredible source of biomarkers, potential drugs acting as paracrine and autocrine factors, and structural entities comprising the extracellular matrix (1, 2). The number of studies dealing with analysis of conditioned media from cells of various tissues/organs in vitro increased in last several years. This increase reflects the technical and instrument platform development and progress in techniques and methodology which make this analysis feasible and also facilitate the recognition of secreted protein clinical/therapeutic potential (3). In cardiovascular research, many of secreted proteins have been investigated to clarify their effect on heart function improvement, regulation, and regeneration (4–6). Regardless of the methodology and techniques used, it is

Fernando Vivanco (ed.), *Heart Proteomics: Methods and Protocols*, Methods in Molecular Biology, vol. 1005,
DOI 10.1007/978-1-62703-386-2_18, © Springer Science+Business Media New York 2013

critical to obtain maximum secreted proteome coverage. The major steps in analysis of proteome secreted by cardiac cells into conditioned medium in vitro are (1) quality sample preparation (cell harvesting, culturing, and conditioning), (2) fractionation of complex protein mixture by efficient analytical separation methods (e.g., 1-DE, 2-DE (7), LC techniques (8)) and gel-free electrophoresis (capillary or free-flow electrophoresis) (9), (3) mass spectrometry protein identification, and (4) evaluation of the results and linking them to general or specific biological functions by using bioinformatics tools.

Here we present optimized protocol for preparation and analysis of conditioned medium of cardiac cells, specifically neonatal rat cardiomyocytes (8). The protocol can be modified, but the general steps described in this chapter can be applied for analysis of other (cardiac) cells as well.

2 Materials

2.1 Equipment and Chemicals for Cell Culture and Conditioned Medium Processing

Centrifuge (for 15- and 50-ml vials; 250 rcf, 4 °C; e.g., 5810 R Eppendorf).

Microscope (Carl Zeiss, Germany).

Bright Line Counting Chamber (Hausser Scientific).

SpeedVac concentrator (SPD2010 Thermo Scientific).

Incubator (37 °C, 5 % CO_2).

Buffer I (pH 7.4): 136 mM NaCl, 20 mM HEPES, 5.55 mM glucose, 5.36 mM KCl, 0.81 mM $MgSO_4$, 0.44 mM KH_2PO_4, 0.34 mM NaH_2PO_4.

1 % Fibronectin, from bovine plasma (Sigma) for dish/flask pretreatment.

0.05 % Trypsin—EDTA (Invitrogen) for cell trypsinization.

Collagenase type 2 (Worthington).

Dulbecco's Modified Eagle's Medium—BMEM/F-12 (Gibco) supplemented with fetal bovine serum (FBS; Sigma), 1 % L-glutamine (Gibco), 15 mM HEPES (Sigma), and 1 % Pen Strep (Sigma).

Pen Strep (10,000 U/ml penicillin, 10,000 μg/ml streptomycin; Sigma).

45-μm sterile filter; 33 mm (Millipore).

PBS (pH 7.4; Quality Biological).

0.4 % trypan blue stain (Gibco).

2.2 Reversed-Phase Liquid Phase Instrument and Materials

ProteomeLab PF 2D instrument (HPRP module; Beckman Coulter, Inc.) (see Note 1).

C18 column (50 mm × 4.6 mm; nonporous 1.5 μm silica particles; Beckman Coulter, Inc.) (see Note 1).

Fraction collector (96-well; FC 204; Gilson, Inc.).

Acetonitrile (ACN; HPLC grade, submicron filtered; Fisher Scientific).

Trifluoroacetic acid (TFA; peptide synthesis grade; Applied Biosystems).

96-well plates (freezable at −80 °C).

Microcentrifuge tubes (1.5 ml, low retention, Fisher Scientific).

Trypsin (sequencing grade modified; V5111; Promega) for protein digestion.

pH Hydrion papers (Micro Essential Laboratory, Inc.).

2.3 Mass Spectrometry Instrumentation and Materials

ThermoFinnigan LTQ Orbitrap with electrospray ionization (Thermo Electron Corporation) (see Note 2).

Acetonitrile (ACN; HPLC grade, submicron filtered; Fisher Scientific).

Formic acid (99 + % for LC/MS applications, Thermo Scientific).

Water (HPLC grade, submicron filtered; Fisher Scientific).

Sorcerer™-SEQUEST® (Sage-N Research, Inc.) (see Note 3).

Scaffold 3/Trans-Proteomic Pipeline (TPP) (Proteome Software, Inc.) (see Note 4).

3 Methods

3.1 Cell Culture

For cell culturing follow all usual sterile and safety requirements. Pre-coat all flasks/dishes used for cell cultures and conditioning with fibronectin. The procedures, buffers, and medium used for various cardiac cell culturing and harvesting might differ. In this section, we describe the preparation of primary cultures of neonatal rat cardiomyocytes (8, 10–12).

1. Perform the excision of the hearts from 2-day-old Sprague–Dawley rats and place the hearts in cold Buffer I.
2. Wash and mince the ventricles, incubate with 0.1 % trypsin overnight, and dissociate the cells using 0.1 % collagenase.
3. Pre-plate cardiomyocytes in two 45 min steps to separate from cardiac fibroblasts.
4. Resuspend the isolated myocytes in culture medium DMEM/F-12 supplemented with 10 % FBS, 1 % L-glutamine, 15 mM HEPES, and 1 % Pen Strep, and plate the myocytes (3×10^6 cells) in 100 mm dishes.
5. Place the myocyte culture to the incubator (37 °C at 5 % CO_2).
6. After 48 h (approx. 80 % cell confluence), wash the myocytes with PBS three or more times (see Note 5) to remove high

content of FBS and replace with 1 % FBS (see Note 6) in DMEM/F-12 supplemented with 1 % L-glutamine, 15 mM HEPES, and 1 % Pen Strep.

7. Place the myocyte in the incubator and condition for a time which you select as optimal by using the steps in Subheading 3.2. We use conditioning for 48 h (see Note 6).

3.2 Optimization of Cell Conditioning Conditions (FBS Concentration and Time of Conditioning)

After cell culturing you need to determine the optimal FBS concentration in conditioned medium and time of conditioning for following cell conditioning step. Although the studies have been published in which serum-free medium was used for cell conditioning (7, 13), the proteins secreted by cells can be different at serum-free conditions, and addition of low concentration of FBS might be necessary. The serum concentration should reflect the minimum concentration (mostly 0–2 %) which will be acceptable for subsequent protein identification by mass spectrometry, but simultaneously the concentration which the cells will still tolerate without lysis, extensive cell death, and/or feature/morphology changes. These optimized conditions should allow distinguishing between secreted proteins and those artificially liberated with cell lysis and/or death (see Note 7).

1. Prepare the cell culture medium used for cell culturing with several different FBS concentrations, e.g., 0, 1, 2, and 10 % (see Note 8).
2. Wash the cells three times with PBS after cultured in medium with 10 % FBS (see Note 5).
3. Split the cell culture into pre-coated dishes (each dish containing approximately the same number of cells; e.g., 1×10^6 cells/10 ml medium) with medium of different FBS concentrations (see step 1) in triplicate (for three time points).
4. Condition the cells for 12, 24, and 48 h. At each time point, measure the total number of the cells and the number of dead cells in media of different FBS concentration by using the general methods for cell counting, e.g., with 0.4 % trypan blue labeling (see Note 9) as follows.
5. Collect the cells by trypsinization (see Note 10), spin down by using 250 rcf for 5 min at 4 °C, remove the supernatant by aspiration, and resuspend the pellet in 2 ml of PBS.
6. Take 50 μl of cell suspension, add 50 μl 0.4 % trypan blue, mix, and let incubate for 10 min.
7. For cell counting, pipette 10 μl of cell suspension with trypan blue into chamber (Bright Line Counting Chamber) and count the total number of cells and number of dead cells in four squares by use of microscope. Calculate the average. Perform the counting in triplicate for each FBS concentration and each time point. Determine the optimal FBS concentration and the time of conditioning (see Note 11).

3.3 Conditioned Medium Collection and Processing

1. After conditioning, collect the conditioned medium (medium with cell-secreted proteins).
2. Centrifuge at 250 rcf for 5 min at 4 °C and filter the supernatant by using 45 μm sterile filter.
3. Concentrate 1.4 ml of conditioned medium in SpeedVac concentrator to a final volume 210 μl (see Note 12).

3.4 RPLC

1. Add 60 μl of 100 % ACN, 30 μl of 10 % TFA to 210 μl of concentrated conditioned medium, gently vortex.
2. Add an appropriate volume of 20 % ACN/1 % TFA to obtain the total protein concentration as 100 μg/100 μl of 20 % ACN/1 % TFA (see Note 13). Gently vortex. Ensure that you have enough sample volume for at least two 100 μl consecutive RPLC run injections.
3. Let incubate for 15 min.
4. Prepare solvent A (0.1 % TFA in water) and solvent B (0.08 % TFA in ACN) for RPLC. Set the gradient run from 0 to 100 % solvent B in 30 min followed by 100 % B for 5 min and 100 % A for 10 min with a flow rate 0.75 ml/min. Set the wavelength on UV detector to 214 nm (see Note 14). Set the fraction collector with collection times from 9 to 24 min with fractions collected in 0.25 min intervals into 96-well plate.
5. Inject 100 μl (100 μg) of conditioned medium sample into reversed-phase C18 column of HPLC instrument.
6. Collect the fractions into 96-well.
7. Repeat **step 5** and collect the fractions into same 96-well, so your total amount of protein collected will be 200 μg (see Notes 15 and 16). For our experimental conditions, we use 100 μg as a maximum protein load per run (repeated twice) rather than one large load to prevent from a clogging of chromatographic column and also obtain a good peak resolution.
8. Combine the fractions based on your chromatogram pattern into 1.5 ml microcentrifuge tubes (Fig. 1, see Note 17).
9. Do not collect the fraction containing albumin (retention time from 15.8 to 16.2 min; see Note 18 and Fig. 1).
10. Dry the fractions in SpeedVac concentrator to dryness.
11. Prepare trypsin solution: Add 50 μl of trypsin resuspension buffer to 20 μg trypsin (Promega), aliquot 5 μl per tube (store at −20 °C). In use, add 195 μl of 20 mM NH_4HCO_3 to 5 μl aliquot (i.e., 10 ng/μl trypsin final concentration).
12. Add 25 μl of trypsin solution to each dried fraction.
13. Check the pH of each fraction by dropping 0.5 μl on pH paper strip to ensure pH value is around 8 (optimal pH for trypsin

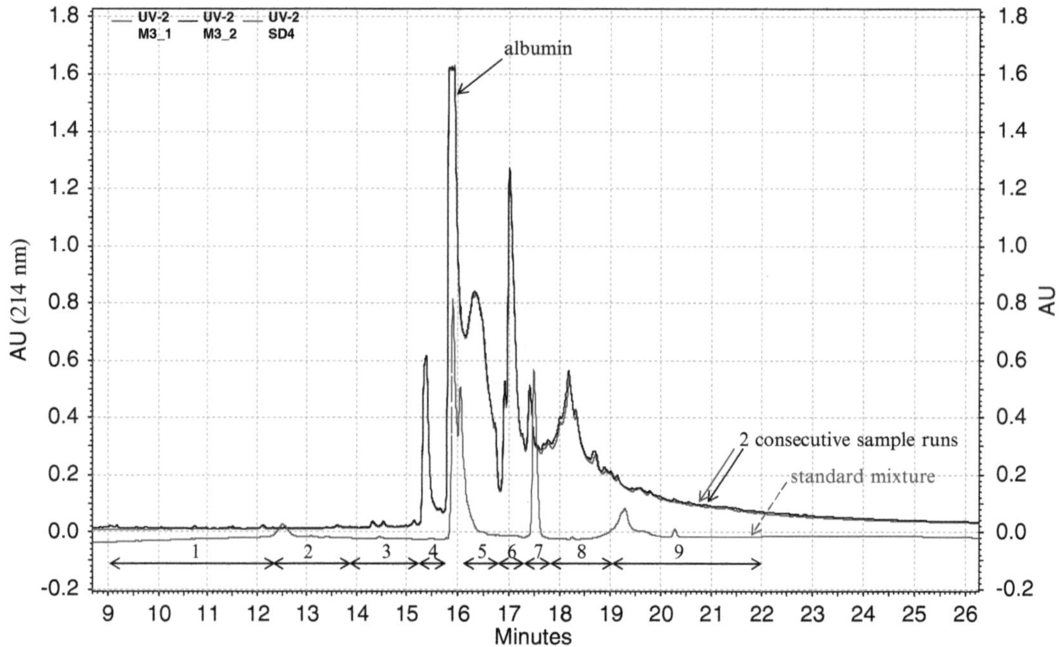

Fig. 1 Example of chromatogram obtained by analysis of conditioned medium of cardiomyocytes by RPLC. Two consecutive runs (100 μg of total protein per run) are shown to demonstrate the reproducibility. The peak with retention time from 15.8 to 16.2 min contains albumin from FBS. The chromatogram of standard protein mixture containing BSA is shown as well. The retention times are marked at the bottom of the figure for nine final fractions. The fraction of albumin was excluded from further mass spectrometry identification

enzymatic activity). If needed, adjust to pH 8 by adding 1–5 μl of 1 M NH_4HCO_3.

14. Incubate at 37 °C overnight.
15. Add 10 μl of 2 % of formic acid, pH value should be between 2 and 3 to stop the trypsin function.
16. Dry the fractions to dryness.

3.5 Mass Spectrometry and Protein Identification

1. Resuspend the peptides from digested dried fractions 1, 2, and 3 in 30 μl of 60 % ACN/0.1 % formic acid (see Note 19). Plate 5 μl of sample on plate for mass spectrometry analysis, each fraction in triplicate. Dry to dryness.
2. Resuspend the peptides from digested dried fractions 4, 5, 6, 7, 8, and 9 in 30 μl of 60 % ACN/0.1 % formic acid. Take 8 μl of this sample solution and dilute with additional 12 μl of 60 % ACN/0.1 % formic acid. Plate 5 μl of sample on plate for mass spectrometry analysis, each fraction in triplicate (see Note 19). Dry to dryness.

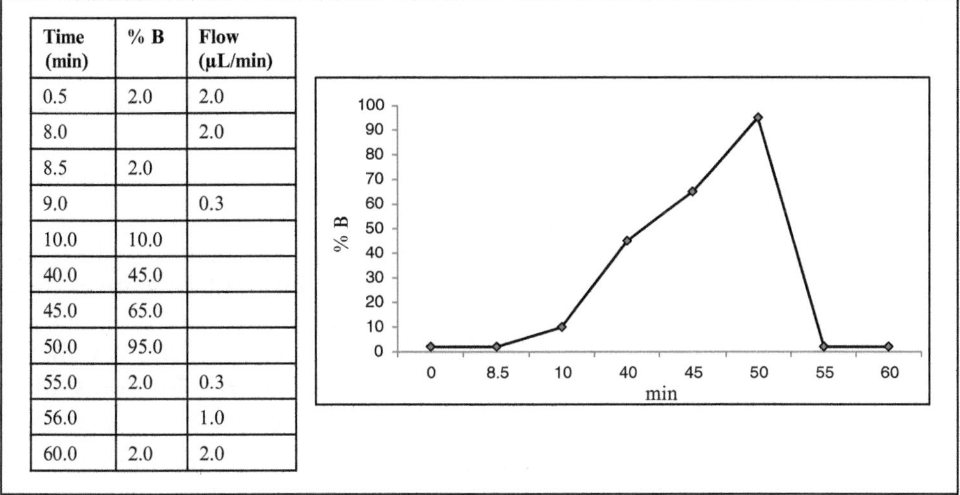

Time (min)	% B	Flow (μL/min)
0.5	2.0	2.0
8.0		2.0
8.5	2.0	
9.0		0.3
10.0	10.0	
40.0	45.0	
45.0	65.0	
50.0	95.0	
55.0	2.0	0.3
56.0		1.0
60.0	2.0	2.0

Fig. 2 Schematically shown the gradient (% of B solvent) and flow rate (μl/min) setting used for 60 min analysis of cardiomyocyte-conditioned medium by nano-LC system connected on-line to LTQ Orbitrap. B is 0.1 % formic acid in 90 % of ACN

3. During mass spectrometry analysis, the samples on plate are resuspended in 6 μl of A:B (20:1) solvent (A is 0.1 % formic acid in water; B is 0.1 % formic acid in 90 % of ACN) with 5 μl injection into C18 reversed-phase column (fused silica 75 μm × 100 mm; 5 μm particle) of nano-LC system (Agilent Technologies). The setting of the gradient and time of analysis depends on your sample and the goal of your project. For analysis of cardiomyocyte-conditioned medium, we use 60 min analysis with gradient shown in Fig. 2. After that, the peptides are on-line eluted into LTQ Orbitrap.
4. After mass spectrometry analysis, submit the raw files into Sorcerer™-SEQUEST® with the search criteria and database depending on your sample analyzed. We use UniProtKB/Swiss-Prot Release 2011_02 of 08-Feb-2011 database (all species; semi-trypsin search: the generated peptides will contain trypsin specificity only at one terminus; two missed cleavages: number of sites that trypsin missed), peptide mass tolerance 50 ppm, fragment mass type: monoisotopic, post-search analysis: using Scaffold and/or TPP (see Note 4), no differential modifications, maximum instances per type: 3.
5. View and evaluate your protein identifications by using Scaffold 3 and/or TPP.
6. For analysis of biological samples, analyze the three biological replicates for each sample.

4 Notes

1. Any HPLC instrument equipped with C18 column and fraction collector.
2. Any tandem mass spectrometry instrument suitable for your sample analysis and goal of your project.
3. We use Sage-N Research Sorcerer platform which includes SEQUEST as a search engine. SEQUEST identifies the sets of tandem mass spectra to peptide sequences that have been generated from databases of protein sequences. Another search engines that can be used are X!Tandem, Mascot, OMSSA, or any of their combination for higher proteome coverage (14).
4. In Sage-N Research Sorcerer platform, you can select Scaffold and/or Trans-Proteomic Pipeline (TPP) for post-search analysis to visualize and validate your MS/MS experiments.
5. Three washes with PBS or three washes with serum-free medium or combination of both should be sufficient to remove 10 % of FBS from cells.
6. FBS concentration in conditioned media and time of cell conditioning under serum deprivation conditions should be optimized for each type of cells individually, see Subheading 3.2.
7. Some cells might be very sensitive to serum deprivation conditions and change the feature/morphology dramatically. In this case, the better option is to keep FBS concentration in conditioned medium the same as during cell culturing (e.g., 10 %) with depletion of high abundant serum proteins in conditioned medium prior to RPLC.
8. The highest FBS concentration should be the same concentration as used for cell culturing, and it serves as control in the optimization process.
9. The cell viability assessment at serum deprivation condition can be done by using various methods, alternatively to simple method described here, e.g., by labeling and incubation of the cells for 15 min with Annexin V-FITC and 7AAD after cell collection by trypsinization, and analysis of labeled cells by flow cytometer.
10. In addition to using trypsin (trypsinization), the cells can be detached from culture dish/flask and dissociated by using other methods (some not involving trypsin but, e.g., papain, accutase, or by using nonenzymatic cell dissociation), but in all cases, you have to ensure that cell membrane damage and cell death are minimal.
11. In our cell counting experiments, the optimum FBS concentration (1 %) and the time for cell conditioning (48 h) were selected based on the highest number of total cells and the

lowest number of dead cells measured. The percentage of dead cells should not extend 10 %.

12. Medium concentrating prevents you from running many RPLC runs to be able to collect enough samples for mass spectrometry protein identification (for at least three mass spectrometry technical replicates and possible reruns).
13. For example, our total protein concentration in collected conditioned medium is 0.32 mg/ml, i.e., 448 μg of total protein amount in 1.4 ml conditioned medium (in 210 μl concentrated medium). Adding 60 μl of 100 % ACN and 30 μl of 10 % TFA to 210 μl of concentrated medium gives us 448 μg total protein concentration in 300 μl. Dilution of 300 μl with addition of 148 μl of 20 % ACN/1 % TFA results in 448 μg of total protein in 448 μl conditioned medium sample. This volume is sufficient for 2–3 consecutive runs on RPLC instrument, 100 μl (100 μg) injection for each run.
14. For detection of peptides and for proteins with aromatic amino acids, the wavelength should be set at 210 and 280 nm, respectively.
15. Since you collect the fractions into same 96-well, your retention time reproducibility is crucial (see Fig. 1).
16. At this point, you can freeze the fractions in the plate (−80 °C) or continue to step 8.
17. For cardiomyocyte-conditioned medium containing 1 % FBS, we mostly combine collected fractions to final nine fractions, see Fig. 1.
18. There are several choices how to avoid the interfering of high abundant serum proteins with low abundant secreted proteins to make mass spectrometry identification of secreted proteins feasible. In our experiments, we separate albumin from other proteins by using RPLC, and we further analyze the fractions by mass spectrometry with exclusion of the fraction with highest albumin content (retention times from 15.8 to 16.2 min). It should be noted that some fractions will still contain certain amount of albumin (especially fractions adjacent to albumin fraction) and you will loose some proteins eluted in same fraction (retention time) as albumin.
19. The volume of resuspension solution depends on concentration of peptides in each individual fraction. This fraction dilution needs to be optimized during mass spectrometry experiments. In our experiments, we usually use sixfold dilution for fractions 1, 2, and 3 and 15-fold dilution for fractions 4, 5, 6, 7, 8, and 9. See Fig. 1 for the fractions and Table 1 for mass spectrometry sample dilution optimization. As well, based on our optimization experiments, we run three technical replicates for each fraction (see Fig. 3).

Table 1
Optimization of RPLC fraction dilution for mass spectrometry analysis

Fraction #_dilution	Intensity (TIC)	# Peptides	# Proteins	
2_6	5.55E+08	309	136	Dilution selected
2_15	4.38E+08	208	93	
2_30	6.40E+08	63	36	
2_60	3.50E+08	34	13	
6_6	1.35E+10	241	57	
6_15	4.97E+09	170	49	Dilution selected
6_30	3.43E+09	153	51	
6_60	2.04E+09	131	37	
8_6	5.78E+09	279	66	
8_15	1.32E+09	252	68	Dilution selected
8_30	1.96E+09	174	56	
8_60	9.59E+08	107	36	

The fractions 2, 6, and 8 obtained from RPLC were diluted 6-, 15-, 30-, and 60-fold and injected into nano-LC system on-line connected to LTQ Orbitrap. The signal intensity (TIC—total ion count) was monitored, and the number of peptides/proteins was calculated. For fraction 2, dilution which yielded in the highest number of peptides/proteins was selected (sixfold). For fraction 6, sixfold dilution resulted in the highest number of peptides/proteins, but since the signal intensity was rather high, we opted for 15-fold dilution as optimal. For fraction 8, we selected 15-fold dilution for which the highest number of proteins was identified

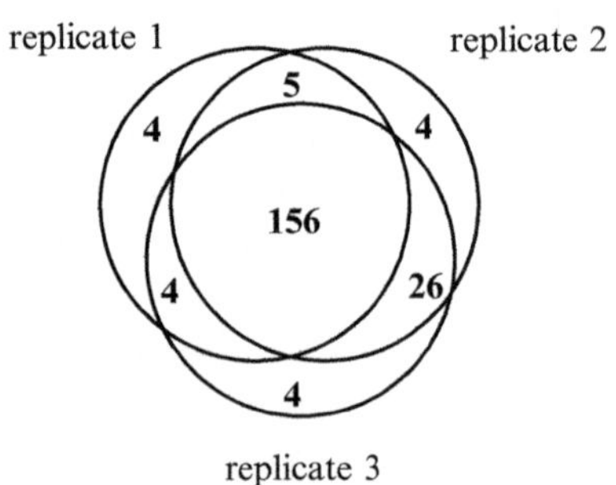

Fig. 3 Number of proteins identified in three mass spectrometry technical replicates. The proteins identified in nine RPLC fractions of cardiomyocyte-conditioned medium sample (three LTQ Orbitrap technical replicates for each fraction) were evaluated and calculated. From total 203 proteins, 156 proteins were identified in all three replicates, additional four proteins were found in each replicate 1, 2, and 3. Additional 5, 26, and 4 proteins overlapped between replicates 1 and 2, replicates 2 and 3, and replicates 1 and 3, respectively. We recommend using three technical replicates, whenever possible, for broader proteome coverage

Acknowledgement

Supported by the National Heart, Lung, and Blood Institute Proteomic Initiative—contract NHLBI-HV-10-05 (2) (JEV), by NHLBI PO1 HL10026 (Glycoconjungates and Cardiovascular Disease) (JEV), by the AHA grant-in-aid (JEV), and by the Institutional Support RVO: 68081715 (MS).

References

1. El-Armouche A, Ouchi N, Tanaka K, Doros G, Wittköpper K, Schulze T, Eschenhagen T, Walsh K, Sam F (2011) Follistatin-like 1 in chronic systolic heart failure: a marker of left ventricular remodeling. Circ Heart Fail 4: 621–627
2. Stastna M, Van Eyk JE (2012) Secreted proteins as a fundamental source for biomarker discovery. Proteomics 12:722–735
3. Stastna M, Van Eyk JE (2012) Investigating the secretome: lessons about the cells that comprise the heart. Circ Cardiovasc Genet 5:8–18
4. Harris BS, Zhang Y, Card L, Rivera LB, Brekken RA, Bradshaw AD (2011) SPARC regulates collagen interaction with cardiac fibroblast cell surfaces. Am J Physiol Heart Circ Physiol 301:H841–H847
5. Barandon L, Casassus F, Leroux L, Moreau C, Allières C, Lamazière JM, Dufourcq P, Couffinhal T, Duplàa C (2011) Secreted frizzled-related protein-1 improves postinfarction scar formation through a modulation of inflammatory response. Arterioscler Thromb Vasc Biol 31:e80–e87
6. Accornero F, van Berlo JH, Benard MJ, Lorenz JN, Carmeliet P, Molkentin JD (2011) Placental growth factor regulates cardiac adaptation and hypertrophy through a paracrine mechanism. Circ Res 109:272–280
7. Dupont A, Corseaux D, Dekeyzer O, Drobecq H, Guihot AL, Susen S, Vincentelli A, Amouyel P, Jude B, Pinet F (2005) The proteome and secretome of human arterial smooth muscle cells. Proteomics 5:585–596
8. Stastna M, Chimentti I, Marban E, Van Eyk JE (2010) Identification and functionality of proteomes secreted by rat cardiac stem cells and neonatal cardiomyocytes. Proteomics 10: 245–253
9. Tunica DG, Yin X, Sidibe A, Stegemann C, Nissum M, Zeng L, Brunet M, Mayr M (2009) Proteomic analysis of the secretome of human umbilical vein endothelial cells using a combination of free-flow electrophoresis and nanoflow LC-MS/MS. Proteomics 9:4991–4996
10. Bursac N, Papadaki M, Cohen RJ, Schoen FJ, Eisenberg SR, Carrier R, Vunjak-Novakovic G, Freed LE (1999) Cardiac muscle tissue engineering: towards an in vitro model for electrophysiological studies. Am J Physiol Heart Circ Physiol 277:H433–H444
11. Hein P, Michel MC, Leineweber K, Wieland T, Wettschureck N, Offermanns S (2005) Biomedical and analytical techniques. Chapter 6. In: Dhein S, Mohr W, Delmar M (eds) Practical methods in cardiovascular research. Springer-Verlag Berlin, Heidelberg
12. Smith RR, Barile L, Cho HC, Leppo MK, Hare JM, Messina E, Giacomello A, Abraham MR, Marbán E (2007) Regenerative potential of cardiosphere-derived cells expanded from percutaneous endomyocardial biopsy specimens. Circulation 115:896–908
13. Duran MC, Martin-Ventura JL, Mas S, Barderas MG, Darde VM, Jensen ON, Egido J, Vivanco F (2007) Characterization of the human atheroma plaque secretome by proteomic analysis. Methods Mol Biol 357:141–150
14. Gundry RL, White MY, Murray CI, Kane LA, Fu Q, Stanley BA, Van Eyk JE (2009) Preparation of proteins and peptides for mass spectrometry analysis in a bottom-up proteomics workflow. Curr Protoc Mol Biol 10: Unit10.25

Chapter 19

Secretome of Human Aortic Valves

Fernando de la Cuesta, Gloria Alvarez-Llamas, Felix Gil-Dones, Verónica M. Darde, Enrique Calvo, Juan Antonio López, Fernando Vivanco, and María G. Barderas

Abstract

One of the major challenges in cardiovascular medicine is to identify candidate biomarker proteins. Between different proteomics studies, the secretome is a valuable tool in the search for biomarkers locally released by a studied tissue and remains particularly important while working with vascular tissues, since the secreted proteins would be probably shed to the blood. In this chapter, we described a method to validate the origin of secreted proteins in human aortic valves, demonstrating their synthesis and release by the tissue and ruling out blood origin.

Key words Secretome, Biomarkers, Aortic valves, Aortic stenosis, nLC–MS/MS, SRM validation

1 Introduction

To date, aortic stenosis (AS) has been considered as a passive process secondary to calcium deposition in the worn-out aortic valves. Indeed, clinicians and researchers have mainly focused on surgical treatments for valvular heart disease, while the underlying mechanisms remained largely unknown (1). In this sense, proteomics has become a particularly suitable platform to analyze in a non-biased way the proteins involved in the pathogenesis of different diseases, aortic stenosis between others (2, 3), and pave the road to the development of potential biomarkers which will help to detect patients at risk of developing degenerative AS. With this in mind, aortic valve secretome emerges as an attractive source to further approach the understanding of AS pathogenic process due to the inherent limitations of plasma and serum, in relation to their complex nature and their great dynamic range of protein concentrations, which obscures detection of low-abundance proteins in favor of major ones (4, 5). In a general way, this kind of sample can be considered as a plasma

Fernando Vivanco (ed.), *Heart Proteomics: Methods and Protocols*, Methods in Molecular Biology, vol. 1005, DOI 10.1007/978-1-62703-386-2_19,

subproteome, and it provides a more direct approximation to the in vivo situation, since it focuses on a subset of proteins released by a tissue under certain biological conditions (6).

2 Materials

2.1 Equipment

1. Milli-Q system (Millipore, Bedford, MA).
2. Laminar flow cabinet Telstar AV-100 (Telstar).
3. Cell incubator (Nuaire).
4. Biophotometer (Eppendorf, AG. Hamburg, Germany).
5. Thermomixer comfort (Eppendorf, AG. Hamburg, Germany).
6. Mini Protean System (Bio-Rad).
7. nano-HPLC (Ultimate 3000, Dionex).
8. PicoTip™ emitter nano-spray needle (New Objective, Woburn, MA).
9. LTQ-Orbitrap XL mass spectrometer (Thermo Fisher, Inc., MA, USA).
10. Proteome Discoverer 1.0 software (Thermo Fisher, Inc., MA, USA).
11. TEMPO nano-LC System (AB Sciex, CA, USA).
12. Autosampler coupled to a modified triple quadrupole (4000 QTRAP LC/MS/MS, AB Sciex, CA, USA).
13. MRMpilot software v1.1 (AB Sciex, CA, USA).
14. SEQUEST™ (version 1.0.43.2, Thermo Fisher Scientific, Inc., MA, USA).
15. Mascot™ (version 2.1, Matrix Science) software.

2.2 Reagents

2.2.1 Tissue Culture

1. PBS: phosphate saline buffer, pH 7.4.
2. Culture medium: 1640 RPMI medium without l-glutamine (Cell Culture Technologies, Invitrus) supplemented with penicillin, streptomycin (Pen/Strep, Lonza), and amphotericin B (Fungizone 50 mg, Bristol-Myers Squibb). l-Lysine 2HCl (U–$^{13}C_6$, 97–99 %) and l-arginine HCl (U–$^{13}C_6$, 97–98 %: Cambridge Isotope Laboratories, Inc., Andover, MA).
3. Petri dish (Cell Star®).
4. Sterile tweezers and surgical blades.

2.2.2 Secretome Analysis

1. Milli-Q water.
2. Centriplus, 3-kDa cutoff (Millipore).
3. Bradford–Lowry reagent: Protein Assay (Bio-Rad).
4. 0.03 M Tris buffer (Bio-Rad).

5. Coomassie Brilliant Blue G-250 (Page Blue staining solution, Fermentas).
6. Dithiothreitol (Sigma Aldrich).
7. Ammonium bicarbonate, 99 % purity (Scharlau).
8. Iodoacetamide (Bio-Rad).
9. Sequencing grade modified porcine trypsin (Promega).
10. Formic acid, 99.5 % purity.
11. Trifluoroacetic acid (HPLC grade).
12. Acetic acid (HPLC grade).
13. C-18 reversed-phase (RP) nano-column (Mediterranea Sea, Teknokroma).
14. Acetonitrile (ACN) (HPLC grade).
15. Pep-Clean spin columns (Pierce).
16. µ-Precolumn Cartridge (Acclaim Pep Map 100 C18, 5 µm, 100 Å; 300 µm i.d. × 5 mm, LC Packings).

3 Methods

3.1 Aortic Valve Tissue Collection, Culture, and Labeling

1. Aortic valve material may come from aortic valve replacement surgery or from autopsies (see Note 1).
2. Wash at least three times the aortic valve in PBS in order to reduce blood contaminants.
3. Transfer the sample to a Petri dish.
4. Cut into pieces of about 1 mm^3.
5. Incubate at 37 °C in lysine–arginine free 1640 RPMI medium supplemented with 5 mg/ml fungizone, 250 mg/ml amikazin, 2 mg/ml L-lysine 2HCl ($U{-}^{13}C_6$, 97–99 %), and 10 mg/ml L-arginine HCl ($U{-}^{13}C_6$, 97–98 %) in a humidified atmosphere of 5 % CO_2.
6. Replace culture medium after 1 h, overnight, and every 4 h in the following 8 h period (a total of four washes).
7. Maintain tissue in culture for additional 96 h prior to collection of final secretome samples.
8. Store at −80 °C until analysis.

3.2 Secretome Analysis by nLC–MS/MS

3.2.1 Concentration and 1-DE Prefractionation

1. Centrifuge at 14,000 × *g* to remove remaining tissue. Transfer supernatant to a Centriplus device.
2. Concentrate supernatants by ultrafiltration at a maximum of 4,000 × *g* in a swinging bucket rotor until volume has been reduced to approximately 100 µl (see Note 2).
3. Determine protein concentration using Bradford method.

4. Load 25 μg of total protein on a 12 % SDS-PAGE gel and run at 80 V until bromophenol blue front has barely passed into the resolving gel (see Note 3).
5. Stain the gel with Coomassie Brilliant Blue G-250 for 1 h. Remove excess dye with bidistilled water.
6. Excise the resulting band, cut into 15–20 pieces, and store in ultrapure water at 4 °C until use.

3.2.2 Proteins Digestion

1. Reduce disulfide bonds of gel pieces with 100 mM dithiothreitol in 50 mM ammonium bicarbonate for 30 min at 37 °C.
2. Alkylate thiol groups 20 min at room temperature (RT) with 550 mM iodoacetamide in 50 mM ammonium bicarbonate.
3. Digest overnight in 50 mM ammonium bicarbonate, 15 % acetonitrile with sequencing grade modified porcine trypsin at a final concentration of 1:50 (trypsin: protein).

3.2.3 Analysis by nLC–MS/MS

1. Vacuum dry on a SpeedVac.
2. Dissolve pellet in 1 % acetic acid (HPLC grade).
3. Inject sample onto the C-18 reversed-phase (RP) nano-column of the Ultimate chromatographer. Separate in a continuous acetonitrile gradient consisting of 0–43 % B in 140 min and 50–90 % B in 1 min with a flow rate of 300 nl/min and analyze eluting peptides in the on-line coupled Orbitrap instrument (see Note 4).
4. Extract tandem mass spectra and deconvolute charge state using Proteome Discoverer software. Analyze MS/MS spectra with SEQUEST and Mascot softwares (see Note 5).
5. Labeled proteins found exclusively in control or stenotic valves may be considered for differential secretion validation analysis by SRM.

(Detailed workflow corresponding with tissue culture, labeling, and nLC–MS/MS can be found in Fig. 1.)

3.3 Selected Reaction Monitoring Validation

3.3.1 Protein Digestion

1. Reduce protein samples with 100 mM dithiothreitol in 50 mM ammonium bicarbonate for 30 min at 37 °C.
2. Alkylate protein samples for 20 min at room temperature (RT) with 550 mM iodoacetamide in 50 mM ammonium bicarbonate.
3. Digest the proteins in 50 mM ammonium bicarbonate, 15 % acetonitrile with sequencing grade modified porcine trypsin at a final concentration of 1:50 (trypsin: protein).

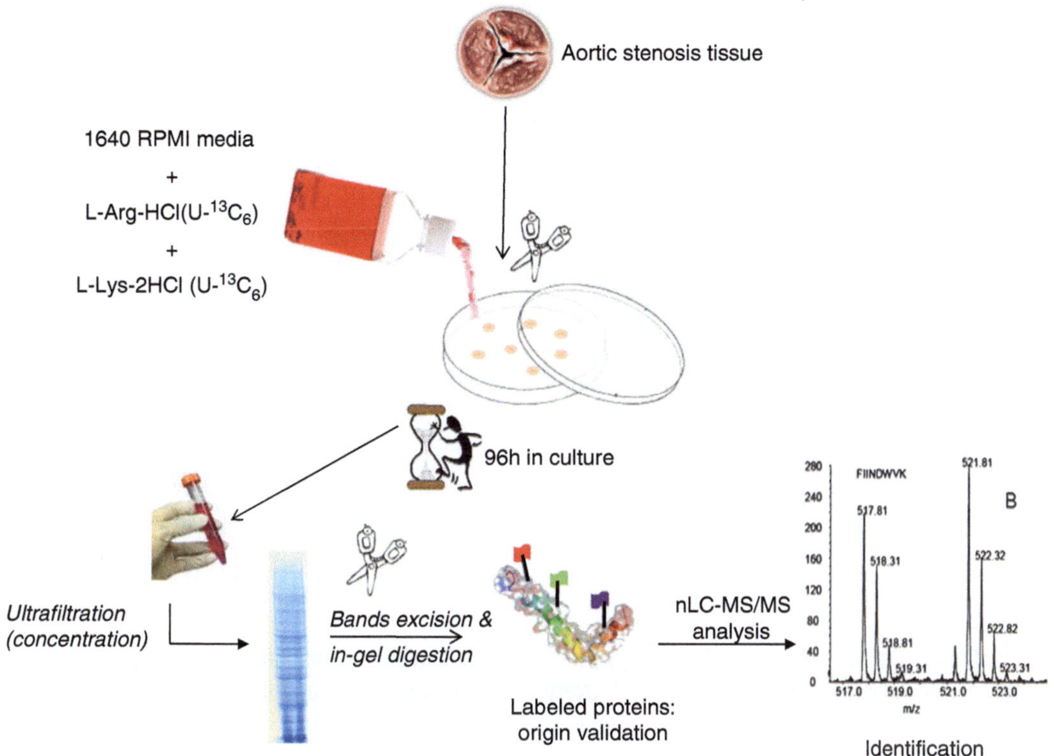

Fig. 1 Schematic representation of human aortic valve secretome analysis

4. Incubate the digestion overnight at 37 °C.
5. Add 2 % formic acid (HPLC grade).
6. Clean the samples with Pep-Clean spin columns according to the manufacturer's instructions.
7. Dry the tryptic digests in SpeedVac.
8. Resuspend in 2 % acetonitrile, 2 % formic acid (FA) prior to MS analysis.

3.3.2 LC–MS/MS

1. Perform three replicate injections (2 μg of protein in 4 μl) per sample in the mobile phase A (2 % ACN/98 % water, 0.1 % FA) at a flow rate of 10 μl/min for 5 min.
2. Load peptides onto a μ-Precolumn Cartridge to preconcentrate and desalt samples.
3. Perform a reversed-phase liquid chromatography (RPLC) using TEMPO nano LC system on a C18 column in a gradient of phase A and phase B (98 % ACN/2 % water, 0.1 % FA).

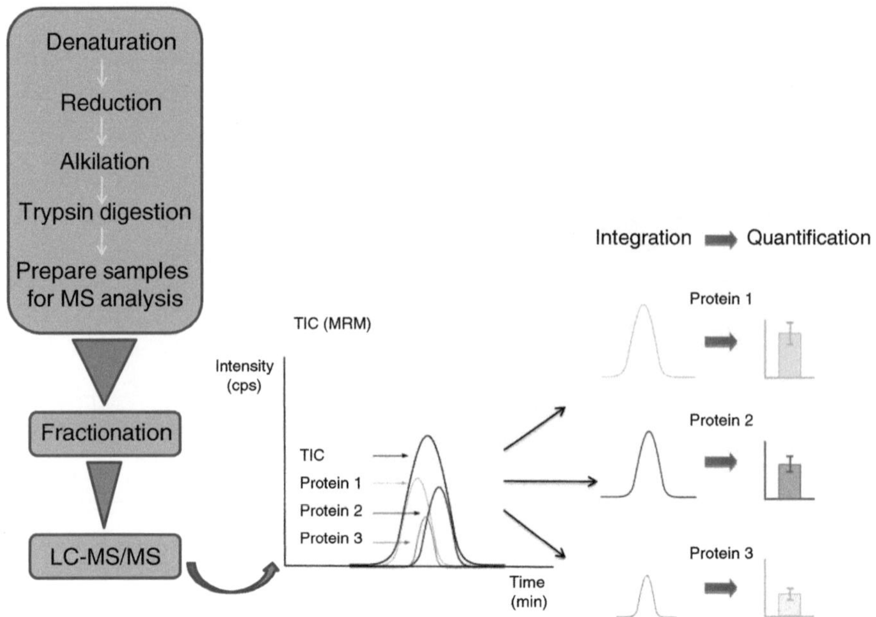

Fig. 2 Schematic representation of SRM analysis for protein validation and quantification. SRM peak areas are integrated to calculate peptide abundances in every chromatographic run and these data are then used to perform subsequent statistical analysis

4. Elute the peptides at a flow rate of 300 nl/min in a continuous acetonitrile gradient: 2–15 % B for 2 min, 15–50 % B for 38 min, 50–90 % B for 2 min, and 90 % B for 3 min.
5. Control both, the TEMPO nano-LC and 4000 QTRAP system by Analyst software v.1.4.2 (see Notes 6 and 7).

Detailed scheme can be found in Fig. 2.

4 Notes

1. All material to be in contact with tissue must to be sterile. Work in a laminar flow hood.
2. Ultrafiltration at 4,000 × *g* takes approximately 2 h to reduce secretome volume from 1 ml to 100 μl.
3. This method provides sample cleaning and the possibility to perform in-gel digestion, which may be more efficient than in-solution one, and it is more reproducible for comparison than 1-DE prefractionation.
4. Resolution of the enhanced FT-resolution spectrum was set to 60,000. Five most intense parent ions were fragmented and MS/MS spectra analyzed during the whole chromatographic run (180 min). Dynamic exclusion was set at 0.5 min.

5. SEQUEST was configured to search human.fasta (version 1.0, 199,762 entries) and Mascot to search the human subset of the Swiss-Prot (version 51.6, 236,102 entries) protein database, both search engines assuming trypsin digestion. Two missed cleavages were allowed and an error of 15 ppm and 0.8 Da set for full MS or MS/MS spectra searches, respectively. Oxidation of methionine and phosphorylation of serine, threonine, or tyrosine were selected as variable modifications together with SILAC +6 modifications at arginine or lysine residues, to search for labeled proteins. All identifications were performed by decoy database search with the false discovery rate (FDR) set at 0.05 using the corresponding filters (Mascot significance threshold, and score versus charge rate for SEQUEST).
6. The mass spectrometer was set to operate in positive ion mode with ion spray voltage of 2,800 V and a nanoflow interface heater temperature of 150 °C.
7. The source and curtain gas were set to 20 and 10 psi, respectively, and nitrogen was applied as both curtain and collision gases.

Acknowledgments

This work was supported by grants from the Instituto de Salud Carlos III (FIS PI070537, PI11/02239, PI11/01401), Fondos Feder, Redes temáticas de Investigación Cooperativa en Salud (RD06/0014/1015), grants from Fundación para la Investigación Sanitaria de Castilla-La Mancha (FISCAM PI2008-08, PI2008-28, PI2008-52).

References

1. Stewart BF, Siscovick D, Lind BK, Gardin JM, Gondiener JS, Smit VE, Kitzman DW, Otto CM (1997) Clinical factors associates with calcific aortic valve disease: cardiovascular health study. J Am Coll Cardiol 29: 630–634
2. Gil-Dones F, Darde VM, Alonso-Orgaz S, Lopez-Almodovar L, Mourino-Alvarez L, Padial LR, Vivanco F, Barderas MG (2012) Inside human aortic stenosis: a proteomic profile of plasma. J Proteomics 75: 1639–1653
3. Martin-Rojas T, Gil-Dones F, Lopez-Almodovar L, Padial LR, Vivanco F, Barderas MG (2012) Proteomic profile of human aortic stenosis: insights into degenerative process. J Proteom Res 11:1537–1550
4. Hanash SM, Pitteri SJ, Faca VM (2008) Mining the plasma proteome for cancer biomarkers. Nature 452:571–579
5. Good DM, Thongbookerd V, Novak J, Bascands JL, Schanstra JP, Coon JJ, Dominiczak A, Mischak H (2007) Body fluid proteomics for biomarkers discovery: lessons from the past hold the key to success in the future. J Proteome Res 6:4549–4555
6. De la Cuesta F, Barderas MG, Calvo E, Zubiri I, Maroto AS, Darde VM, Martin-Rojas T, Gil-Dones F, Posada-Ayala M, Tejerina T, Lopez JA, Vivanco F, Alvarez-Llamas G (2011) Secretome analysis of atherosclerotic and non-atherosclerotic arteries reveals dynamic extracellular remodeling during pathogenesis. J Proteomics 10(4):M110.003517

Chapter 20

A Comparative Study of Immunodepletion and Equalization Methods for Aortic Stenosis Human Plasma

Felix Gil-Dones, Verónica M. Darde, Fernando Vivanco, and María G. Barderas

Abstract

Calcified aortic valve disease is a slowly progressive disorder that ranges from mild valve thickening with no obstruction of blood flow, known as aortic sclerosis, to severe calcification with impaired leaflet motion or aortic stenosis. Until now, aortic stenosis (AS) was thought to result from aging and "wear and tear" of the aortic valve, but nowadays, it is known that it presents the same risk factors as atherosclerosis and cardiovascular diseases.

A proteomic analysis of plasma could permit to identify the changes in protein expression induced by AS in this biological sample. However, the characterization of human plasma proteome is a very complicated task, due to the wide dynamic range of concentration that separates the most abundant proteins and the less common ones (10–12 orders of magnitude). For this reason, plasma analysis requires pre-fractionation methods, and several such techniques are currently used to deplete albumin and other abundant plasma proteins.

In this work we describe two different and optimized protocols to decrease the plasma proteome complexity for proteomic analysis. With this, comprehensive and systematic characterization of the plasma proteome in the healthy and diseased aortic stenosis (AS) state will greatly facilitate the development of "useful" biomarkers for early disease detection, clinical diagnosis, and therapy.

Key words Aortic stenosis, Human plasma, Proteomic analysis, Immunoaffinity chromatography, Combinatorial peptide ligand library, Equalization, Two-dimensional gel electrophoresis

1 Introduction

Calcific aortic stenosis (AS) has become the most common cardiac valve disease in developed countries, affecting 2–3 % of the population older than 65 years. For this reason, aortic valve replacement is warranted when the valvular stenosis is hemodynamically significant, becoming the most common worldwide cause of aortic valve surgery (1–3).

Several lines of evidence have demonstrated that degenerative AS like atherosclerosis and other cardiovascular diseases are active process in which inflammation plays a key role. As such, preventative

Fernando Vivanco (ed.), *Heart Proteomics: Methods and Protocols*, Methods in Molecular Biology, vol. 1005, DOI 10.1007/978-1-62703-386-2_20, © Springer Science+Business Media New York 2013 approaches similar to those used in coronary artery disease (CAD) may be also applicable to AS (4, 5). However, recent studies reported no reductions in AS progression using statins, which significantly benefit patients with atherosclerosis. Hence, a deeper understanding of the pathophysiology of this disease is required to identify appropriate preventive measures.

Proteomics has emerged as a particularly suitable platform for the non-biased analysis of proteins involved in the pathogenesis of various diseases, such as AS. In this chapter, we focused our attention on the proteomics analysis of plasma, one of the most important clinical samples in terms of diagnosis and prognosis. This is mainly due to sample collection and processing being both simple, noninvasive technique, and inexpensive (6). In addition, as blood circulates through every organ and tissue of the body, it should contain valuable information pertaining to the physiological and pathological state of the organism (7). A proteomic analysis of plasma will permit to know and identify the changes in protein expression induced by AS in this tissue.

Proteomics has unraveled important questions in the biology of cardiovascular disease and holds even greater promise for the development of novel diagnostic and prognostic biomarkers. This approach may establish early detection strategies and monitor responses to therapies. Despite of this, human plasma has not been used often in clinical proteomics, because the wide dynamic range of proteins present in plasma (10–12 orders) poses a significant challenge for proteomic analysis, and the most abundant proteins tend to mask those that are less abundant (8). This means that around 20 proteins, including albumin and immunoglobulins, comprise approximately 99 % of total protein content in plasma, and thus, potential candidate proteins present at low concentrations may not be detected using currently available proteomic techniques (Fig. 1). For this reason, plasma proteomic analysis requires pre-fractionation steps, and several techniques have been developed including binding to solid-phase libraries, enrichment of low-molecular-weight proteins, and liquid chromatography. The most common method is immunodepletion, which is used extensively to remove specific highly abundant proteins through the action of specific antibodies (9–14). This approach involves binding the selected antibodies to a chromatographic support. When plasma proteins are in contact with the antibody-decorated beads, the high-abundance proteins are retained and the low-abundance proteins are eluted for using them in downstream analysis schemes (Fig. 2a). However, this methodology also presents some disadvantages as the relative high cost of antibodies and dilution of the sample, and nowadays, there has been a concern regarding whether less-abundant plasma proteins are removed along with albumin and other commonly depleted proteins by "nonspecific" binding (15, 16).

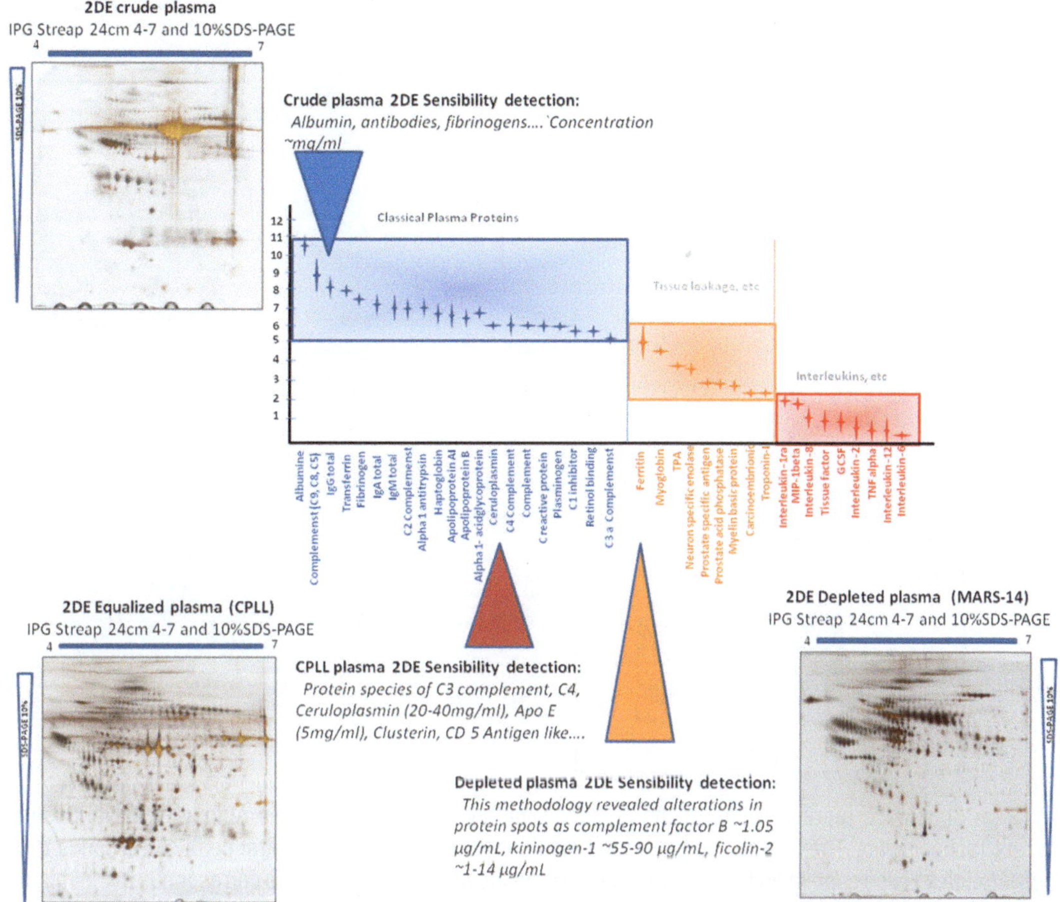

Fig. 1 Human plasma proteome: the large dynamic range of protein concentrations in the human proteome represents a significant experimental challenge as technologies must be sensitive across nearly 12 orders of magnitude (a one trillion-fold range) for comprehensive analysis and the development of biomarkers. Estimated concentration of plasma proteins (based in Anderson and Anderson (7)) Also we show different spot maps obtained from crude and prefractioned plasma

Saturation protein binding to a combinatorial peptide ligand libraries (CPLL, ProteoMiner, Bio-Rad) has also recently been proposed as an alternative approach to analyze the "low-abundance proteome" in association with mass spectrometry (MS) (16–18). CPLL technology uses a combinatorial library of hexapeptides bound to a chromatographic support, an alternative approach that should overcome most, if not all, of the disadvantages of the immunodepletion while still effectively equalizing the high-abundance proteins. The principle of this strategy is that the most abundant plasma proteins saturate all the corresponding hexapeptides, eliminating the excess and effectively decreasing their concentration (18, 19). Thus, the less-abundant plasma proteins are better represented and more accessible to differential protein expression analysis.

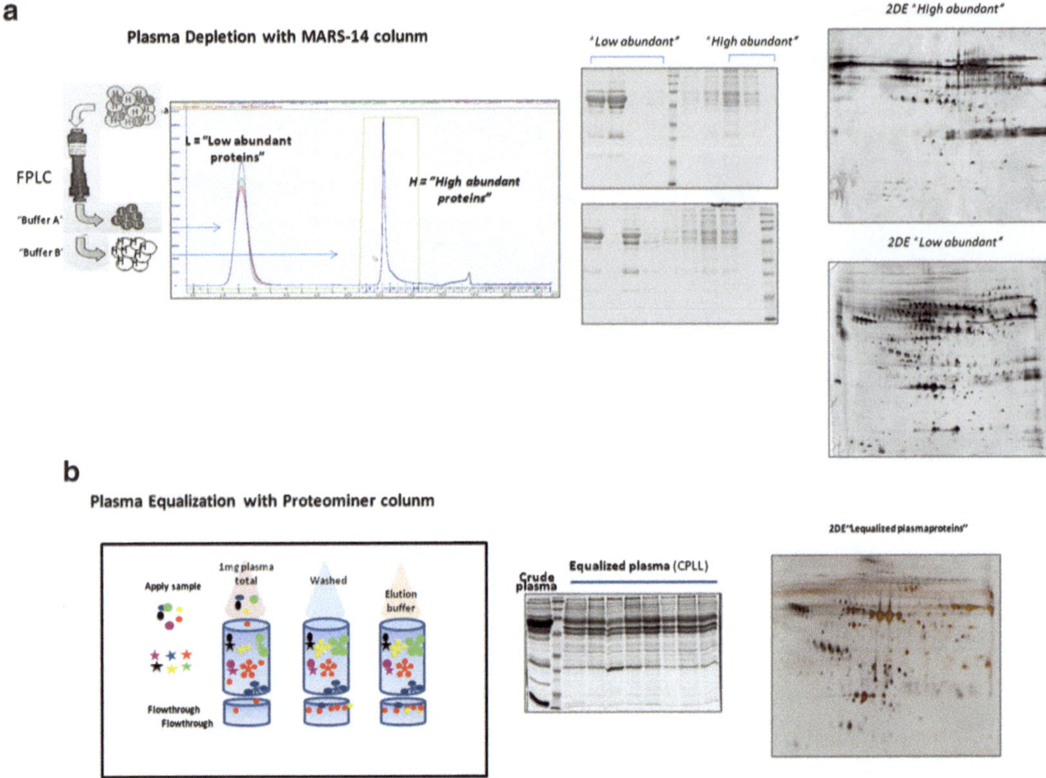

Fig. 2 (**a**) Schematic representation of the depletion method using the "Multiple Affinity Removal System" (Agilent Technologies). Three peaks appear in the resulting chromatograms. The first peak corresponds to the flow-through fraction containing the low-abundance proteins, while the second one corresponds to the retained fraction. There is a third small peak with no protein content, that might reflect changes in buffers from B to A. Similar chromatograms are obtained when overlaying several chromatographic runs, confirming the high reproducibility of this method. A 1D-SDS and 2DE gels also showed the reproducibility of this technique. Removal of the 14 most abundant plasma proteins (retained fraction) allows the detection of many low-abundance proteins (flow-through fraction) using high-resolution 2-D gels (first dimension, 24 cm IPG, pH 4–7 range; second dimension 9–16 % acrylamide gradient). (**b**) Schematic representation (binding, washed, and elution) of the plasma equalization method using the "CPLL" (ProteoMiner, Bio-Rad). 1D-SDS page show the reproducibility of this technique, each lane was loaded with the elutes of different samples. Also we show a representative plasma equalized 2DE image

In this chapter, we describe two different procedures to decrease the complexity and the dynamic range of human plasma. The first one allows to remove the 14 most abundant plasma proteins, which constitute over 99.5 % of total protein content, using a commercially available affinity system (Human 14 Multiple Affinity Removal System, Agilent Technologies) (20).

The second alternative was designed to reduce the complexity of plasma proteome avoiding losing most abundant proteins. Both procedures allow raising the concentration of low-abundance proteins, and they make possible to perform differential analysis of plasma or sera samples from healthy people (controls) and diseased patients by comparison of the expression profiles of their low-abundance plasma proteins (21).

2 Materials

2.1 Equipment for Immunodepletion (MARS-14) and Combinatorial Peptide Ligand Libraries (CPLL, ProteoMiner)

1. EDTA glass tubes (Becton Dickinson, Franklin Lakes, NJ, USA)
2. Milli-Q system (Millipore, Bedford, MA, USA)
3. Shaker (Promax 1020, Heidolph Instruments GmbH & Co. Schwabach, Germany)
4. Vortex mixer (Promax Reax Top, Heidolph Instruments GmbH & Co. Schwabach, Germany)
5. Thermomixer compact (Eppendorf AG. Hamburg, Germany)
6. HITACHI U-1100 spectrophotometer (Hitachi, Ltd. Tokyo, Japan)
7. Freezemobile 12SL Freeze Dryer (The VirTis Company Inc. Gardiner, NY, USA)
8. Allegra X-12R centrifuge (Beckman Coulter, Inc. Fullerton, CA, USA)
9. Micro Tweezers MTW7 (FONTAX Technologie SA. Lausanne, Switzerland)
10. FPLC system (Agilent Technologies 1200 series)
11. Agilent Human 14 Multiple Affinity Removal Column 4.6 mm × 50 mm (Agilent Technologies, Inc. Palo Alto, CA, USA)
12. Vivaspin 5 kDa cutoff spin concentrators (Vivascience AG. Hannover, Germany)
13. Spin filters, 0.22 μm cellulose acetate (Agilent Technologies, Inc. Palo Alto, CA, USA)
14. ProteoMiner kit (Catalog # 163-3000): Contain 10-spin column, each containing 500 μl bead slurry, wash buffer, elution reagent, and rehydration reagent
15. Collection tubes: 20 capless 2 ml

2.2 Reagents (MARS-14 and CPLL)

1. Buffer A for loading, washing, and equilibrating Multiple Affinity Removal Column, and buffer B for elution of bound proteins from Multiple Affinity Removal Column purchased from Agilent Technologies, Inc. (Palo Alto, CA, USA) (buffer A 1 l ref 5185-5987; buffer B 1 l ref 5185-5988). Both buffers are not supplied with the MARS-14 column.

2. Spin columns each containing 500 μl bead slurry (20 % beads, 20 % v/v aqueous EtOH, 0,5 % v/v acetonitrile). The columns are supplied with the ProteoMiner kit (Catalog # 163-3000).
3. Wash buffer for CPLL: 50 ml PBS buffer (150 mM NaCl, 10 mM NaH_2PO_4, pH 7,4) is also supplied with the ProteoMiner kit (Catalog # 163-3000).
4. Elution reagent: Five vials, lyophilized urea CHAPS (4 M urea, 1 % CHAPS), supplied with the ProteoMiner kit (Catalog # 163-3000) (see Note 1).
5. Rehydration reagent: 5 ml, 5 % acetic acid, supplied with the ProteoMiner kit (Catalog # 163-3000).

2.3 Solutions

Ammonium bicarbonate solution (25 mM): Dissolve 99 mg of ammonium bicarbonate in double-distilled water (50 ml final volume). Prepare just prior to use.

3 Methods

3.1 Depletion of High-Abundance Proteins (MARS 14)

To analyze the low-abundance proteins of plasma and serum using proteomic tools, removal of high-abundance proteins is required (7), and different methods for removal of albumin have been described through years (9–15). The classical albumin depletion strategy involves the use of the hydrophobic dye Cibacron blue, a chlorotriazine dye that has high affinity for albumin (22). Another classical affinity medium is the Protein A/G, which is used for the removal of the immunoglobulins (23). Among these methods, it was recently found that antibody-based methods are more effective than those based on ion-exchange chromatography (19, 20). In 2003, Piepper et al. (20) developed a strategy to generate multicomponent immunoaffinity subtraction chromatography matrices capable of specifically depleting albumin and other high-abundance plasma proteins. A similar approach has been applied by Agilent Technologies to produce the "Agilent Multiple Affinity Removal System," which comprises an affinity column, with binding capacity for 20 μl of human serum or plasma (~1–1.4 mg), and optimized mobile phases (Buffers A and B). The affinity column is packed with immobilized affinity-purified polyclonal antibody resins for removing albumin, IgG, α1-antitrypsin, IgA, transferrin, haptoglobin, α2-macroglobulin, α1-acid glycoprotein, apo AI, apo AII, IgM, transthyretin, C3, and 92–99 % fibrinogens, with high specificity (Fig. 2). The ratio of immobilized antibodies coupled to the affinity column reflects the ratio of target proteins in human serum or plasma. We describe an optimized protocol to deplete 14 most abundant plasma proteins to enable proteomic analysis (2DE, 2D-DIGE, iTRAQ, LC-Orbitrap, etc.)

3.1.1 Sample Preparation and Chromatographic Steps

1. To avoid column saturation, dilute up to 15–20 μl of human plasma five times in buffer A (15 μl of plasma and 60 μl of buffer) and spun through a 0.22 μm spin filter tube at maximum speed (about 16,000 × *g*) on table microfuge for 1 min (see Note 2).

   Alternatively, if enough plasma is available, add 400 μl of buffer A to 100 μl of plasma. Distribute the 500 μl in five aliquots of 100 μl and use 90 μl for the analysis (this corresponds to 18 μl of plasma).
2. Then, the sample is ready for injection into an FPLC System (Agilent 1200 series, ÄKTApurifier, or similar) coupled to the "Multiple Affinity Removal Column" (Agilent Technologies).
3. Buffer A and buffer B must be set up as the only mobile phases. The chromatographic steps must be performed according to the instruction manual supplied with the column.
4. Inject 75–100 μl of the diluted plasma with buffer A at a flow rate of 0.25 ml/min for 9 min.
5. Monitor the absorbance at 280 nm and the conductivity, which reflects the linear concentration increase for buffer B (% gradient).
6. Collect flow-through fraction when the absorbance is >0.02 absorbance units (it usually appears between 1.5 and 4.5 min) (Fig. 1) and store collected fractions at −20 °C if not analyzed immediately.
7. Elute bound proteins from the column with buffer B at a flow rate of 1 ml/min for 3.5 min. Collect the elution fraction when the absorbance is >0.02 absorbance units. Store collected fractions at −20 °C if not analyzed immediately.
8. Regenerate column by equilibrating it with buffer A for 7.5 min at a flow rate of 1 ml/min, with a total run cycle of 20 min (see Note 4).
9. Control the pressure limit of the column (120 bar) (see Note 3).

   Similar chromatograms are obtained when overlaying several chromatographic runs (Fig. 2a). The great reproducibility of this method allows the combination of flow-through fractions after several chromatographic cycles if high amounts of protein are necessary for downstream applications. Normally, 100–300 μg of non-retained proteins, in 1 ml (two chromatographic fractions, see Fig. 1), are recovered after every injection (see Note 5). The flow-through fractions can be also used for SELDI-TOF analysis, LC-MS/MS analysis, etc.
10. Transfer the flow-through fractions (4 × 1 ml) into 5 kDa cut-off spin concentrators and centrifuge at 3,000 × *g* in a swinging-bucket rotor for 30 min (the volume is reduced to about 0.5 ml).

11. Equilibrate fractions in a 25 mM ammonium bicarbonate solution by addition of 3–4 ml, and concentrate them again to about 0.5 ml.
12. Quantify the protein content using the Bradford method and pipet the amount of protein required for further applications (see Note 6).
13. Freeze aliquots at −80 °C and lyophilize them for 15–18 h.

3.2 Equalization of High-Abundance Proteins and Enrichment of Low-Abundance Plasma Proteins (CPLL, ProteoMiner)

The ProteoMiner technology is a novel tool to decrease the complexity and dynamic range of biological samples, especially plasma proteome. ProteoMiner technology uses a combinatorial library of hexapeptides ligands bound to a chromatographic support. Essentially, the high-abundance proteins saturate their high-affinity ligands and excess protein is washed away, and the medium and low-abundance proteins are concentrated on their specific affinity ligands without altering their concentrations and allowing to perform quantitative proteomic analysis (16). This protocol has been optimized for plasma and serum samples with protein concentration of ≥50 mg/ml. The ratio of protein to beads is crucial for optimal performance.

This procedure presents different steps: column preparation, sample binding, sample wash, and elution.

3.2.1 Column Preparation

1. Remove the top cap and snap off the bottom cap from each of the spin columns you will be using.
2. Remove the storage solution: Place the column in a capless collection tube and centrifuge at 1,000 × *g* for 2 min. Discard collected material.
3. Replace the bottom cap and add 1 ml double-distilled water and rotate the column end-to-end several times over a 5 min period.
4. Place the column without cap in a capless collection tube and centrifuge 1,000 × *g* for 2 min to remove any remaining material. Discard collected material.
5. Repeat steps 3 and 4 steps using 1 ml wash buffer in place of double-distilled water.
6. Centrifuge again at 1,000 × *g* for 1 min to remove any remaining material.
7. Replace the bottom cap on spin column. The column is now ready for sample binding.

3.2.2 Sample Binding

Centrifuge samples at 10,000 × *g* for 10 min to clarify (see Note 7). 900 μl plasma and 100 μl plasma preparation buffer to spin column and replace bottom cap and rotate the column in order to eliminate all debris end-to-end for 2 h at room temperature (see Note 8).

3.2.3 Sample Wash

1. After sample binding, remove bottom caps and centrifuge at 1,000 × *g* for 2 min. Discard collected material and centrifuge again at 1,000 × *g* for 1 min to remove any remaining material.
2. Add 1 ml of wash buffer to column, and incubate by rotation end-to-end for 5 min.
3. Remove caps, and centrifuge the column in a capless collection tube at 1,000 × *g* for 2 min. Discard collected material and centrifuge again 1,000 × *g* for 1 min to remove any remaining material.
4. To avoid any unspecific union to the hexapeptide library ligands, repeat steps 2 and 3 two more times.

3.2.4 Elution

1. After all wash buffer has been removed, replace the bottom cap, add 1 ml double-distilled water, attach top cap, and incubate in a rotating wheel for 1 min.
2. Centrifuge 1,000 × *g* for 2 min without caps in a capless collection tube. Discard collected material.
3. Centrifuge again at 1,000 × *g* for 1 min to remove any remaining material. Discard collected material.
4. Attach bottom cap to the column (see Note 8). Add 100 μl of rehydrated elution reagent (refer to Subheading 2.3) to the column and replace the cap (see Note 9). Lightly vortex for 5 s.
5. Incubate column at room temperature; lightly vortex several times over a period of 15 min.
6. Remove the caps, place in a clean collection tube labeled E1 and centrifuge at 1,000 × *g* for 2 min. This solution contains the equalized proteins. Do not discard.
7. Repeat steps 4–7 two more times (see Note 10: Elutions may be pooled or analyzed individually). Store-eluted proteins at −20 °C or proceeds with downstream analysis (2DE, DIGE, iTRAQ, LC-MS/MS, etc.).

4 Notes

1. Solutions containing urea must not be heated over 37 °C, in order to prevent urea decomposition into isocyanate, which may cause protein modification by carbamylation.
2. Columns of higher capacity are available from Agilent Technologies (4.6 × 100 mm; 1.66 ml), for 30–40 μl of human serum/plasma, or the most recent one (Agilent High Capacity Multiple Affinity Removal System for the Depletion of High-Abundant Proteins from Human Proteomic Samples. Ref.

5188-5921), with at least twice loading capacity and 300 injections in the FPLC with high reproducibility. This column is made with a novel attachment process for antibodies, and the same buffers and protocols are used.

3. Do not load the column directly with crude serum or plasma. It is essential to dilute it in order to prevent clogging of the column. If the inlet frit is clogged, replace both, the inlet and outlet frits, simultaneously. Clogged inlet frits may increase backpressure, which affects negatively to column lifetime. Remove particulate materials that are sometimes present in serum or plasma by quick spin using the 0.22 μm spin filters.
4. When not in use, store the column after equilibration with buffer A at 2–8 °C, in a refrigerator to minimize loss in column capacity. Be sure that the end caps are tightly sealed. Do not expose column to organic solvents or reducing agents (which affect the structure of the antibodies). Do not freeze the column.
5. After the first elution cycle of a plasma sample, the binding capacity of the column may drop a 10–25 %. This is a phenomenon frequently observed, known as "first cycle effect." It is probably caused by high-avidity binding of some plasma proteins to a fraction of the immobilized antibodies (10, 20).
6. Protein estimation is very important, and it depends on the next applications. For example, the amount of protein that can be loaded to a single IPG strip depends on its length. Usually 50–500 μg of proteins are applied to 17 or 24 cm IPG strips.

Notes ProteoMiner

7. Be careful to avoid the bottom aggregate proteins and top lipid layer when recovering your sample. It is "highly" recommended that at least 1 ml of sample (protein concentration ~50 mg/ml) is added to the column, as lower volumes or protein concentrations may not achieve optimal results.
8. With plasma or serum samples, clumping may occur after 1 h of binding; this is expected and will not negatively impact your sample preparation.
9. Take caution to ensure the bottom cap is tightly attached.
10. Different eluted samples may be pooled or analyzed individually. In addition, there is a sequential elution kit that uses four different elution reagents (Bio-Rad cat. 163-3002), allowing higher sample fractionation that could be necessary in some analysis or studies.

Acknowledgements

We thank the Proteomics Unit of the HNP for assistance and Carmen Bermudez for her technical support. This work was supported by grants from the Instituto de Salud Carlos III (FIS PI070537, PI11/02239), grants from Fondos Feder-Redes temáticas de Investigación Cooperativa en Salud (RD06/0014/1015), and grants from Fundación para la Investigación Sanitaria de Castilla-La Mancha (FISCAM PI2008-08, PI2008-28, PI2008-52).

References

1. Iung B, Baron G, Butchart E, Delahaye F, Gohlke-Bärwolf C, Levang OW et al (2003) A prospective survey of patients with valvular heart disease in Europe: the Euro Heart Survey on Valvular Heart Disease. Eur Heart J 24: 1231–1243
2. Lindroos M, Kupari M, Heikkila J, Tilvis R (1993) Prevalence of aortic valve abnormalities in the elderly: an echocardiographic study of a random population sample. J Am Coll Cardiol 21:1220–1225
3. Pomerance A (1972) Pathogenesis of aortic stenosis and its relation to age. Br Heart J 34: 569–574
4. Otto CM, Kuusisto J, Reichenbach DD, Gown AM, O'Brien KD (1994) Characterization of the early lesion of `degenerative' valvular aortic stenosis: histologic and immunohistochemical studies. Circulation 90:844–853
5. Rajamannan NM (2010) Mechanisms of aortic valve calcification: the LDL-density-radius theory: a translation from cell signalling to physiology. Am J Physiol Heart Circ Physiol 298:H5–H15
6. Veenstra TD, Conrads TP, Hood BL, Avellino AM, Ellenbogen RG, Morrison RS (2005) Biomarkers: mining the biofluid proteome. Mol Cell Proteomics 4:409–418
7. Anderson NL, Anderson NG (2002) The human plasma proteome: history, character, and diagnostic prospects. Mol Cell Proteomics 11:845–867
8. Anderson NL (2005) Candidate-based proteomics in the search for biomarkers of cardiovascular disease. J Physiol 563:23–60
9. Zolotarjova J, Martosella G, Nicol J, Bailey B, Boyes E, Barrett WC (2005) Differences among techniques for high abundant protein depletion. Proteomics 5:3304–3313
10. Wang YY, Cheng P, Chan DW (2003) A simple affinity spin tube filter method for removing high-abundant common proteins or enriching low-abundant biomarkers for serum proteomic analysis. Proteomics 3:243–248
11. Travis J, Bowen J, Tewksbury D, Johnson D, Pannell R (1976) Isolation of albumin from whole human plasma and fractionation of albumin-depleted plasma. Biochem J 157: 301–306
12. Travis J, Pannell R (1973) Selective removal of albumin from plasma by affinity chromatography. Clin Chim Acta 49:49–52
13. Leatherbarrow RJ, Dean PD (1980) Studies on the mechanism of binding of serum albumin to immobilized Cibacron Blue F3GA. Biochem J 189.27–34
14. Seam N, Gonzales DA, Kern SJ, Hortin GL, Hoehn GT, Suffredini AF (2007) Quality control of serum albumin depletion for proteomic analysis. Clin Chem 53:1915–1920
15. Granger J, Siddiqui J, Copeland S, Remick D (2005) Albumin depletion of human plasma also removes low abundance proteins including the cytokines. Proteomics 5:4713–4718
16. Righetti PG, Boschetti E, Lomas L, Citterio A (2006) Protein equalizer technology: the quest for a "democratic proteome". Proteomics 6:3980–3992
17. Sihlbom C, Kanmert I, von Bahr H, Davidsson P (2008) Evaluation of the combination of bead technology with SELDI-TOF-MS and 2-D DIGE for detection of plasma proteins. J Proteome Res 9:4191–4198
18. Paulus A, Freeby S, Academia K, Thulassiraman V (2009) Accessing low-abundance proteins in serum and plasma with a novel, simple enrichment and depletion method. Tech note 5632. Sample preparation. BioRad, Life science Group 6000. James Watson Drive, Hercules, CA.
19. Bjorhall K, Miliotis T, Davidsson P (2005) Comparison of different depletion strategies for

improved resolution in proteomic analysis of human serum samples. Proteomics 5:307–317

20. Pieper R, Su Q, Gatlin CL, Huang ST, Anderson NL, Steiner S (2003) Multi-component immunoaffinity subtraction chromatography: an innovative step towards a comprehensive survey of the human plasma proteome. Proteomics 3:422–432
21. Gil-Dones F, Darde VM, Alonso Orgaz S, Lopez-Almodovar LF, Mourino-Alvarez L, Padial LR, Vivanco F, Barderas MG (2012) Inside human aortic stenosis: a proteomic analysis of plasma. J Proteomics 75:1639–1653
22. Gianazza E, Arnaud P (1982) A general method for fractionation of plasma proteins. Dye-ligand affinity chromatography on immobilized Cibacron blue F3-GA. Biochem J 201:129–136
23. Greenough C, Jenkins RE, Kiterringham NR, Pirmohamed M, Park BK, Pennington SR (2004) A method for the rapid depletion of albumin and immunoglobulin from human plasma. Proteomics 4:3107–3111

Index

A

B

C

Fernando Vivanco (ed.), *Heart Proteomics: Methods and Protocols*, Methods in Molecular Biology, vol. 1005, DOI 10.1007/978-1-62703-386-2, © Springer Science+Business Media New York 2013

H

I

J

K

L

M

N

O

P

Q

R

S

T

U

V

W

X

Z

MIX
Papier aus verantwortungsvollen Quellen
Paper from responsible sources
FSC® C105338

If you have any concerns about our products,
you can contact us on
ProductSafety@springernature.com

In case Publisher is established outside the EU,
the EU authorized representative is:
Springer Nature Customer Service Center GmbH
Europaplatz 3, 69115 Heidelberg, Germany

Printed by Libri Plureos GmbH
in Hamburg, Germany